Serono Symposia USA
Norwell, Massachusetts

Springer Science+Business Media, LLC

PROCEEDINGS IN THE SERONO SYMPOSIA USA SERIES

CELL DEATH IN REPRODUCTIVE PHYSIOLOGY
Edited by Jonathan L. Tilly, Jerome F. Strauss III,
and Martin Tenniswood

*INHIBIN, ACTIVIN, AND FOLLISTATIN: Regulatory Functions
in System and Cell Biology. A Serono Symposia S.A. Publication*
Edited by Toshihiro Aono, Hiromu Sugino, and Wylie W. Vale

PERIMENOPAUSE
Edited by Rogerio A. Lobo

*GROWTH FACTORS AND WOUND HEALING: Basic Science and
Potential Clinical Applications*
Edited by Thomas R. Ziegler, Glenn F. Pierce, and David N. Herndon

POLYCYSTIC OVARY SYNDROME
Edited by R. Jeffrey Chang

IDEA TO PRODUCT: The Process
Edited by Nancy J. Alexander and Anne Colston Wentz

BOVINE SPONGIFORM ENCEPHALOPATHY: The BSE Dilemma
Edited by Clarence J. Gibbs, Jr.

GROWTH HORMONE SECRETAGOGUES
Edited by Barry B. Bercu and Richard F. Walker

CELLULAR AND MOLECULAR REGULATION OF TESTICULAR CELLS
Edited by Claude Desjardins

GENETIC MODELS OF IMMUNE AND INFLAMMATORY DISEASES
Edited by Abul K. Abbas and Richard A. Flavell

MOLECULAR AND CELLULAR ASPECTS OF PERIIMPLANTATION PROCESSES
Edited by S.K. Dey

*THE SOMATOTROPHIC AXIS AND THE REPRODUCTIVE PROCESS
IN HEALTH AND DISEASE*
Edited by Eli Y. Adashi and Michael O. Thorner

GHRH, GH, AND IGF-I: Basic and Clinical Advances
Edited by Marc R. Blackman, S. Mitchell Harman, Jesse Roth,
and Jay R. Shapiro

IMMUNOBIOLOGY OF REPRODUCTION
Edited by Joan S. Hunt

FUNCTION OF SOMATIC CELLS IN THE TESTIS
Edited by Andrzej Bartke

GLYCOPROTEIN HORMONES: Structure, Function and Clinical Implications
Edited by Joyce W. Lustbader, David Puett, and Raymond W. Ruddon

Continued after Index

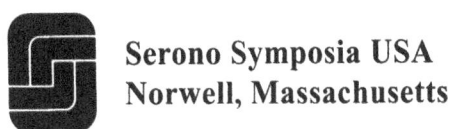

Serono Symposia USA
Norwell, Massachusetts

Jonathan L. Tilly   Jerome F. Strauss III
Martin Tenniswood
Editors

Cell Death in Reproductive Physiology

With 76 Figures

Springer

Jonathan L. Tilly, Ph.D.
Vincent Center for Reproductive Biology
Massachusetts General Hospital
Boston, MA 02114
USA

Jerome F. Strauss III, M.D., Ph.D.
Center for Research on Reproduction
  and Women's Health
University of Pennsylvania School of Medicine
Philadelphia, PA 19104
USA

Martin Tenniswood, Ph.D.
W. Alton Jones Cell Science Center
Lake Placid, NY 12946
USA

Proceedings of the International Symposium on Cell Death in Reproductive Physiology, sponsored by Serono Symposia USA, Inc., held April 11 to 14, 1996, in Chicago, Illinois.

For information on previous volumes, contact Serono Symposia, USA, Inc.

Library of Congress Cataloging-in-Publication Data
Cell death in reproductive physiology/edited by Jonathan L. Tilly,
    Jerome F. Strauss III, and Martin Tenniswood.
        p. cm.
    At head of title: Serono Symposium USA.
    Includes bibliographical references and index.
    ISBN 978-0-387-98344-8   ISBN 978-1-4612-1944-6 (eBook)
    DOI 10.1007/978-1-4612-1944-6
    1. Generative organs—Physiology—Congresses. 2. Apoptosis—
Congresses. I. Tilly, Jonathan L., 1962– . II. Strauss, Jerome
F. (Jerome Frank), 1947– . III. Tenniswood, M. (Martin)
IV. Serono Symposia, USA. V. International Symposium on Cell Death
in Reproductive Physiology (1996: Chicago, Ill.)
    [DNLM: 1. Apoptosis—congresses. 2. Reproduction—physiology—
congresses. QH 671 C3937 1997]
QP251.C36   1997
612.6—dc21   97-36129

Printed on acid-free paper.

Production coordinated by Chernow Editorial Services, Inc., and managed by Francine McNeill; manufacturing supervised by Jacqui Ashri.
Typeset by KP Company, Brooklyn, NY.

9 8 7 6 5 4 3 2 1

ISBN 978-0-387-98344-8

SYMPOSIUM ON CELL DEATH IN REPRODUCTIVE PHYSIOLOGY

Scientific Committee

Jonathan L. Tilly, Ph.D., Chair
Massachusetts General Hospital
Harvard Medical School
Boston, Massachusetts

Jerome F. Strauss III, M.D., Ph.D.
University of Pennsylvania
School of Medicine
Philadelphia, Pennsylvania

Martin Tenniswood, Ph.D.
W. Alton Jones Cell Science Center
Lake Placid, New York

Organizing Secretary

Leslie Nies
Serono Symposia USA, Inc.
100 Longwater Circle
Norwell, Massachusetts

These proceedings are dedicated to the memory of Bruce C. Moulton, Ph.D., an outstanding scientist, a devoted teacher, and a loyal friend. Bruce's contribution contained in these proceedings, based on his exciting lecture presented at the symposium, was one of the last scientific discourses that he prepared. His unique insight into female reproductive physiology, as clearly exemplified by the intriguing hypotheses raised in his chapter concerning the molecular and genetic events surrounding decidualization in the uterus during embryo implantation, will be truly missed, in general, by the scientific community, and in particular, by those people fortunate enough to have been personally acquainted with Bruce during his long and productive research career.

Preface

The regulation of cell death in various reproductive tissues, as in other major organ systems of the body, has become a focal point of research activity in many laboratories over the past few years. As such, the need for a "formal" meeting to highlight recent work in this field, as well as to integrate knowledge from other sources (such as investigators working on cell death in cancer and immune function) in the broad context of identifying conserved pathways that coordinate life-and-death decisions in diverse cell types, became apparent. Therefore, the goals of the Scientific Committee of the International Symposium on Cell Death in Reproductive Physiology, sponsored by Serono Symposia USA, were already predetermined by this need. Simply stated, we sought to bring together for the first time a select cohort of reproductive biologists and cell death researchers, many but not all chosen based on their pioneering efforts in elucidating the fundamental aspects of apoptosis in reproductive and nonreproductive tissues, as a means to review the current status of the field, foster new ideas, and promote scientific collaborations. In the ensuing chapters of this book, summaries of work discussed at the meeting are presented to emphasize both the diversity and the similarities in the occurrence and regulation of apoptosis in tissues of the male and female reproductive systems. Before the reader begins examination of the more formal chapters prepared by each of these scientists, a brief overview is provided in Chapter 1 as a means to introduce the unifying themes discussed throughout the conference, as well as to stimulate thought on future directions of research in this quickly expanding field.

This conference was brought to fruition through the dedicated work of many individuals. I would like to first thank my esteemed co-chairs, Jerome F. Strauss III and Martin Tenniswood, for their assistance in organizing the meeting and editing these proceedings. I am also most appreciative of the efforts of the staff of Serono Symposia USA, without whom this conference would have not taken place. I would like to offer special thanks to Leslie Nies (President) and Judy Donahue (Editor) for their tremendous input throughout the planning stages of the symposium, to Judy Donahue for her incredible patience and dedication in preparing these chapters for publication, and to Dianne Ferreira (Conference Coordinator) for assistance with site preparation and travel arrangements for the invited participants. Lastly and most importantly, I am indebted to the outstanding cohort of speakers who took time out of their busy schedules to attend the meeting, give exciting lectures, and provide thought-provoking chapters that serve as the basis of these proceedings. As I scan the chapters and think back to the symposium, the goals that were set by the Scientific Committee were certainly met, and accordingly I must agree with a recent review of the meeting by one of the participants (De Felici M. How life may come from death: lessons from reproductive biology. Cell Death Differ 1996; 3:439–41) that the conference was an "undoubted success."

JONATHAN L. TILLY

Contents

Contributors

DORIT AHARONI, Department of Molecular Cell Biology, The Weizmann Institute of Science, Rehovot, Israel.

KAMIL C. AKCALI, Department of Cell Biology, Neurobiology and Anatomy, University of Cincinnati College of Medicine, Cincinnati, Ohio, USA.

ALESSANDRA AMENDOLA, Department of Biology, University of Rome "Tor Vergata," Rome, Italy.

ABRAHAM AMSTERDAM, Department of Molecular Cell Biology, The Weizmann Institute of Science, Rehovot, Israel.

RAYMOND F. ATEN, Department of Obstetrics and Gynecology, Yale University School of Medicine, New Haven, Connecticut, USA.

HAROLD R. BEHRMAN, Department of Obstetrics and Gynecology, Yale University School of Medicine, New Haven, Connecticut, USA.

HÅKAN BILLIG, Department of Physiology, University of Göteborg, Göteborg, Sweden.

MINA J. BISSELL, Life Sciences Division, Ernest Orlando Lawrence Berkeley National Laboratory, University of California, Berkeley, California, USA.

DAVID L. BOONE, Reproductive Biology Unit, Departments of Obstetrics and Gynecology and Physiology, University of Ottawa, and The Loeb Medical Research Institute, Ottawa Civic Hospital, Ottawa, Ontario, Canada.

CARL D. BORTNER, Molecular Endocrinology Group, The Laboratory of Integrative Biology, National Institute of Environmental Health Sciences, Research Triangle Park, North Carolina, USA.

WILLIAM S. BRANHAM, Division of Reproductive and Developmental Toxicology, National Center for Toxicological Research, Jefferson, Arkansas, USA.

MAREN BRECKWOLDT, Department of Molecular Cell Biology, The Weizmann Institute of Science, Rehovot, Israel.

THOMAS L. BROWN, Department of Cell Biology, Lerner Research Institute, Cleveland Clinic Foundation, Cleveland, Ohio, USA.

ROBERT F. CASPER, Division of Reproductive Sciences, Toronto Hospital Research Institute, and Department of Obstetrics and Gynecology, University of Toronto, Toronto, Ontario, Canada.

JOHN A. CIDLOWSKI, Molecular Endocrinology Group, The Laboratory of Integrative Biology, National Institute of Environmental Health Sciences, Research Triangle Park, North Carolina, USA.

ADA DANTES, Department of Molecular Cell Biology, The Weizmann Institute of Science, Rehovot, Israel.

MASSIMO DE FELICI, Department of Public Health and Cell Biology, University of Rome "Tor Vergata," Rome, Italy.

VIOLETTA DELGADO, Department of Obstetrics and Gynecology, University of Pennsylvania, Philadelphia, Pennsylvania, USA.

PATRICIA K. DONAHOE, Pediatric Surgical Research Laboratories, Massachusetts General Hospital, Boston, Massachusetts, USA.

LEO DUNKEL, Children's Hospital, University of Helsinki, Helsinki, Finland.

KRISTA ERKKILÄ, Children's Hospital, University of Helsinki, Helsinki, Finland.

MARIA GRAZIA FARRACE, Department of Biology, University of Rome "Tor Vergata," Rome, Italy.

EMMA E. FURTH, Department of Pathology and Laboratory Medicine, University of Pennsylvania, Philadelphia, Pennsylvania, USA.

SEAN GUENETTE, W. Alton Jones Cell Science Center, Lake Placid, New York, USA.

TINA L. GUMIENNY, Department of Genetics, State University of New York at Stony Brook, Stony Brook, New York, USA.

DEBORA L. HAMERNIK, Department of Physiology, University of Arizona, Tucson, Arizona, USA.

WILLIAM J. HENDRY III, Department of Biological Sciences, Wichita State University, Wichita, Kansas, USA.

MICHAEL O. HENGARTNER, Cold Spring Harbor Laboratory, Cold Spring Harbor, New York, USA.

PATRICIA B. HOYER, Department of Physiology, University of Arizona, Tucson, Arizona, USA.

ALAN L. JOHNSON, Department of Biological Sciences, University of Notre Dame, Notre Dame, Indiana, USA.

ANDREA JURISICOVA, Division of Reproductive Sciences, Toronto Hospital Research Institute, and Department of Zoology, University of Toronto, Toronto, Ontario, Canada.

SOHAIB A. KHAN, Department of Cell Biology, Neurobiology and Anatomy, University of Cincinnati College of Medicine, Cincinnati, Ohio, USA.

PINAR KODAMAN, Department of Obstetrics and Gynecology, Yale University School of Medicine, New Haven, Connecticut, USA.

JOHNATHON LAKINS, Department of Biochemistry, University of Ottawa, Ottawa, Ontario, Canada; and W. Alton Jones Cell Science Center, Lake Placid, New York, USA.

WENDELL W. LEAVITT, Department of Biological Sciences, Wichita State University, Wichita, Kansas, USA.

HANQIN LEI, Department of Obstetrics and Gynecology, University of Pennsylvania, Philadelphia, Pennsylvania, USA.

CARLOS A. MOLINA, Department of Obstetrics and Gynecology, New Jersey Medical School, Newark, New Jersey, USA.

COLM MORRISSEY, University College Dublin, Dublin, Ireland; and W. Alton Jones Cell Science Center, Lake Placid, New York, USA.

JOAN MOTZ, Department of Obstetrics and Gynecology, University of Cincinnati College of Medicine, Cincinnati, Ohio, USA.

BRUCE C. MOULTON, Department of Obstetrics and Gynecology, University of Cincinnati College of Medicine, Cincinnati, Ohio, USA.

SELVARAJ NATARAJAGOUNDER, Department of Molecular Cell Biology, The Weizmann Institute of Science, Rehovot, Israel.

THOMAS F. OGLE, Department of Physiology and Endocrinology, Medical College of Georgia, Augusta, Georgia, USA.

JACINTHA O'SULLIVAN, University College Dublin, Dublin, Ireland; and W. Alton Jones Cell Science Center, Lake Placid, New York, USA.

LAURIE G. PAAVOLA, Department of Anatomy and Cell Biology, Temple University School of Medicine, Philadelphia, Pennsylvania, USA.

JOHN J. PELUSO, Department of Obstetrics and Gynecology, University of Connecticut Health Center, Hartford, Connecticut, USA.

MAURIZIO PESCE, Department of Public Health and Cell Biology, University of Rome "Tor Vergata," Rome, Italy.

MAURO PIACENTINI, Department of Biology, University of Rome "Tor Vergata," Rome, Italy.

PAOLO RINAUDO, Department of Obstetrics and Gynecology, Yale University School of Medicine, New Haven, Connecticut, USA.

BO R. RUEDA, Department of Physiology, University of Arizona, Tucson, Arizona, USA.

DAVID W. SCHOMBERG, Departments of Obstetrics and Gynecology and Cell Biology, Duke University Medical Center, Durham, North Carolina, USA.

DANIEL M. SHEEHAN, Division of Reproductive and Developmental Toxicology, National Center for Toxicological Research, Jefferson, Arkansas, USA.

SRIKALA SRIDHAR, Department of Biochemistry, University of Ottawa, Ottawa, Ontario, Canada.

JEROME F. STRAUSS III, Departments of Obstetrics and Gynecology and Pathology and Laboratory Medicine, University of Pennsylvania, Philadelphia, Pennsylvania, USA.

HAILUN TANG, W. Alton Jones Cell Science Center, Lake Placid, New York, USA.

SEPPO TASKINEN, Children's Hospital, University of Helsinki, Helsinki, Finland.

MARTIN TENNISWOOD, W. Alton Jones Cell Science Center, Lake Placid, New York, USA.

JONATHAN L. TILLY, Vincent Center for Reproductive Biology, Department of Gynecology and Obstetrics, Massachusetts General Hospital and Harvard Medical School, Boston, Massachusetts, USA.

KIM I. TILLY, Department of Gynecology and Obstetrics, Massachusetts General Hospital, Boston, Massachusetts, USA.

BENJAMIN K. TSANG, Reproductive Biology Unit, Departments of Obstetrics and Gynecology and Physiology, University of Ottawa, and The Loeb Medical Research Institute, Ottawa Civic Hospital, Ottawa, Ontario, Canada.

FELIPE VADILLO-ORTEGA, The National Institute of Nutrition, Mexico City, Mexico.

SUSSANAH L. VARMUZA, Department of Zoology, University of Toronto, Toronto, Ontario, Canada.

TONGWEN WANG, Pediatric Surgical Research Laboratories, Massachusetts General Hospital, Boston, Massachusetts, USA.

ZHENGQI WANG, Department of Chemistry, Clarkson University, Potsdam, New York, and W. Alton Jones Cell Science Center, Lake Placid, New York, USA.

SAKARI WIKSTRÖM, Children's Hospital, University of Helsinki, Helsinki, Finland.

PING ZHAN, Department of Chemistry, Clarkson University, Potsdam, New York, and W. Alton Jones Cell Science Center, Lake Placid, New York, USA.

XINGLONG ZHENG, Department of Biological Sciences, Wichita State University, Wichita, Kansas, USA.

TONY G. ZREIK, Department of Obstetrics and Gynecology, Yale University School of Medicine, New Haven, Connecticut, USA.

1

Dissecting the Functions, Mechanisms, and Genes of Physiological Cell Death in Reproductive Tissues: An Overview

JONATHAN L. TILLY AND KIM I. TILLY

Physiological cell death, a process referred to most often as apoptosis or programmed cell death, is fundamental to almost every aspect of normal tissue development, function, and homeostasis in multicellular organisms (1–3). In reproductive physiology, cell death plays a fundamental role in phenotypic sex determination during embryogenesis, and maintains important functions throughout postnatal life in the gonads, reproductive tract, and accessory reproductive tissues of both males and females. For instance, controlled cellular deletion has been characterized during Müllerian duct regression in males (Chapter 6) (4) and Wolffian duct regression in females (5), ovarian germ cell endowment and depletion (Chapters 2, 3, and 8–11) (6–8), spermatogenesis (Chapter 16) (9), early blastocyst development (Chapter 4) and embryo implantation (Chapter 5), parturition (Chapter 7), and luteolysis (Chapters 13 and 14). Furthermore, physiological cell death serves to balance cellular replication and maintain general homeostasis or cyclicity of nongonadal reproductive tissues such as the prostate (Chapter 17) (10), uterus (11, 12), and breast (Chapters 12 and 17) (13).

Despite this wide diversity of tissues, recent data suggest that there are many common features associated with the regulation of cell death throughout the male and female reproductive systems. One of the most obvious is the apparent requirement by cells of these tissues for a cell-type specific survival factor(s), generally in the form of a trophic hormone or growth factor. In the ovary and testis, important roles for pituitary—and placental/trophectodermal—derived gonadotropins (Chapters 8–11) (6–8, 14–16), as well as locally produced cytokines (Chapter 3) (6–8, 17), growth factors (Chapters 3, 8–11) (6–8, 18–22), neuropeptides (23), and steroids (Chapter 11) (22), in suppression of germ cell or somatic cell death have already been documented. Similarly, the prostate appears to require androgen (Chapter 17) (10) and the breast lactational hormones (Chapter 17) (13) for a maintenance of cellular mass. It should be noted, however, that there are also several instances of hormone-induced cell death, with one of the most well-characterized being the regression of the Müllerian duct triggered by Müllerian-inhibiting substance in males during early fetal development (Chapter 6) (4). Additionally, transforming growth factor-β has been reported to induce apoptosis in cells of the prostate (24), uterus (25), and ovary (26), and engagement of the FAS antigen (a membrane receptor) by antibodies that mimic the natural ligand to activate downstream signaling events triggers apoptosis in human granulosa-lutein cells (27) and rat granulosa cells (28).

A second "universal" feature is the apparent dependency of many, if not all, of these tissues on interactions of their different cell types with each other (mesenchymal or stromal cell-epithelial cell interaction, epithelial cell-germ cell interaction; [Chapters 16 and 17] [6–8]) and with their surrounding substratum (extracellular matrix proteins, cadherins; [Chapters 11, 12, and 17] [29, 30]) for survival to be achieved. With all of these extrinsic factors to account for, one then wonders about the complexity of intracellular events that must be set in motion to decipher the plethora of external signals impinging upon the cell at any given time. Only recently have reproductive biologists begun to tease apart the cascade of "second messengers" (ceramide, kinases, phosphatases, reactive oxygen species, ions) and transcription factors (p53, nur77, c-myc, NF-κB, retinoic acid receptor, ICER) involved in transducing extracellular signals within cells of reproductive tissues (Chapters 9 and 15) (6–8, 15, 31–36). Much of this work, which has primarily relied on comparative analogies with data derived from studies of cell lines (Chapter 18) (1–3, 37–41), is still in the early stages of analysis but will clearly contribute to our overall understanding of how cells respond to external signaling events and consequently either activate or repress apoptosis.

It is known that many of these early cytoplasmic signals apparently end up at the nucleus, as reflected by hormone (Chapter 8) (6–8, 15, 31, 42–44)— and extracellular matrix protein (Chapter 12) (29, 30)—induced changes in expression of genes encoding a cohort of cell death regulatory proteins conserved both functionally and structurally through evolution. Among the rapidly growing list of such proteins, evidence derived from gene expression analyses implicates members of the *ced-9/bcl-2* (Chapter 8) (30, 42, 45–47) and *ced-3*/interleukin-1β-converting enzyme (*Ice*) (Chapter 12) (29, 43, 48, 49) gene families, as well as the p53 tumor suppressor protein (31, 32), as primary determinants of cell survival or death in reproductive tissues (6–8). Additionally, data have been put forth to support a role for an alteration in the cellular reduction-oxidation (redox) state, possibly resulting from disruptions in mitochondrial function (50), as an enhancer of death susceptibility (51–53). Indeed, recent experiments have shown that antioxidants are potent inhibitors of cell death in hormone-deprived ovarian cells (15), paving the way for more work on the mechanisms underlying the role of reactive oxygen species and mitochondria in the decision making process that leads to apoptosis induction. To support studies of changes in expression of endogenous cell death genes, recent analyses of "gene knock-out" mice have provided concrete proof that some of these gene products live up to their expected roles in directing cell fate in certain reproductive tissues (for instance, the BCL-2 death-repressor and the BAX death-inducer proteins in the ovary; [45–47]). However, this may not hold true in all reproductive tissues for putative cell death regulators. For example, ablation of BAX actually increases male germ cell death (46, 47), and loss-of-function of a member of the *ced-3/Ice* family of death-inducing proteases may in fact reduce survival rates of female germ cells (49).

Lastly, a retrospective analysis of data generated from earlier toxicological assessments has confirmed that several man-made chemicals, such as polycyclic aromatic hydrocarbons, cause female sterility by initiating apoptosis in oocytes of primordial follicles (54–56). These findings have been recently supported and extended by investigations of the actions of other environmental toxicants on apoptosis and cell death gene expression in the female gonad (56–58). Thus, the occurrence of physiological, and possibly gene-directed, cell death in tissues following exposure to these nonphysiological stimuli reinforces the idea that not all cell deaths, despite similarities in the mechanisms by which the death is ultimately accomplished, are "genetically programmed." In some cases, however, other pathways of cellular programming that indirectly, but ultimately, lead to alterations in cell fate can be initiated by prior exposure of developing tissues to "endocrine disrupters" (Chapter 20). Consequently, the study of cell death will not only require evaluations of cells collected during discrete points during normal development (for example, studies reporting maturation-related changes in apoptotic endonuclease activity in ovarian granulosa cells; [Chapter19]), but will also be dependent upon interpretation of these experimental data in the context of prior conditions that may have permanently altered the internal programming of specific cell lineages. In closing, we as reproductive biologists have learned a great deal by close examination of data derived from studies of gene-directed cell death in nonmammalian species and nonreproductive tissues. Considering the progress made just over the past few years, the future holds tremendous promise as we continue to characterize the roles and regulation of physiological cell death in development and function of the male and female reproductive systems.

References

1. Wyllie AH. The genetic regulation of apoptosis. Curr Opin Genet Develop 1995; 5:97–104.
2. Stellar H. Mechanisms and genes of cellular suicide. Science 1995;267:1445–9.
3. Yang E, Korsmeyer SJ. Molecular thanatopsis: a discourse on the BCL-2 family and cell death. Blood 1996;88:386–401.

4. Price MJ, Donahoe PK, Ito Y, Hendren WH. Programmed cell death in the Müllerian duct induced by Müllerian inhibiting substance. Am J Anat 1977;149: 353–76.
5. Djehiche B, Segalen J, Chambon Y. Ultrastructure of Müllerian and Wolffian ducts of fetal rabbit in vivo and in organ culture. Tissue Cell 1994;26:323–32.
6. Tilly JL. Apoptosis and ovarian function. Rev Reprod 1996;1:162–72.
7. Tilly JL, Tilly KI, Perez GI. The genes of cell death and cellular susceptibility to apoptosis in the ovary: a hypothesis. Cell Death Differ 1997;4:180–7.
8. Martimbeau S, Tilly JL. Physiological cell death in endocrine-dependent tissues: an ovarian perspective. Clin Endocrinol 1997;46:241–54.
9. Dunkel L, Hirvonen V, Erkkilä K. Clinical aspects of male germ cell apoptosis during testis development and spermatogenesis. Cell Death Differ 1997;4: 171–9.
10. Banerjee PP, Banerjee S, Tilly KI, Tilly JL, Brown TR, Zirkin BR. Lobe specific apoptotic cell death in rat prostate after androgen ablation by castration. Endocrinology 1995;136:4368–76.
11. Hopwood D, Levison DA. Atrophy and apoptosis in the cyclical human endometrium. J Pathol 1975;119:159–66.
12. Tao X-J, Tilly KI, Maravei DV, Shifren JL, Krawjewski S, Reed JC, Tilly JL, Isaacson KB. Differential expression of members of the *bcl-2* gene family in proliferative and secretory human endometrium: glandular epithelial cell apoptosis is associated with increased expression of *bax*. J Clin Endocrinol Metab 1997;82:(In Press).
13. Walker NI, Bennett RE, Kerr JFR. Cell death by apoptosis during involution of the lactating breast in mice and rats. Am J Anat 1989;185:19–32.
14. Chun S-Y, Billig H, Tilly JL, Furuta I, Tsafriri A, Hsueh AJW. Gonadotropin suppression of apoptosis in cultured preovulatory follicles: mediatory role of endogenous insulin-like growth factor-I. Endocrinology 1994;135:1845–53.
15. Tilly JL, Tilly KI. Inhibitors of oxidative stress mimic the ability of follicle-stimulating hormone to suppress apoptosis in cultured rat ovarian follicles. Endocrinology 1995;136:242–52.
16. Dharmarajan AM, Goodman SB, Tilly KI, Tilly JL. Apoptosis during functional corpus luteum regression: evidence of a role for chorionic gonadotropin in promoting luteal cell survival. Endocr J (Endocrine) 1994;2:295–303.
17. Hughes Jr FM, Fong Y-Y, Gorospe WC. Interleukin-6 stimulates apoptosis in FSH-stimulated rat granulosa cells in vitro: development and utilization of an in vitro model. Endocrine 1996;2:997–1002
18. Pesce M, Farrace MG, Piacentini M, Dolci S, De Felici M. Stem cell factor and leukemia inhibitory factor promote primordial germ cell survival by suppressing programmed cell death (apoptosis). Development 1993;118:1089–94.
19. De Miguel M, De Boer-Brouwer M, Paniagua R, van den Hurk R, De Rooj DG, van Dissel-Emiliani FMF. Leukemia inhibitory factor and ciliary neurotropic factor promote the survival of Sertoli cells and gonocytes in a coculture system. Endocrinology 1996;137:1885–93.
20. Martimbeau S, Manganaro TK, Donahoe PK, Tilly JL. Preliminary characterization of an in vitro model to elucidate the regulation of germ cell apoptosis in the intact fetal rat ovary. J Soc Gynecol Invest 1996;3(Supplement):216A.
21. Tilly JL, Billig H, Kowalski KI, Hsueh AJW. Epidermal growth factor and ba-

sic fibroblast growth factor suppress the spontaneous onset of apoptosis in cultured rat ovarian granulosa cells and follicles by a tyrosine kinase-dependent mechanism. Mol Endocrinol 1992;6:1942–50.

22. Luciano AM, Pappalardo A, Ray C, Peluso JJ. Epidermal growth factor inhibits large granulosa cell apoptosis by stimulating progesterone synthesis and regulating the distribution of intracellular free calcium. Biol Reprod 1994;51:646–54.

23. Flaws JA, DeSanti A, Tilly KI, Javid RO, Kugu K, Johnson AL, Hirshfield AN, Tilly JL. Vasoactive intestinal peptide-mediated suppression of apoptosis in the ovary: potential mechanisms of action and evidence of a conserved anti-atretogenic role through evolution. Endocrinology 1995;136:4351–9.

24. Ilio KY, Sensibar JA, Lee C. Effects of TGF-β1, TGF-α, and EGF on cell proliferation and cell death in rat ventral prostatic epithelial cells in culture. J Androl 1995;16:482–90.

25. Moulton BC. Transforming growth factor-β stimulates endometrial stromal apoptosis in vitro. Endocrinology 1994;134:1055–60.

26. Tilly JL, Flaws JA, DeSanti A, Kugu K, Rubin JS, Hirshfield AN. Role of intrafollicular growth factors in maturation and atresia of rat ovarian follicles. Biol Reprod 1995;52(Supplement):159.

27. Quirk SM, Cowan RG, Joshi SG, Henrikson KP. Fas antigen-mediated apoptosis in human granulosa/luteal cells. Biol Reprod 1995;52:279–87.

28. Hakuno N, Koji T, Yano T, Kobayashi N, Tsutsumi O, Taketani Y, Nakane PK. Fas/APO-1/CD95 system as a mediator of granulosa cell apoptosis in ovarian follicle atresia. Endocrinology 1996;137:1938–48.

29. Boudreau N, Sympson CJ, Werb Z, Bissell MJ. Suppression of ICE and apoptosis in mammary epithelial cells by extracellular matrix. Science 1995;267:891–93.

30. Pullan S, Wilson J, Metcalfe A, Edwards GM, Goberdham N, Tilly JL, Hickman JA, Dive C, Strueli CH. Requirement of basement membrane for the suppression of programmed cell death in mammary epithelium. J Cell Sci 1996;109:631–42.

31. Tilly KI, Banerjee S, Banerjee PP, Tilly JL. Expression of the p53 and the Wilms' tumor suppressor genes in the rat ovary: gonadotropin repression in vivo and immunohistochemical localization of nuclear p53 protein to apoptotic granulosa cells of atretic follicles. Endocrinology 1995;136:1394–402.

32. Keren-Tal I, Suh B-S, Dantes A, Lindner S, Oren M, Amsterdam A. Involvement of p53 expression in cAMP-mediated apoptosis in immortalized granulosa cells. Exp Cell Res 1995;218:283–95.

33. Fraser HM, Lunn SF, Cowen GM, Illingworth PJ. Induced luteal regression in the primate: evidence for apoptosis and changes in c-myc protein. J Endocrinol 1995;147:131–7.

34. Richards JS. Hormonal control of gene expression in the ovary. Endocr Rev 1994;15:725–51.

35. Koshimizu U, Watanabe M, Nakatsuji N. Retinoic acid is a potent growth activator of mouse primordial germ cells in vitro. Develop Biol 1995;168:683–5.

36. Flaws JA, Hirshfield AN, Tilly JL, DeSanti AM, Davis MA. Activation of mitogen-activated protein kinases during follicular atresia and survival. Biol Reprod 1996;54(Supplement):87.

37. Evan GI, Wyllie AH, Gilbert CS, Littlewood TD, Land H, Brooks M, Waters CM, Penn LZ, Hancock DC. Induction of apoptosis in fibroblasts by c-myc protein. Cell 1992;69:119–28.

38. Yonish-Rouach E, Resnitzky D, Lotem J, Sachs L, Kimchi A, Oren M. Wild-type p53 induces apoptosis of myeloid leukaemic cells that is inhibited by interleukin-6. Nature 1991;352:345.

39. Yang Y, Vacchio MS, Ashwell JD. 9-cis-Retinoic acid inhibits activation-driven T-cell apoptosis: implications for retinoid X receptor involvement in thymocyte development. Proc Natl Acad Sci USA 1993;90:6170–4.

40. Xia Z, Dickens M, Raingeaud J, Davis RJ, Greenberg ME. Opposing effects of ERK and JNK-p38 MAP kinases on apoptosis. Science 1995;270:1326–31.

41. Verheij M, Bose R, Lin XH, Yao B, Jarvis WD, Grant S, Birrer MJ, Szabo E, Zon LI, Kyriakis JM, Haimovitz-Friedman A, Fuks Z, Kolesnick RN. Requirement for ceramide-initiated SAPK/JNK signalling in stress-induced apoptosis. Nature 1996;380:75–9.

42. Tilly JL, Tilly KI, Kenton ML, Johnson AL. Expression of members of the *bcl-2* gene family in the immature rat ovary: equine chorionic gonadotropin-mediated inhibition of granulosa cell apoptosis is associated with decreased *bax* and constitutive *bcl-2* and *bcl-x$_{LONG}$* messenger ribonucleic acid levels. Endocrinology 1995;136:232–41.

43. Flaws JA, Kugu K, Trbovich AM, DeSanti A, Tilly KI, Hirshfield AN, Tilly JL. Interleukin-1β-converting enzyme-related proteases (IRPs) and mammalian cell death: dissociation of IRP-induced oligonucleosomal endonuclease activity from morphological apoptosis in granulosa cells of the ovarian follicle. Endocrinology 1995;136:5042–53.

44. Martimbeau S, Tao XJ, Tilly KI, Tilly JL. Enhanced expression of the *dad-1* death repressor gene in rat ovarian granulosa cells during gonadotropin-promoted follicular survival. Proceedings from the 10th International Congress of Endocrinology, San Francisco, CA 1996;p 751.

45. Ratts VS, Flaws JA, Kolp R, Sorenson CM, Tilly JL. Ablation of *bcl-2* gene expression decreases the numbers of oocytes and primordial follicles established in the post-natal female mouse gonad. Endocrinology 1995;136:3665–8.

46. Knudson CM, Tung KSK, Tourtellotte WG, Brown GAJ, Korsmeyer SJ. Bax-deficient mice with lymphoid hyperplasia and male germ cell death. Science 1995;270:96–9.

47. Knudson CM, Tung KSK, Flaws JA, Brown GAJ, Tilly JL, Korsmeyer SJ. Oocyte survival but spermatocyte cell death in BAX-deficient mice. Proceedings from the International Symposium on Cell Death in Reproductive Physiology, Chicago, IL 1996;p 33.

48. Maravei DV, Trbovich AM, Perez GI, Tilly KI, Talanian RV, Banach D, Wong WW, Tilly JL. Cleavage of cytoskeletal proteins by caspases during ovarian cell death: evidence that cell-free systems do not always mimic apoptotic events in intact cells. Cell Death Differ 1997;4:(In Press).

49. Flaws JA, Wang S, Miura M, Tilly JL, Yuan J. Reduced ovarian follicle endowment and delayed activation of primordial follicle growth in mice lacking the CED-3/ICE homolog, ICH-3. Proceedings from the International Symposium on Cell Death in Reproductive Physiology, Chicago, IL 1996;p 30.

50. Henkart PA, Grinstein S. Apoptosis: mitochondria resurrected? J Exp Med 1996;183:1293–5.

51. Hockenberry DM, Oltvai ZN, Yin X-M, Milliman CL, Korsmeyer SJ. Bcl-2 functions in an antioxidant pathway to prevent apoptosis. Cell 1993;75:241–51.

52. Steinman HM. The Bcl-2 oncoprotein functions as a pro-oxidant. J Biol Chem 1995;270:3487–90.
53. Buttke TM, Sandstrom PA. Oxidative stress as a mediator of apoptosis. Immunol Today 1994;15:7–10.
54. Mattison DR. Morphology of oocyte and follicle destruction by polycyclic aromatic hydrocarbons. Toxicol Appl Pharmacol 1980;53:249–59.
55. Perez GI, Knudson CM, Brown GAJ, Korsmeyer SJ, Tilly JL. Resistance of BAX-deficient mouse oocytes to apoptosis induced by 7,12-dimethylbenz(a)anthracene (DMBA) in vitro. Toxicologist 1997;36(Supplement):250.
56. Tilly JL, Perez GI. Mechanisms and genes of physiological cell death: a new direction for toxicological risk assessments? In: Sipes IG, McQueen CA, Gandolfi AJ, eds. Comprehensive toxicology. Oxford (England):Elsevier Press 1997;10:389–95.
57. Springer LN, McAsey ME, Flaws JA, Tilly JL, Sipes IG, Hoyer PB. Involvement of apoptosis in 4-vinylcyclohexene diepoxide-induced ovotoxicity in rats. Toxicol Appl Pharmacol 1996;139:394–401.
58. Springer LN, Tilly JL, Sipes IG, Hoyer PB. Enhanced expression of *bax* in small preantral ovarian follicles during 4-vinylcyclohexene diepoxide-induced ovotoxicity in the rat. Toxicol Appl Pharmacol 1996;139:402–10.

2

C. elegans as a Model System for Germ Cell Death

TINA L. GUMIENNY AND MICHAEL O. HENGARTNER

Programmed cell death plays a vital role in the life of an organism. The developing organism is shaped not only by its cells, but also by the cells it removes through programmed cell death. Programmed cell death also helps regulate homeostasis in a growing number of tissues; the disruption of this pathway can prevent the culling of potentially dangerous cells (facilitating the production of cancer and autoimmune disease) or remove cells that should not die (leading to degenerative diseases and other pathologies) (1). Programmed cell death is particularly abundant in the mammalian germline. For example, over 99.9% of all primary germ cells in women fail to ovulate, instead dying by a process called atresia, which has recently been recognized to be apoptotic (2, 3). Programmed cell deaths also occur in the germlines of other species, both vertebrate and invertebrate (reviewed in 3). The broad conservation of germ cell death suggests that the molecular mechanisms controlling this process may also be similar across evolutionary lines.

This chapter reviews germ cell death in the small nematode, *Caenorhabditis elegans*, and discusses the possibility of using this species as a model system to understand germ cell apoptosis in humans.

Programmed Cell Death in *C. elegans*

The nematode *C. elegans* has firmly established itself as a model system in which to study programmed cell death (Fig. 2.1) (reviewed in 4). During the development of the animal, 131 of the 1090 somatic cells created reproducibly undergo programmed cell death, and the identity and timing of each death is known (5). Genetic dissection of the *C. elegans* cell death pathway has revealed a number of mutations that affect this process, defining at least thirteen genes that can be ordered in a multistep genetic pathway (Fig. 2.2). The execution of the death sentence is carried out through the activity of *ced-3* (cell death abnormal) and *ced-4*, and is stayed by *ced-9*. If either *ced-3*

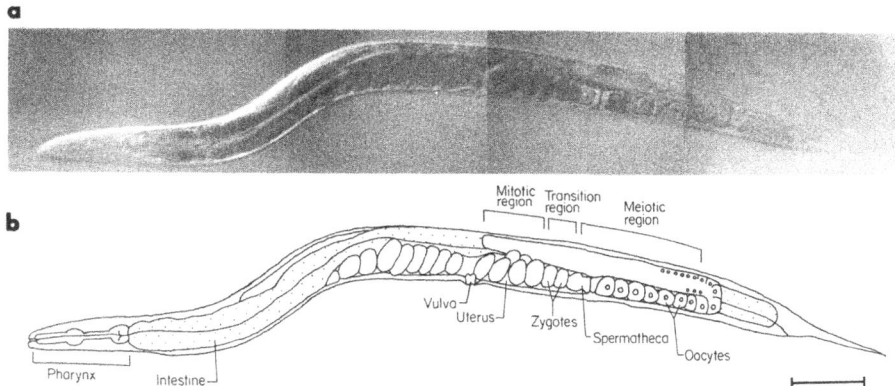

FIGURE 2.1. *C. elegans* adult hermaphrodite. Anterior is at left, dorsal on top. (a) Composite photograph. (b) Diagram of (a). Bar = 100μm.

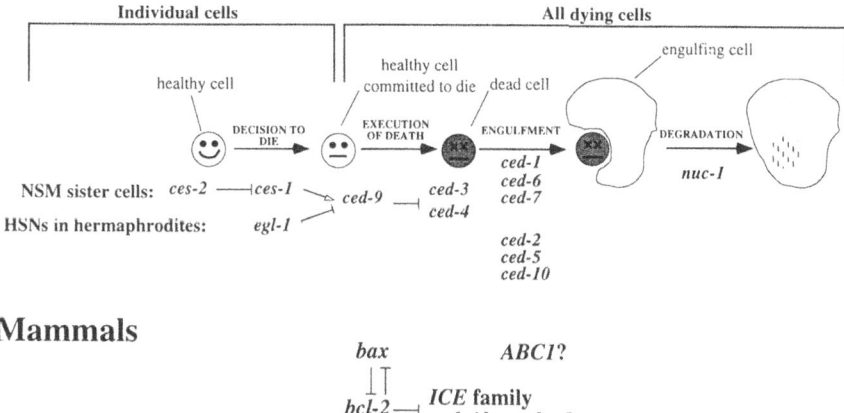

FIGURE 2.2. The programmed cell death pathway in *C. elegans* and mammals. Mutations in at least 13 genes define four steps in the process of programmed cell death in *C. elegans*. The first stage is cell specific: genes defining this step determine whether certain cells should live or die. The last three stages are executed in all normal programmed cell deaths. The mammalian cell death genes that are (and may be) conserved in *C. elegans* are shown. Gene names are explained in the text. HSN = hermaphrodite specific neuron. NSM = neurosecretory motoneuron. Genetic interactions are represented as ———▶, positive regulation; ———|, negative regulation.

or *ced-4* function is lost, cells do not die. When the function of *ced-9* is lost, many cells that normally survive undergo programmed cell death. At least six genes are required for the efficient engulfment of dying cells, and an endonuclease activity degrades the DNA of engulfed corpses in the final step (6, 7). Several additional genes act upstream of the general cell death pathway. These genes affect the decisions of small numbers of cells to live or die and are thought to be involved in the cell type-specific control of the death program.

Programmed cell death in *C. elegans* bears the characteristics marking mammalian apoptotic cell death, suggesting a similar mechanism of death. Indeed, the mammalian death-repressor, Bcl-2, has its homolog in the cell survival gene product, CED-9. The death-inducing interleukin-1-beta converting enzyme (ICE) protease family is represented by CED-3, and an ATP-binding cassette (ABC) transporter required to phagocytose corpses in mammals may be homologous to the *C. elegans* engulfment gene product, CED-7 (Fig. 2.2) (8–10).

The *C. elegans* Germline

The germ line is the most important tissue in *C. elegans*, both quantitatively and qualitatively. Adult worms contain 1000 to 2000 germ cells (11–13). Thus, adult hermaphrodites have, on average, more cells in their germ line than in all their somatic tissues combined (adult *C. elegans* hermaphrodites have 959 somatic cell nuclei, while adult males have 1031) (23). Unlike somatic development, the number of germ cells and the pattern of germ cell divisions are not fixed. The number of germ cells increases continuously throughout the animal's larval development and adult life.

The development of the gonad and germline is well documented (11, 13, 14). Only four gonadal cells are present in larvae after hatching (Fig. 2.3). Two of these cells (Z1 and Z4) will divide during larval development to produce the somatic gonad (uterus, spermatheca, and sheath cells), whereas the two other cells (Z2 and Z3) are the germline precursors. During the early larval stages, all germ cells divide mitotically. Starting in the third larval stage, some germ cells leave the mitotic cycle and enter meiosis. In hermaphrodites, the first few dozen cells to enter meiosis undergo spermatogenesis. By the time the animals reach adulthood, production has switched over to oogenesis, and all subsequent germ cells become oocytes. By contrast, males produce only sperm. Development of the germline is coordinated with the development of the somatic gonad such that animals become sexually mature right after the molt from the fourth larval stage to adulthood (11, 14).

In adult *C. elegans* hermaphrodites, the ovaries (or more precisely the ovotestes) consist of two symmetrical, U-shaped tubes, which are linked by a common uterus (Fig. 2.3) (11). Each tube can be divided into a proximal (near the uterus) and distal (away from the uterus) arm. Mitotic germ cells

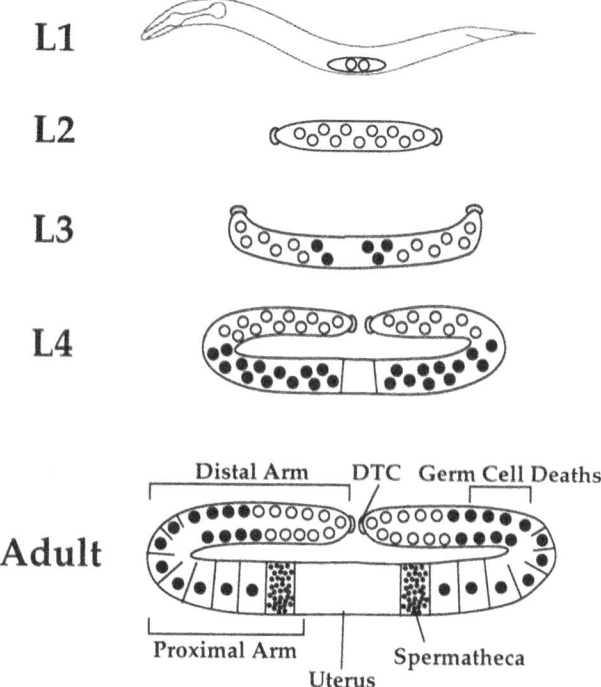

L1

L2

L3

L4

Adult

Distal Arm   DTC   Germ Cell Deaths

Proximal Arm

Uterus

Spermatheca

FIGURE 2.3. Schematic of gonad morphology and cell type distribution at consecutive stages of *C. elegans* hermaphrodite development. Mitotic germ nuclei are represented by large open circles, meiotic nuclei by large closed circles. Sperm are indicated by small open circles and are located within the spermatheca. Larval stages are indicated by L1–L4. DTC = distal tip cell.

are present in the distal end of each tube. These cells are kept in the mitotic state through a signal emanating from the distal tip cell (DTC), a somatic cell capping the distal tip of each tube. As dividing germ cells are pushed further away from the DTC signal, they enter meiosis and progressively mature as they migrate towards the spermatheca and uterus. Thus, each gonad contains germ cells at various stages of differentiation, progressing from mitosis at the distal end to mature gametes at the proximal end.

An interesting peculiarity of the *C. elegans* germline is that the mitotic and early meiotic nuclei are joined in a single large syncytial cell. Maturing spermatocytes and oocytes bud off from this syncytium, with the oocytes incorporating a large amount of the common cytoplasm during their cellularization (becoming 250 times larger than the syncytial compartments) (11). Oocytes usually start to enlarge and cellularize at the bend located between the two gonad arms (Fig. 2.3). The proximal arm consists of a single row of mature oocytes, which become fertilized when they pass through the spermatheca (where the mature sperm is stored) to enter the uterus.

C. elegans as a Model System for Germ Cell Death

A number of features make the nematode an attractive model system in which to study programmed cell death in the germline. The entire nematode germline (in living or fixed animals) is easily visualized under high-power light microscopy. The animals are inexpensive and easily maintained, have a short generation time, have the ability to produce self-progeny or to outcross with males, have large broods, are readily mutagenized, and should have a fully sequenced genome by 1998 (15–17). These features allow the rapid identification and isolation of genes of interest. In addition, the germline of C. elegans, when compared to the soma, possesses an important mammalian cell feature: germ cell fates are not rigid, as they can be influenced by endogenous gene products as well as by supplies of sperm and food (18). Protein, DNA, or RNA can be introduced into the germline by microinjection and the effects can be seen on germ cells or the resultant progeny (19, 20).

Features of Germ Cell Death

Germ cell corpses, visualized using Nomarski optics and electron microscopy, resemble those seen during the development of wild-type animals (Fig. 2.4 and unpublished observations). A germline nucleus undergoing programmed cell death cellularizes away from the syncytium, contains less cytoplasm than neighboring nuclear compartments, and becomes more refractile. DAPI and Hoechst 33342 stainings indicate that chromatin condensation occurs in these germ cells. Under normal growth conditions, germ cell death is restricted to adult hermaphrodites, and is never seen in larvae or wild-type males. However, corpses are consistently seen throughout the egg-bearing time, suggesting that germ cell death occurs continuously in adult hermaphrodites (unpublished observations). Germ cell deaths occur exclusively in the meiotic region (at or just after pachytene arrest) of the gonad, and the corpses are normally rapidly engulfed and degraded (usually before they reach the oocyte-cellularization region at the gonad bend) (21).

Programmed Cell Death Genes Regulate Germline Death

Analyses of germ cell death in the various cell death mutants indicate that the genes functioning in somatic deaths also affect germ cell death, suggesting that the same molecular program is used to eliminate cells in both the soma and the germline (7, 21, 22, unpublished observations).

For example, ced-3 and ced-4 are required for cell death in the soma (Fig. 2.2) (21). Strong loss-of-function mutations in ced-3 and ced-4 prevent all

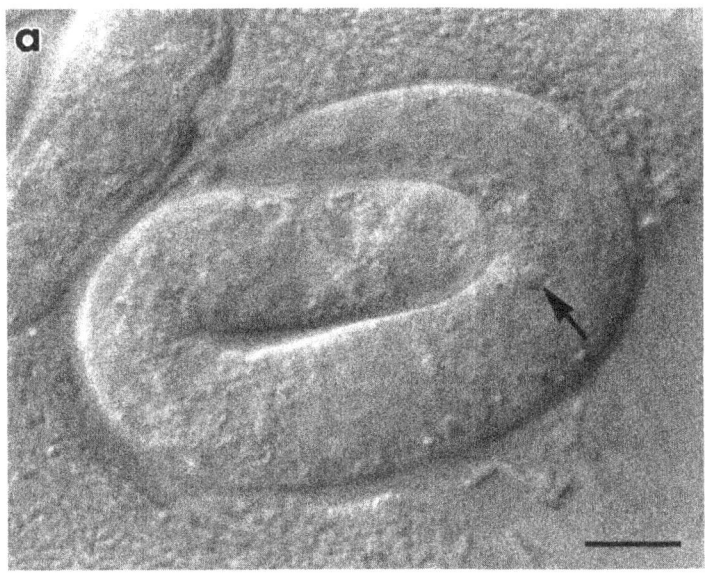

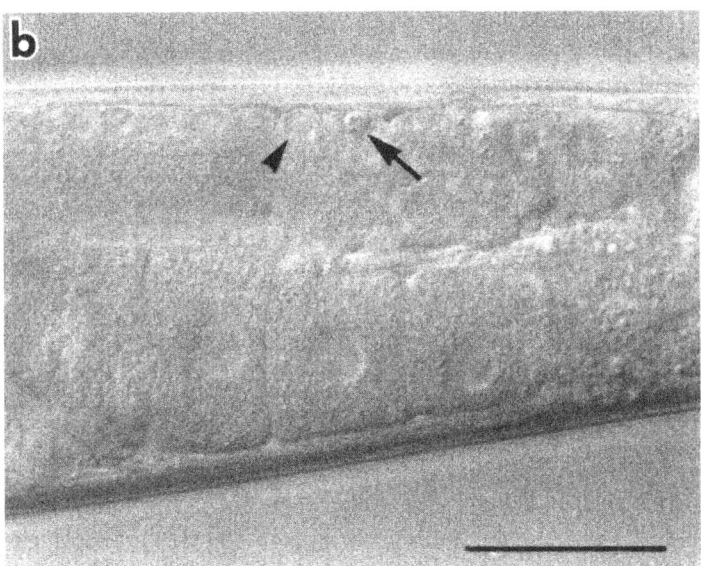

FIGURE 2.4. Cell death in *C. elegans* observed under differential interference contrast optics. Corpses are indicated by arrows. (a) Somatic cell death in the head of an embryo about to hatch. This corpse has condensed and rounded up, becoming more refractile. Bar = 10μm. (b). Germ cell death in an adult hermaphrodite. Note that this corpse has cellularized away from the common cytoplasm, condensed, and become more refractile than surrounding germ cell nuclei (arrowhead). Bar = 100μm.

developmental cell deaths in *C. elegans* (21). Almost no germ cell deaths are observed in strong loss-of-function *ced-3* or *ced-4* mutant animals (unpublished observations), indicating that both genes are also required for programmed death of germ cells.

By comparison *ced-9* protects cells that should live from apoptotic death. Mutations in *ced-9* that reduce or eliminate its function allow excessive cell death to occur (22). Null mutations confer maternal-effect lethality, where homozygous mutant offspring of a heterozygous parent survive, but self-progeny of a homozygous mutant parent die as embryos. Interestingly, homozygous mutants of a heterozygous mother may survive, but they lay fewer embryos, suggesting that *ced-9* function affects not only viability, but also fertility. Indeed, germ cell death is increased in *ced-9(lf)* animals, suggesting that *ced-9* has a protective function in the germline, as it does in the soma (unpublished observations). The increased death in the germline likely explains the low number of embryos laid by *ced-9* mutants.

Efficient engulfment of apoptotic corpses in *C. elegans* requires the function of at least six genes: *ced-1*, *ced-2*, *ced-5*, *ced-6*, *ced-7*, and *ced-10* (15). In animals mutant for any one of these genes, many dying cells fail to be recognized and persist as apoptotic bodies for hours or even days. Many more germ cell corpses can be observed in adult hermaphrodites mutant for any of these genes, suggesting that these genes also function to promote the engulfment of dying germ cells by the surrounding somatic sheath cells (unpublished observations). Older animals can accumulate over 70 corpses in one arm (48 to 60 hours after the L4/adult molt). Old corpses lose the smooth appearance of newly formed corpses, swell, and slowly disintegrate, indicating that these germ cell corpses are undergoing "secondary necrosis," a process that has previously been observed in mammalian apoptotic bodies that escape ingestion (23). Cell debris also builds up before the spermatheca and in the uterus, possibly the result of cell collapse as corpses are forced into the spermatheca by the progression of oocytes. Alternatively, corpses may not be able to withstand the increase in cell volume and may lyse.

The gene *nuc-1* (nuclease-1) is required for proper degradation of apoptotic corpses in the last step of the *C. elegans* cell death pathway. The *nuc-1* mutants lack the major endonuclease present in *C. elegans* (24). This nuclease functions in the intestinal lumen to digest bacterial DNA (*C. elegans* feeds on bacteria) and in the engulfing cells to degrade somatic corpse DNA. In a wild-type animal, germ cell corpses are engulfed by the sheath cell and degraded through the action of *nuc-1*. In *nuc-1* mutants, the genomic DNA of apoptotic cells persists within sheath cell vacuoles. Engulfment is required for the *nuc-1*-dependent digestion of DNA: in an engulfment mutant (which is wild-type for the nuclease), germ cell corpses are not phagocytosed and their DNA is not degraded (unpublished observations).

In addition to the genes that function in all cell deaths, a number of genes have been identified that only affect the death of small numbers of cells. For example, the genes *ces-1* and *ces-2* (cell death specification) affect the deaths of specific pharyngeal cells that undergo programmed cell death in wild-type animals (25). Similarly, dominant mutations in the *egl-1* (egg laying defective) gene cause the two hermaphrodite specific neurons (HSNs) to die in hermaphrodites (these two neurons show sexual dimorphism and normally die in wild-type males) (19). These genes might be required for individual cell types to decide whether they should live or die (i.e., they might function in the cell type-specific regulation of the ubiquitously-present death machinery). No differences in germline corpse numbers are observed between any of these cell-type specific mutants and the wild type, indicating that these genes have no influence on germ cell death (unpublished observations). This suggests that the cell death machinery in the germline may be controlled by another, as yet unidentified, cell death regulator.

Germ Cell Death Occurs Only With Oogenesis

Under normal growth conditions, germline programmed cell death occurs only in adult hermaphrodites. Analysis of germ cell death in mutants with aberrant sexual development or identity suggests that these deaths are specifically associated with oogenesis (unpublished observations and 26). For example, mutants in which the differentiation of germ cells into sperm and oocytes is prevented have no programmed cell death in the germline, suggesting that gamete production is required for germ cell death. In masculinized hermaphrodites that produce sperm but no oocytes, no germ cell corpses are seen. If the hermaphrodite germline is feminized, all differentiated germ cells are oocytes and corpses are observed in both virgin and mated animals. A feminized male animal will produce oocytes and will also accumulate corpses. These data suggest that neither the sex chromosome complement nor the somatic phenotype of the animal is the deciding factor for the induction of cell death. Rather, germ cell death requires the differentiation of germ cells into oocytes. However, full maturation of oocytes is not required for death as mutations that disrupt late events in oogenesis do not affect germ cell death (for instance, animals producing aberrant oocytes also have germ cell deaths).

Why should germ cell death be specifically associated with oogenesis? Germ cell death is not likely to occur solely to eliminate aberrant germ cells. DAPI staining should reveal such events (for instance, mitotic catastrophe), yet all pachytene-arrested nuclei (except for already dying ones) look alike when stained. Furthermore, in animals in which germ cell death has been stymied (e.g., *ced-3* and *ced-4* mutants), there is no apparent decrease in fertility or observable increase in embryonic lethality, suggesting that the germ cells that would have died have the potential to form normal, functional oocytes (unpublished observations). Rather, all syncytial germ cells may not only be presumptive oocytes, but also nurse cells. Since the thousands of germ cell nuclei cannot be accommodated as oocytes, the extra ones are eliminated through programmed cell death. Consistent with this model, corpses can be observed in the meiotic part of the gonad soon after oogenesis begins, but few to no corpses are found in wild-type larvae and males, neither of which make oocytes (larvae are not mature enough to generate oocytes; males have no cell death because the germ cells mature into smaller sperm, which provide no size problem) (18).

The Process of Germ Cell Death Is Conserved Among Nematode Species

Nematode germ cell death is not a phenomenon specific to *C. elegans*. Three related species (*C. vulgaris*, *C. remanei*, and *C. briggsae*), which are as diverged from *C. elegans* as cats are from mice, also have germline corpses, indicating that germline programmed cell death is evolutionarily conserved among nematodes (unpublished observations and 27). *Pelodera strongyloides*, a more distant gonochoristic nematode relative, also has programmed cell death in its germline. In all of these cases, germ cell deaths are associated with the presence of oocytes (28). In addition, cell deaths have been reported in the mitotic germ line of the nematode *Panagrellus redivivus* (28). This conservation of cell death among nematodes, taken with observations of cell death in the ovaries of all studied higher animals, indicates that programmed cell death may be a common phenomenon in the germline of all metazoans (Chapter 3) (3).

Cell Death Accounts for Over One-Half of all Germ Cell Fates

One virgin hermaphrodite produces about 150 to 160 sperm in each of its two gonad arms, limiting its egg production to about 300 (15). Scores of corpses are retained in engulfment mutant germlines, indicating that death is a common and frequent fate for germ cells. Even so, it is likely that some germ cell deaths are recognized and engulfed, since many somatic corpses are properly phagocytosed in engulfment mutants (7). Indeed, some double mutants have nearly twice the germ cell corpse count as single mutants (over 100 corpses in a single arm at one time point), suggesting that about one-half of all germ cells that could have become oocytes die instead (unpublished observations). The kinetics of *C. elegans* germ cell death also support this theory, in that more germ cell corpses are produced than eggs laid over the reproductive life of an animal (unpublished observations and 29). The conclusion that cell death is a major fate for nematode germ cells may be an evolutionary reflection of the massive amount of cell death occurring in mammalian ovaries, where greater than 90% of all postnatal germ cells die (3).

Conclusion

C. elegans has proved to be a powerful tool in the genetic dissection of programmed cell death in somatic tissues. Growing evidence suggests that survival of cells in the *C. elegans* reproductive system (as well as in vertebrate reproductive systems) is regulated by many of the same genes shown to control survival of somatic cells. In *C. elegans*, the cell death program in the germline is dependent upon oocyte production, and accounts for over one-half of all germ cell outcomes. Significantly, the genetic program regulating cell death is not only conserved between soma and germline, it is conserved between mammals and nematodes. These findings indicate that knowledge obtained about *C. elegans* germ cell death may further our understanding of mammalian atresia, and justify the use of this organism as a model system in which to specifically study cell death in reproductive physiology.

Acknowledgments. We would like to thank Serono Symposia USA Inc. and the meeting's scientific committee—Jon Tilly, Jerome Strauss III, and Martin Tenniswood—for bringing this meeting and publication to fruition. Some nematode strains used in this research were provided by the *Caenorhabditis* Genetics Center, which is funded by the NIH National Center for Research Resources (NCRR). MOH is a Rita Allen Foundation scholar. Work in the authors' laboratory is supported by PHS grant R01-GM52540.

References

1. Thompson CB. Apoptosis in the pathogenesis and treatment of disease. Science 1995;267:1456–62.
2. Baker TG. Radiosensitivity of mammalian oocytes with particular reference to the human female. Am J Obstet Gynecol 1971;110:746–61.
3. Tilly JL. Apoptosis and ovarian function. Rev Reprod 1996;1:162–72.
4. Horvitz HR, Shaham S, Hengartner MO. The genetics of programmed cell death in the nematode *Caenorhabditis elegans*. Cold Spring Harb Symp Quant Biol 1994; 59:377–85.
5. Sulston JE, Horvitz HR. Post-embryonic cell lineages of the nematode, *Caenorhabditis elegans*. Dev Biol 1977;56:110–56.
6. Hedgecock EM, Sulston JE, Thomson JN. Mutations affecting programmed cell deaths in the nematode *Caenorhabditis elegans*. Science 1983;220:1277–9.
7. Ellis RE, Jacobson DM, Horvitz HR. Genes required for the engulfment of cell corpses during programmed cell death in *Caenorhabditis elegans*. Genetics 1991;129:79–94.
8. Hengartner MO, Horvitz HR. *C. elegans* cell survival gene *ced-9* encodes a functional homolog of the mammalian proto-oncogene *bcl-2*. Cell 1994;76:665–76.
9. Yuan J, Shaham S, Ledoux S, Ellis HM, Horvitz HR. The *C. elegans* cell death gene *ced-3* encodes a protein similar to mammalian interleukin-1 beta-converting enzyme. Cell 1993;75:641–52.

10. Luciani M-F, Chimini G. The ATP binding cassette transporter ABC1, is required for the engulfment of corpses generated by apoptotic cell death. EMBO J 1996;15:226–35.
11. Hirsh D, Oppenheim D, Klass M. Development of the reproductive system of *Caenorhabditis elegans*. Dev Biol 1976;49:200–19.
12. Klass M, Wolf N, Hirsh D. Development of the male reproductive system and sexual transformation in the nematode *Caenorhabditis elegans*. Dev Biol 1976;52:1–18.
13. Kimble JE, White JG. On the control of germ cell development in *Caenorhabditis elegans*. Dev Biol 1981;81:208–19.
14. Kimble J, Hirsh D. The postembryonic cell lineages of the hermaphrodite and male gonads in *Caenorhabditis elegans*. Dev Biol 1979;70:396–417.
15. Wood WB. Introduction to *C. elegans* biology. In: Wood WB, ed. The Nematode *Caenorhabditis elegans*. Cold Spring Harbor, NY: Cold Spring Harbor Laboratory Press, 1988;1–16.
16. Waterston R, Sulston J. The genome of *Caenorhabditis elegans*. Proc Natl Acad Sci USA 1995;92:10836–40.
17. Brenner S. The genetics of *Caenorhabditis elegans*. Genetics 1974;77:71–94.
18. Kimble J, Ward S. Germ-line development and fertilization. In: Wood WB, ed. The Nematode *Caenorhabditis elegans*. Cold Spring Harbor, NY: Cold Spring Harbor Laboratory Press, 1988;191–214.
19. Stinchcomb DT, Shaw JE, Carr SH, Hirsh D. Extrachromosomal DNA transformation of *Caenorhabditis elegans*. Mol Cell Biol 1985;5:3484–96.
20. Mello CC, Kramer JM, Stinchcomb DT, Ambros V. Efficient gene transfer in *C. elegans*: extrachromosomal maintenance and integration of transforming sequences. EMBO J 1991;10:3959–70.
21. Ellis HM, Horvitz HR. Genetic control of programmed cell death in the nematode *C. elegans*. Cell 1986;44:817–29.
22. Hengartner MO, Ellis RE, Horvitz HR. *Caenorhabditis elegans* gene *ced-9* protects cells from programmed cell death. Nature 1992;356:494–9.
23. Wyllie AH, Kerr JFR, Currie AR. Cell death: the significance of apoptosis. Int Rev Cytol 1980;68:251–306.
24. Hevelone J, Hartman PS. An endonuclease from *Caenorhabditis elegans*: partial purification and characterization. Biochem Genet 1988;26:447–61.
25. Ellis RE, Horvitz HR. Two *C. elegans* genes control the programmed deaths of specific cells in the pharynx. Development 1991;112:591–603.
26. Kuwabara PE, Kimble J. Molecular genetics of sex determination in *C. elegans*. Trends in Genetics 1992;8:164–8.
27. Fitch DHA, Bugaj-Gaweda B, Emmons SW. 18S ribosomal RNA gene phylogeny for some *Rhabditae* related to *Caenorhabditis*. Mol Biol Evolution 1995;12:346–58.
28. Sternberg PW, Horvitz HR. Gonadal cell lineages of the nematode *Panagrellus redivivus* and implications for evolution by the modification of cell lineage. Dev Biol 1981;88:147–66.
29. Gems D, Riddle DL. Longevity in *Caenorhabditis elegans* reduced by mating but not gamete production. Nature 1996;379:723–5.

3

Stem Cell Factor Regulation of Apoptosis in Mouse Primordial Germ Cells

MAURIZIO PESCE, MARIA GRAZIA FARRACE, ALESSANDRA AMENDOLA, MAURO PIACENTINI, AND MASSIMO DE FELICI

In most mammals extensive degeneration of germ cells occurs during the embryonic stages of oogenesis before follicle formation (1, 2). In the mouse embryo early morphological studies have shown two distinct periods of germ cell death affecting proliferating oogonia (12–13 days post coitum [dpc]) and oocytes at the zygotene/pachytene stage (from 16.5 dpc through birth) (3, 4). Following such episodes of degeneration the total number of germ cells decreases from about 30,000 on day 14 of fetal life to about 6000 to 10,000 at birth seven days later (2, 5, 6). In the present paper, we report the results of studies carried out in our laboratory designed to elucidate the mechanisms of this massive cell death which were obscure until now. Given the current views on the physiological cell death that affects various cell types during development, we have investigated: first if germ cell degeneration requires the activation of an intrinsic cellular program, second if such a program displays the sequence of morphological and biochemical events characteristic of programmed cell death (apoptosis), and third if activator(s) or inhibitor(s) of this death program could be identified.

In the mouse embryo, between 10.5 and 12.5 dpc primordial germ cells (PGCs) migrate into the gonadal ridges. After active proliferation, by 14.5 dpc female germ cells (now called oocytes) enter the prophase of meiosis and pass, in a couple of days, through leptotene and zygotene stages into the pachytene stage. We have investigated the characteristics of cell death in germ cells between 11.5 dpc and 1 week after birth using new methods employed to identify and quantify apoptotic cell degeneration in various cell types both in situ and in isolated cell populations.

Identification and Quantification of Degenerating Germ Cells in Female Embryos From 12.5 DPC Through Birth

The fragmentation of nuclear DNA that occurs in apoptotic cells can be detected histochemically by labeling the newly formed free 3'-ends of the DNA fragments by a technique termed TUNEL (7). Since preliminary experiments had shown that TUNEL histochemistry was able to detect apoptotic cells in embryonic tissues where programmed cell death normally occurs, we elected to use this method on tissue sections of ovaries to identify and quantify apoptotic germ cells from 12.5 dpc to 1 week after birth. About 100 germ cells from several random areas of an ovary (at least three ovaries for each age) were examined and the frequency of TUNEL positive cells was noted. The results, summarized in Table 3.1, showed that labeled cells were rare (< 1%) until 14 to 15.5 dpc. Afterwards, their number increased and was maximal at 18.5 dpc (about 5% to 8%) (Fig. 3.1A). Surprisingly, although TUNEL positive oocytes at the zygotene/pachytene or late pachytene stage often showed an apparently normal nuclear morphology, oocytes in advanced stages of degeneration (probably "secondary necrosis"), frequently observed in every section, were generally negative for TUNEL staining (Fig. 3.1B). Likewise, no TUNEL staining was noted in the nuclei of somatic cells of the ovary.

Since apoptotic cells generally express high levels of tissue transglutaminase (tTGase) (8), we immunolocalized this enzyme in sections of ovaries from 12.5 dpc through birth. No tTGase-immunostained germ cells were detected in the ovaries examined; however, some tTGase-positive PGCs (identified by alkaline phosphatase [APase] staining) were seen in extragonadal sites in embryos at 12.5 to 13.5 dpc (Table 3.1) (9).

In a subsequent series of experiments we analyzed cell populations isolated from 11.5–16.5-dpc gonads according to techniques currently used in our laboratory and described elsewhere (10, 11). Cells attached to poly-L-lysine coated slides were stained for both APase activity and nuclear morphology using the nuclear dye, Hoechst 33342. Germ cells were classified according to their morphology as normal, apoptotic (reduced cell size, small nucleus, pyknotic chromatin), or necrotic (swollen or disintegrating cytoplasm, karyolysis) by two independent observers (Fig. 3.2). Although the percentage of apoptotic cells was higher and more variable than those estimated in tissue sections by TUNEL staining, the overall pattern was similar using the two approaches (Table 3.1). It should be noted that peripheral chromatin caps and apoptotic bodies typical of apoptosis were observed in 11.5 to 12.5 dpc PGCs only. Cell suspensions from 15.5–16.5-dpc ovaries were further analyzed with a viability flow cytometric test consisting of double staining with ethidium bromide, which mostly stains DNA of dead cells, and acridine orange, which beside dead cells strongly stains the nucleic acids of living cells and poorly stains those of apoptotic cells (12). This assay re-

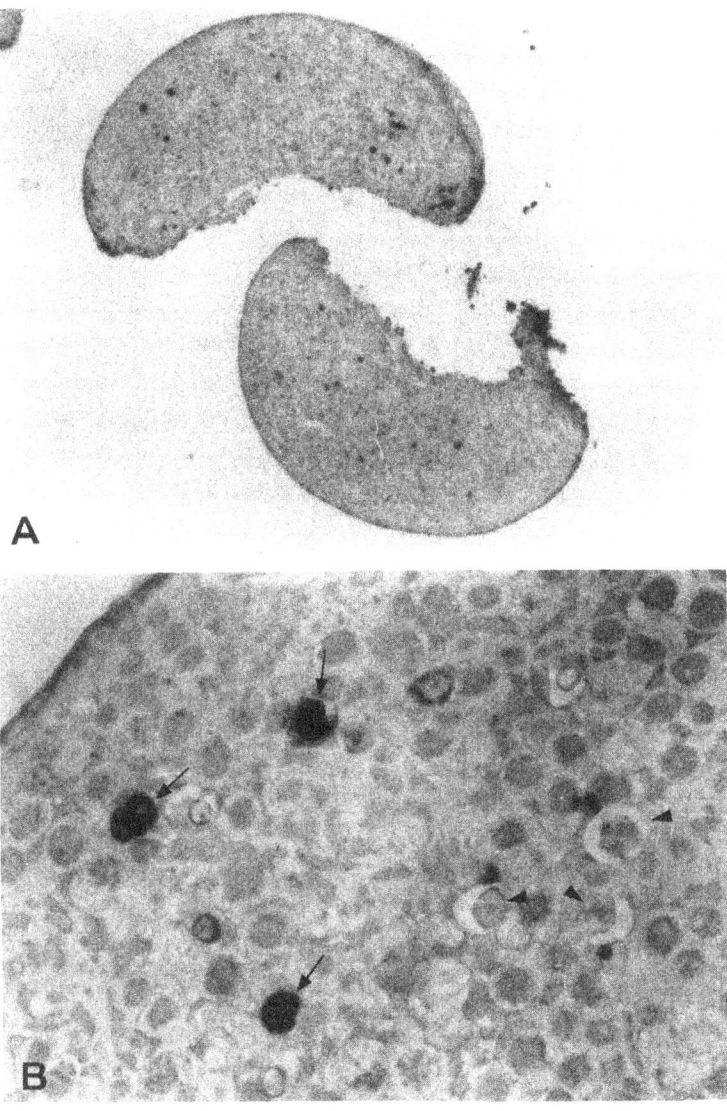

FIGURE 3.1. Paraffin sections of ovaries of 18.5 dpc mouse stained with the TUNEL technique (7). (A) Magnification approximately 220X; (B) Magnification approximately 1500X. Arrows and arrowheads indicate TUNEL-positive and degenerating TUNEL-negative oocytes, respectively.

vealed the presence of a large fraction of apoptotic cells (varying from 20 to 30%) (Fig. 3.3). Cell sorting followed by microscopic observation of cells stained for APase and Hoechst 33342 showed that the "apoptotic fractions" mainly contained somatic cells and about 25% to 35% oocytes (5% to 10% of the whole

TABLE 3.1. Quantification of apoptotic germ cells during pre- and perinatal mouse oogenesis. Tissue sections were labeled by TUNEL or tTGase immunohistochemistry, whereas germ cells isolated from gonads of mice of the same ages were double stained for Apase and with Hoechst 33342.

AGE	TUNEL	tTGase	Hoechst
11.5 dpc[a]	nd[b]	nd	2–3%
12.5 dpc	<1%	±	2–3%
15.5 dpc	1–2%	-[c]	4–5%
16.5 dpc	2–5%	-	5–8%
18.5 dpc	5–8%	-	8–15%
1 dpp[d]	4–5%	nd	nd
7 dpp	1–2%	nd	nd

[a]dpc = days post-coitum
[b]nd = not determined
[c]- = not detected
[d]dpp = days post-partum

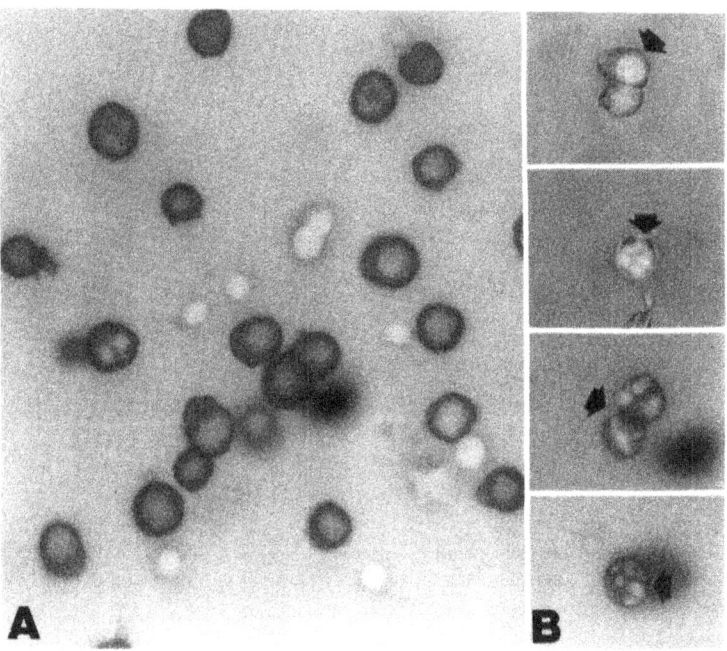

FIGURE 3.2. Photomicrographs of 11.5 dpc PGCs double stained for alkaline phospatase activity and with Hoechst 33342. (A) Normal PGC morphology; (B) Examples of the apoptotic morphology in PGCs in culture. Magnification, 500X.

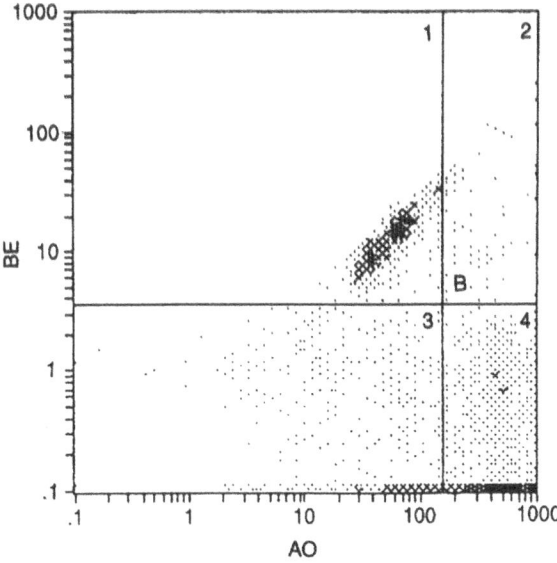

FIGURE 3.3. Example of FACS analysis of oocytes freshly collected from 15.5-dpc ovaries and double stained with acridine organce (AO) and ethidium bromide (BE) to identify and quantify necrotic (quadrant 1), apoptotic (quadrant 3), and viable (quadrant 4) cells (see ref. 12).

cell population) with an apparently normal nucleus (mostly at the zygotene/pachytene stage) or with features typical of apoptotic cells.

Taken together these results indicate that germ cells showing some features typical of apoptotic cells (DNA fragmentation, pyknotic nucleus, reduced acridine orange incorporation) are present in pre- and perinatal ovaries, mostly among oocytes at the zygotene/pachytene and late pachytene stages. These periods correspond with the waves of germ cell degeneration described in early morphological studies. Since even small numbers of apoptotic cells may be indicative of a high level of cell death (see, for instance, ref. 7), the estimated incidence of this process in situ by TUNEL (0.5% to 5%) and in isolated germ cells by morphological and FACS analysis (2% to 15%) in different periods of germ cell development, is consistent with the more than 50% germ cell death that can be calculated to occur during pre- and perinatal oogenesis. Therefore, the programmed nature of germ cell death by apoptosis seems very probable. In accord with our data, Coucouvanis et al. (13), using a flow cytometric analysis that detects the reduced DNA content of apoptotic cells, recently came to the same conclusion. Similarly, internucleosomal clevage of DNA detected in mouse ovaries between 15.5 dpc and day 0 postpartum (14) confirmed our results. However, it is to be noted that in oocytes apoptosis may be "atypical," lacking one or more of the features that characterize classical apoptosis (15). Apoptotic oocytes,

unlike proliferating PGCs, do not appear to show classical chromatin mar-
gination, do not express high levels of tTGase (see also next paragraph) and
seem to be retained in the ovary with their degraded nuclei before becoming
TUNEL negative and undergoing "secondary necrosis". Moreover, preliminary results suggest that some genes associated with apoptosis in other cell
types, such as *p53, c-myc,* and *bcl-2,* are not expressed, at least at levels
detectable by immunohistochemistry, in pre- and perinatal ovaries.

Germ Cell Death in In Vitro Cell Culture

A crucial factor in the biochemical analysis of programmed cell death and
its regulation during oogenesis is selection of a model in which the susceptible cell population(s) can be studied and manipulated easily and reproducibly. Our previous studies showed that PGCs obtained from 11.5-
and 12.5-dpc embryos undergo extensive loss of viability (as determined
by the erythrosin B exclusion test) when cultured for 1 day in the absence
of somatic cells (viability about 20% to 30%). On the other hand, under
the same culture conditions the viability of 15.5–16.5-dpc germ cells is
much higher (around 60% to 65%) (16, 17). We have used the in vitro culture system to characterize the sequence of events leading to PGC degeneration and to identify compounds able to influence this process. To this
end, PGCs were double labeled with the fluorescent nuclear dye, Hoechst
33342, and for APase activity. The results showed that while at the beginning of culture very few 11.5-dpc PGCs showed apoptotic features, after
5 to 6 h of culture a significant percentage of PGCs (about 10% to 20%)
exibited an apoptotic morphology, with a further increase at 16 to 18 h
(25% to 30%) (Fig. 3.4). At this time, many cellular fragments of PGCs,
as well as PGCs showing karyolysis and disintegrated cytoplasm, were
observed. These were considered to be cells undergoing "secondary necrosis."
Similar results were obtained with 12.5-dpc PGCs (data not shown). Electron microscopic observations with scanning and transmission electron
microscopes confirmed the morphological characteristics of apoptosis in
degenerating PGCs in culture (18). Moreover, TUNEL histochemistry (data
not shown), FACS analysis of DNA content with propidium iodide (PI)
according to the method reported by Nicoletti et al. (19) (Fig. 3.5), and the
formation of DNA "ladders" of low molecular weight nucleosomal fragments (9), proved that apoptosis in PGCs is accompanied by DNA degradation. We also observed that although PGCs at the beginning of culture
were negative for tTGase immunostaining, almost 60% showed strong
staining after 4 to 5 h (9).

We next studied the extent of apoptosis in 15.5–16.5-dpc oocytes that, as
reported above, maintain relatively good viability in culture. So far only
oocytes double labeled for APase and with Hoechst 33342, or labeled for
tTGase have been studied. Results from these studies indicate that oocytes in
leptotene or leptotene/zygotene stages are quite resistant to apoptotic degen-

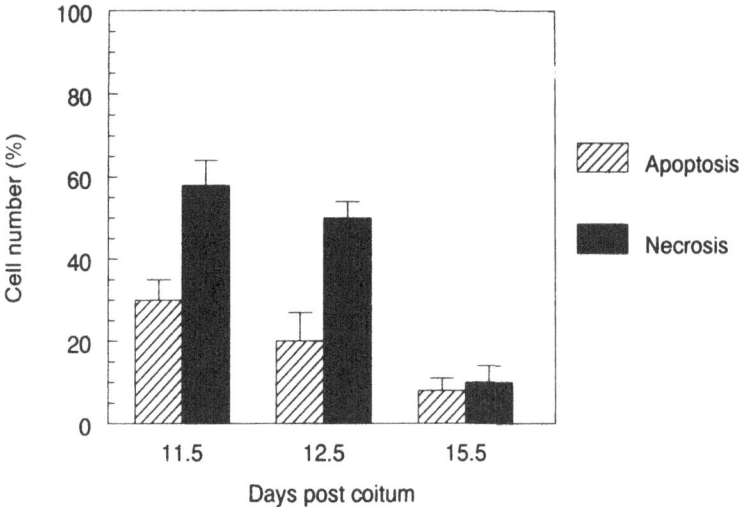

FIGURE 3.4. Morphological analysis of germ cells obtained from embryos of various ages and cultured for 16–18 h under the conditions reported in ref. 9. Cells were double stained for Apase and with Hoechst 33342, and then classified as apoptotic or necrotic according to the criteria described in the text. For each sample all attached cells were examined for a total of approximately 500 cells (at least three samples for each of several ovaries studied).

eration (Fig. 3.4) and do not express tTGase in culture (9). However, massive apoptosis occurs as they reach the pachytene stage after 2 to 3 days of culture. Degenerating pachytene nuclei were smaller than normal with very condensed chromatin (data not shown).

The Role of Stem Cell Factor in Germ Cell Survival

Studies performed with cells of various types in culture demonstrate that apoptosis can be initiated by depriving them of appropriate growth factors. In proliferating cells, it has been demonstrated that the presence of both mitogens to drive cell proliferation and survival factors to prevent apoptosis are necessary to maintain cell viability (20). The recent finding that stem cell factor (SCF) is essential for the survival and/or proliferation of 10.5 to 11.5 dpc mouse PGCs in culture (21, 22, 23), prompted us to examine the effect of this cytokine on apoptosis in cultured germ cells. Results presented in Figure 3.6 show the effect of 100 ng/mL SCF on PGCs over a 24 h culture period. It appears that in the presence of SCF the proportion of cells showing an apoptotic or necrotic morphology was significantly reduced. This effect was confirmed by flow cytometric analysis of 12.5 dpc PGCs double stained with ethidium bromide and acridine orange (Fig. 3.7). In addition, we found that in the presence of SCF the expression of tTGase in cultured PGCs was delayed by several hours (9). Lastly, SCF decreased DNA frag-

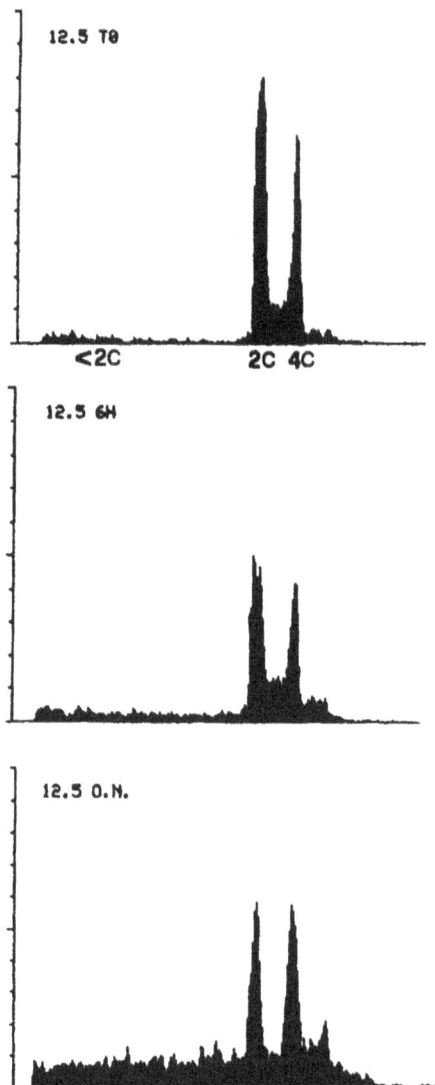

FIGURE 3.5. Example of FACS analysis with propidium iodide staining of DNA in 12.5-dpc PGCs in culture (19). A hypodiploid peak typical of apoptotic nuclei is barely detectable at the beginning of culture (TO), clearly visible after 6 h of culture (6H) and markedly increased after 16–18 h (O.N.).

mentation, as evaluated by either gel electrophoresis (9) or FACS analysis of PI-stained nuclei (data not shown).

Evaluation of nuclear morphology performed with 15.5–16.5-dpc oocytes over 1 to 4 days of culture showed that SCF almost completely prevented the wave of degeneration occurring in oocytes at the pachytene stage and favored the meiotic progression of most oocytes to the diplotene stage.

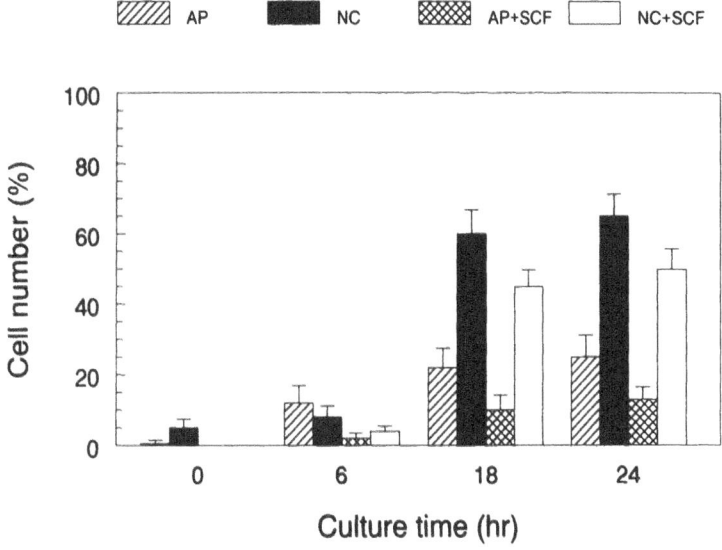

FIGURE 3.6. The effect of 100 ng/mL SCF on the occurrence of apoptotic and necrotic degeneration in 11.5-dpc PGCs after various culture times. Evaluation of cellular morphology was performed as described in legend X to Figure 3.4.

In an effort to elucidate the way(s) by which the presence or absence of SCF may influence the activation of the apoptotic program in PGCs, we found evidence that: first, under the culture conditions employed, SCF does not directly stimulate the proliferation of 11.5-dpc PGCs (24), and second, interaction of the membrane-associated form of SCF with its receptor, c-kit, acts to mediate or consolidate PGC adhesion to somatic cells (25).

This latter finding, coupled with the inhibitory effect of SCF on PGC apoptosis reported above, suggests that adhesive interactions between PGCs and somatic cells via SCF/c-kit are critical for germ cell survival in the embryo. In fact, under normal conditions, apoptosis is seldom detectable in PGCs within gonadal ridges (see previous section). However, apoptosis due to the absence of SCF/c-kit interactions can explain the lack of PGCs in dominant white-spotting (W[e]) or steel (Sl[d]) mutations, as well as the degeneration of PGCs that fail to reach the gonadal ridges. In addition, it is possible that germ cell tumours in vivo (26) and embryonic stem cell-like cells in vitro (27, 28) originate from PGCs when apoptosis fails and they are exposed to signals that alter the normal program of germ cell differentiation.

Is There a Role for *BCL-2* in Germ Cell Apoptosis?

There is emerging evidence that, at the molecular level, growth factor deprivation may lead cells to apoptosis by down regulation of survival genes (i.e., *bcl-2*) and/or up regulation of death genes (i.e., *bax*). Only limited data are

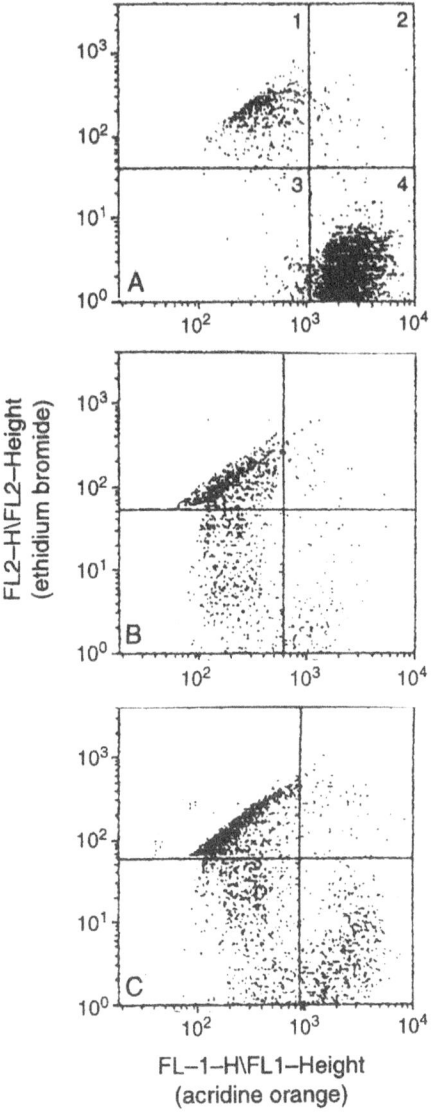

FIGURE 3.7. The effect of 100 ng/mL mouse SCF on the occurrence of apoptotic and necrotic degeneration is 12.5-dpc PGCs after 16–18h of culture as evaluated by the FACS analysis described in legend to Figure 3.3. (A) Beginning of culture; (B) After 16–18 h; (C) After 16–18 h of culture in the presence of SCF.

available regarding the regulation of cellular BCL-2 protein levels. In some cells the expression of the *bcl-2* gene is stimulated by signals that prevent their entry into apoptosis. In particular, Carson et al. (29), showed that SCF prevents apoptosis of human natural killer cells through upregulation of *bcl-2*. Moreover, Ratts et al. (14) reported that ablation of *bcl-2* gene expression

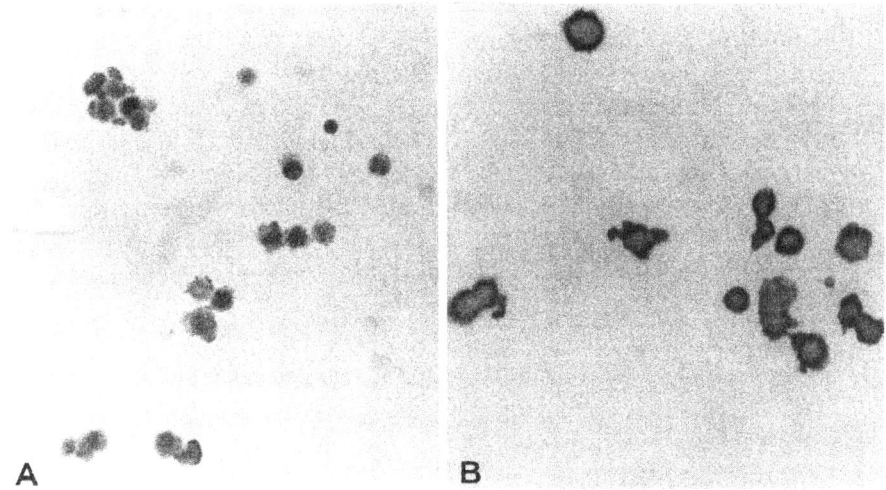

FIGURE 3.8. Expression of BCL-2 in 12.5-dpc PGCs in culture. Cells were not immunostained at the beginning of culture (A) but became strongly positive after 16–18 h of culure (B). In all experiments the 4C11 antibody (Santa Cruz Biotechnology, Inc.) against mouse BCL-2 p25 protein was employed. Magnification, 500X.

decreases the number of oocytes and primordial follicles endowed in mouse ovaries. Using RT-PCR (data not shown), immunohistochemistry, and Western blotting (Figs. 3.8 and 3.9), we did not obtain evidence that addition of

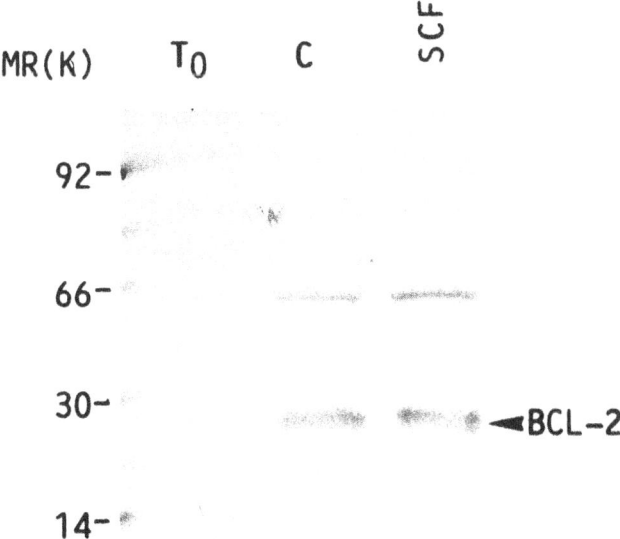

FIGURE 3.9. Western blot analysis of BCL-2 expression in germ cells freshly collected (To) or cultured for 16–18 h without (C) or with 100 ng/ml SCF. The antibody used was the same as that indicated in the legend to Figure 3.8.

soluble SCF to the culture medium results in upregulation of *bcl-2* in 11.5–12.5-dpc PGCs. Interestingly, in some experiments we found that although BCL-2 was not detectable in freshly isolated germ cells, significant amounts of the protein were expressed in both 12.5-dpc and 15.5-dpc germ cells with or without SCF after 16 to 18 h of culture (Figs. 3.8 and 3.9). This suggests that under the stress of the culture conditions PGCs increase production of BCL-2. However, the increased level of BCL-2 appears to be insufficient to save early PGCs, but is adequate to prevent apoptosis in oocytes. Additional experiments are currently in progress to clarify these intriguing results.

Acknowledgments. We wish to thank Dr. Fabrizio Poccia for FACS analyses and Prof. Gregorio Siracusa for critical reading of this manuscript. Financial support from M.U.R.S.T. (40% and 60%), A.I.R.C. to M.P. and C.N.R. (Progetto Finalizzato FAT.MA, grant no. 95.00912.PF41) is gratefully acknowledged.

References

1. Gondos B. Oogonia and oocytes in mammals. In: Jones R, ed. The vertebrate ovary. New York; Plenum Press, 1978;83–120.
2. Peters H, McNatty K. Atresia. In: Peters H, McNatty K, eds. The ovary. London: Granada Publishing, 1980;98–106.
3. Borum, K. Oogenesis in the mouse. A study of the meiotic prophase. Exp Cell Res 1961;24:495–507.
4. Bakken AH, McClanahan M. Patterns of RNA synthesis in early meiotic prophase oocytes from fetal mouse ovaries. Chromosoma 1978;67:21–40.
5. Tam P, Snow MHL. Proliferation and migration of primordial germ cells duriong compensatory growth in mouse embryos. J Embr Exp Morph 1981;64:133–47.
6. Borgoyne PS, Baker TG. Oocyte depletion in XO mice and their sibs from 12 to 200 days post partum. J Repr Fert 1981;61:207–12.
7. Gavrieli Y, Sherman Y, Shmuel A, Ben-Sasson A. Identification of programmed cell death in situ via specific labeling of nuclear DNA fragmentation. J Cell Biol 1992;119:493–501.
8. Fesus L, Davies P, Piacentini M. Apoptosis: molecular mechanisms in cell death. Eur J Cell Biol 1991;56:170–7.
9. Pesce M, Farrace MG, Dolci S, Piacentini M, De Felici M. Stem cell factor and leukemia inhibitory factor promote primordial germ cell survival by suppressing programmed cell death (apoptosis). Development 1993;118:1089–94.
10. De Felici M, McLaren A. Isolation of mouse primordial germ cells. Exp Cell Res 1982;142:476–82.
11. Pesce M, De Felici M. Purification of mouse primordial germ cells by MiniMACS magnetic separation system. Dev Biol 1995;170:722–25.
12. Gougeon ML, Garcia S, Guetard D, Olivier R, Dauguet C, Montagnier L. Apoptosis as a mechanism of cell death in peripheral lymphocytes from HIV-1-infected individuals. In: Janossy G, Autran B, Miedema F, eds. Immunodeficency in HIV and AIDS. Basel: Karger, 1992;115–26.

13. Coucouvanis EC, Sherwood SW, Carswell-Crumpton C, Spack E, Jones PP. Evidence that the mechanism of prenatal germ cell death in the mouse is apoptosis. Exp Cell Res 1993;209:238–46.
14. Ratts V, Flaws JA, Kolp R, Sorenson C, Tilly JL. Ablation of *bcl-2* gene expression decreases the numbers of oocytes and primordial follicles established in the postnatal female mouse gonad. Endocrinology 1995;136:3665–9.
15. Zakeri Z, Bursch W, Tenniswood M, Locksin RA. Cell death: programmed, apoptosis, necrosis, or other? Cell Death Differ 1995;2:87–96.
16. De Felici M, McLaren A. In vitro culture of mouse primordial germ cells. Exp Cell Res 1983;144:417–27.
17. Wabik-Sliz B, McLaren A. Culture of mouse germ cells isolated from fetal gonads. Exp Cell Res 1984;154:530–6.
18. Pesce M, De Felici M. Apoptosis in mouse primordial germ cells: a study by transmission and scanning electron microscope. Anat Embryol 1994;189:435–40.
19. Nicoletti I, Miglorati G, Pagliacci MC, Grignani F, Riccardi C. A rapid and simple method for measuring thymocyte apoptosis by propidium iodide staining and flow cytometry. J Immunol Meth 1991;139:271–9.
20. Evan GI, Wyllie AH, Gilbert CS, Littewood TD, Land H, Brooks M, et al. Induction of apoptosis in fibroblasts by *c-myc* protein. Cell 1992;69:119–28.
21. Dolci S, Williams DE, Ernst MK, Resnick JL, Brannan CL, Fock LF, et al. Requirement for mast cell growth factor for primordial germ cell survival in culture. Nature 1991;352:809–11.
22. Matsui Y, Toksoz D, Nishikawa S, Nishikawa SL, Williams D, Zsebo K, Hogan B. Effect of *Steel* factor and leukemia inhibitory factor on murine primordial germ cells in culture. Nature 1991;353:750–2.
23. Godin I, Deed R, Cooke J, Zsebo K, Dexter M, Wylie CC. Effect of the *Steel* gene product on mouse primordial germ cells in culture. Nature 1991;352:807–9.
24. Dolci S, Pesce M, De Felici M. Combined action of stem cell factor, leukemia inhibitory factor and cAMP on in vitro proliferation of mouse primordial germ cells. Mol Repr Dev 1993;35:134–9.
25. De Felici M, Pesce M. Interactions between migratory primordial germ cells and cellular substrates in the mouse. In: Marsh J, Goode J, eds. Germline development. London: John Wiley & Sons Ltd, 1994;140–53.
26. Stevens LC. Origin of testicular teratomas from primordial germ cells in mice. J Natl Cancer Inst 1967;37:859–61.
27. Resnick JL, Bixler LS, Cheng L, Donovan PJ. Long term proliferation of mouse primordial germ cells in culture. Nature 1992;359:550–1.
28. Matsui Y, Zsebo K, Hogan B. Derivation of pluripotential embryonic stem cells from murine primordial embryonic germ cells in culture. Cell 1992;70:841–7.
29. Carson WE, Haldar S, Baiocchi RA, Croce CM, Caligiuri MA. The c-kit ligand suppresses apoptosis of human natural killer cells through upregulation of *bcl-2*. Proc Natl Acad Sci USA 1994;91:7553–7.

4

Developmental Consequences of Programmed Cell Death in Human Preimplantation Embryos

Andrea Jurisicova, Sussanah L. Varmuza, and Robert F. Casper

For many couples the failure to achieve and maintain pregnancy remains a major problem. An alternative solution for these couples is in-vitro fertilization (IVF). When the first IVF baby was born, no one expected that this unconventional method of conception would become such a widely used treatment for infertility. From the report of clinical results of assisted reproductive technology procedures in the USA and Canada for 1991, it is clear that of the 21,083 oocyte retrievals, 87.1% led to a successful fertilization and preembryo transfer. However, only 4,017 clinical pregnancies (19.1%) were achieved (1). The overall birth rate per IVF retrieval does not exceed 15.25%.

To increase the chances of favorable IVF outcome, controlled ovarian hyperstimulaton and superovulation of patients is used. A higher number of retrieved oocytes increases the probability of obtaining several embryos. The best embryos, following morphological evaluation, are subsequently transferred to the uterus. Nonetheless, a significant proportion of transferred embryos (70% or more) fail to implant and, at most, only 10% will give rise to a full-term infant (2). Moreover, even in the general population, approximately 25% to 40% of conceptions may be lost before they are clinically diagnosed as a pregnancy (2).

Development of the Human Preimplantation Embryo

In the human, the first visible sign of fertilization (e.g., the appearance of two pronuclei in the cytoplasm of the human oocyte) can be observed 18 to 22 h after insemination. From that point, the embryo starts to cleave approximately twice every 24 h. At the 16-cell stage, mammalian embryos undergo a process called compaction, whereby the blastomeres flatten, maximizing cell-cell contact and minimizing intracellular spaces, to form a morula (3).

The diameter of the embryo is unchanged from that of the zygote because growth is restricted by the zona pellucida (approximately 120 μm). The first embryonic differentiation event (e.g., the formation of the blastocyst) is observed 5 days after insemination in vitro, and full blastocoele expansion can be seen on day 6 to day 7. With more pressure inside the embryo, the zona pellucida decreases in width, allowing the embryo to expand in size. At this stage, the human embryo consists of approximately 100 cells (4) and contains two different cell types: trophectoderm, which will give rise to the placenta, and the inner cell mass (ICM), which will form the embryo proper. The next stage in development occurs when the blastocyst hatches out of the zona pellucida, allowing further expansion, growth and interaction of the embryo with the uterine endometrium. The implantation process is thought to start in vivo on day 7 after ovulation and to be complete by about day 14 (3).

Progression of fertilized mammalian oocytes through cleavage, blastocyst formation and implantation is dependent on the successful implementation of the genetic and developmental program contained within the oocyte and the embryo itself, and on the successful interaction of the preimplantation embryo with its environment (5). Very little is known about the pattern of early gene expression. In the human, as in other mammals, zygote development and the first two cleavage divisions depend on, and are probably controlled by, maternal RNA transcribed and accumulated during oogenesis. Activation of the embryonic genome can be first observed between the four- to eight-cell stage in the human (6, 7) or at the two-cell stage in the mouse (8). The onset of gene expression coincides with major changes in the proteins synthesized by the mammalian embryo, some of which are transcription-dependent (6).

Human early embryos cultured in vitro display a remarkably high rate of spontaneous cleavage arrest (9). Using various culture media in conjunction with very stringent quality-control procedures, it has been shown that only 17% to 30% of spare human preimplantation embryos cleave regularly to the 6-day-old blastocyst stage in vitro (4), despite the fact that 60% to 70% of the fertilized zygotes develop to the four- to six-cell stage (day 2) in most IVF centers. The highest incidence of developmental arrest occurs between the four- to eight-cell stages, but also after compaction of morulae. Another morphological anomaly observed in many human embryos is cellular fragmentation. Less than 50% of embryos cleave regularly into equal-sized blastomeres without fragmentation. The remaining embryos often contain variable-sized blastomeres with multiple cellular fragments. Subsequent in vitro development of these fragmented embryos is impaired, often leading to cleavage arrest and embryo degeneration. In addition, the survival rate after cryopreservation is decreased. Severely fragmented human embryos have limited developmental potential and rarely result in a viable pregnancy (10, 11).

Our knowledge of the etiology and underlying mechanism of embryo fragmentation is limited. In mammals, embryo fragmentation can be observed in several species (12) and seems to be a naturally occurring phenomenon observed in embryos conceived in vivo as well as in vitro from both stimulated and unstimulated cycles (13). The morphologic appearance of fragments in early embryos, and previous reports describing nuclear abnormalities (14), led us to hypothesize (15) and later confirm (16) that fragmentation in these embryos is the consequence of programmed cell death (PCD).

Can Early Embryo Fragmentation Be a Result of Activated PCD?

A total of 229 human embryos, arrested at different stages of development ranging from the 2-cell stage to uncompacted morulae, were studied using combined nuclear staining with DAPI and analysis of integrity of chromatin following terminal transferase-mediated DNA end-labeling (TUNEL) (16). In this sample, 203 embryos showed various degrees of fragmentation whereas 26 apparently normal embryos had no visible cellular fragmentation. The normal, unfragmented embryos had normal looking nuclei as judged by both a diffuse DAPI staining pattern and a negative TUNEL reaction. In contrast, 153 of the 203 fragmented embryos (75.4%) displayed hallmarks of apoptosis with or without the presence of some normal nuclei. Evidence of apoptosis was confirmed by transmission and scanning electron microscopy (Fig. 4.1). We observed several undegraded cell corpses with dense cytoplasm, multiple cellular fragments which contained normal appearing cytoplasmic organelles, and dense masses resembling condensed chromatin. Despite condensed cytoplasm in these cell corpses, intact mitochondria and several other organelles were visible. Corpses were not phagocytosed and were always found in the intercellular space within the zona pellucida. Some of the cellular fragments showed secondary necrotic changes with disrupted cellular membranes and swollen cytoplasmic organelles. Necrotic changes were also evident in some nonapoptotic blastomeres. These observations were consistent with the generalized DAPI/TUNEL signal characteristic of necrotic cells observed in some embryos.

This study concentrated on early cleavage stage embryos that arrest and fail to develop to the blastocyst stage in vitro. The distinguishing feature of these embryos was excessive blastomere fragmentation, which was easily visible under a dissection microscope. Embryos were processed 24 h after the last cleavage and had reached variable stages of development. Therefore, embryos with asynchronously dying blastomeres were sampled at different times with respect to the first apoptotic process that produced the original fragments. This may account for the different categories of embryos observed in our population. Embryos with a few fragments but no condensed DAPI/TUNEL signals may represent an earlier stage of apoptosis than those with both fragments and condensed DAPI/TUNEL signals. Embryos showing evidence of necrotic changes may be more advanced still,

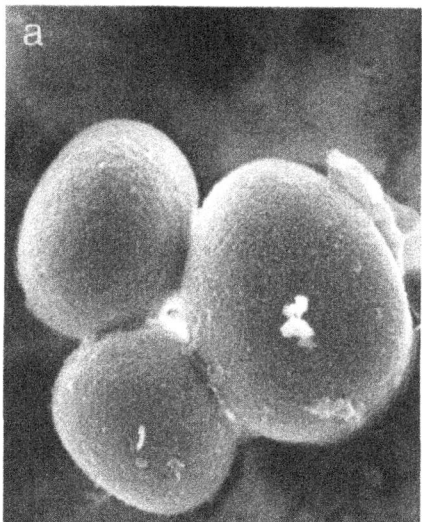
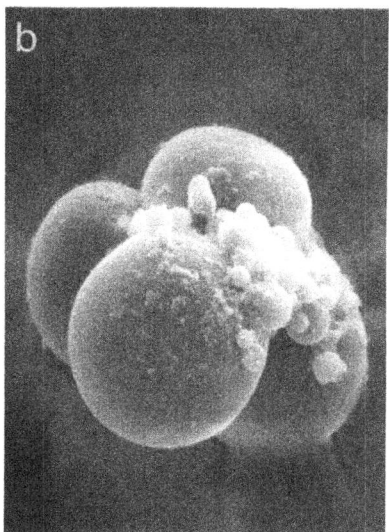
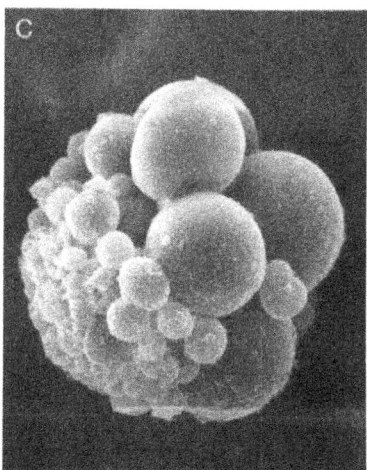

FIGURE 4.1. Scanning electron micrographs of human embryos at days 2 and 3 after fertilization (original magnification, 1200 x). (a) Three-cell human embryo at day 2 with blastomeres of regular size and no fragmentation. Blastomere on left completed second cellular division, whereas blastomere on right has not yet cleaved. (b) Four-cell embryo with a small proportion of cellular fragmentation. (c) Day 3 human embryo at approximately the eight-cell stage with several normal appearing blastomeres and excessive cellular fragmentation. Fragments are comparably smaller in size than regular blastomeres and fill almost 50% of the total volume of embryo within the zona pellucida.

although we cannot rule out the possibility that some cells/embryos die through arrest-mediated necrosis rather than apoptosis.

Blastomeres probably have no phagocytic capability since we found cell corpses in the intercellular space, none of which were engulfed by other cells. Alternatively, blastomere corpses may not promote phagocytotosis, possibly due to an inability to express cell surface molecules (apogens) responsible for recognition of apoptotic cells (17). Thus, within developing preimplantation embryos, cell corpses and fragments that are not phagocytosed effectively may undergo secondary necrosis, which in turn may trigger arrest and subsequent necrosis of surrounding blastomeres. The population of embryos in the study described above represented those that failed to reach the blastocyst stage. Fragmented embryos that managed to reach the blastocyst stage may have the ability to deal with cell corpses more effectively than their arrested counterparts. Trophoectoderm, the first differentiated cell type that arises in the embryo during blastocyst formation, was reported to possess highly effective phagocytic activity (18).

These observations support the hypothesis that PCD and the resultant apoptosis is responsible for a significant proportion of fragmented human embryos, and reinforces the original morphologic description of abnormal embryos reported by Hardy et al. (14) and Winston et al. (19).

PCD as a Part of Normal Development in the Mammalian Blastocyst

Several articles have described the morphologic appearance of dead cells in mammalian embryos at the blastocyst stage. However, it is not known how many cells can die without compromising normal development. Analysis of ultrathin sections of blastocysts of various mammalian species revealed the presence of cells with clumped, condensed chromatin, swelling of the endoplasmic reticulum, multiple cellular fragments and corpses (20, 21). The presence of cell death was observed in blastocysts from different mammalian species obtained in vivo, as well as in embryos produced in vitro. Although not specifically mentioned in these reports, the morphological description, as well as the published photographs, show typical hallmarks of apoptosis.

The rate of human blastocyst formation in vitro is at best 40%. Upon transfer, only a subset of these blastocysts (about 40%) are able to implant (13). We speculate that an activated cell death program in many cells during human blastocyst formation will lead to embryo demise. Cell number and cell death in human blastocysts (evaluated by nuclear staining only) was previously studied by Hardy et al. (4), but no conclusion was reached regarding the physiological range of cell depletion. We have further explored the range of cell death in human blastocysts to define how many cells die during development and how the extent of cell death may effect future developmental potential of the embryo.

We analyzed 66 human embryos that developed to the expanded blastocyst stage by day 6 ($n = 27$) and day 7 ($n = 39$) postfertilization. Out of these 66 blastocysts, 11 originated from oocytes with only one pronucleus and eight from oocytes with three pronuclei (due to failure of second polar body extrusion). The remaining embryos developed from normally fertilized oocytes.

The mean ($\pm$ SEM) number of cells at the day 6 blastocyst stage was 59 ± 4, with only a few cells displaying signs of apoptosis (seven on average). By day 7, with further development to the fully expanded or hatched blastocyst stage, the mean cell number increased to 80 ± 8. Interestingly, the number of dying cells in the same period doubled to 15 ± 2. Occasionally, we also noticed necrotic cells that most likely represented arrested blastomeres from earlier cleavage divisions. In most of the embryos, cell death was random. Dying cells were scattered in both ICM and trophectoderm. However, we identified a subpopulation of embryos (21%) in which the majority of ICM cells had extensive condensation of chromatin and fragmented DNA. The resulting and complete depletion of ICM cells could lead clinically to a spontaneous abortion, classified as a "blighted ovum."

Furthermore, we observed that about 24% of the blastocysts had an increased cell death index, which represented the proportion of dead cells to total number of cells per embryo (%). In the majority of blastocysts this index did not exceed 10% and 23% on day 6 and 7, respectively. In a subpopulation of embryos (16/66), the cell death index increased markedly (up to 56%) indicating widespread activation of PCD.

PCD observed during blastocyst formation has been suggested as the first apoptotic event during mammalian development (22). However the extent of cell death was never clearly identified. In our opinion, the mammalian blastocyst is the first embryonic stage when PCD is present as a normal component of development, responsible for cellular remodeling. However, in some human embryos, one of the two embryonic lineages (ICM) is preferentially eliminated leaving only precursor cells of the placenta. Such a blastocyst may initiate implantation, but will result in a placenta with an empty embryonic sac. A second scenario is that too many cells activate their PCD cascade, compromising further blastocyst development in both cell lineages. These blastocysts most likely would be unable to initiate implantation and would not be recognized as a pregnancy.

Certain murine embryonic carcinoma cell lines with different developmental potential will selectively die, or survive and participate in further development, when deposited into the blastocoele cavity. Cells that exhibit solely embryonic potential are more resistant to apoptosis than cells with the capability of forming trophectoderm. These results led to the speculation that PCD within blastocysts may be designed to eliminate redundant ICM cells with trophectodermal potential (23). Furthermore, in an attempt to isolate factors responsible for this selective killing by the blastocoele fluid, extracelluar hydrogen peroxide generated via polyamine oxidation was suggested as a mediator of PCD in the blastocyst (24).

Possible Triggers of PCD During Mammalian Embryo Development

Fragmentation and death of human embryos through apoptosis suggests a natural "preprogrammed" response to external stimuli or internal defects. Are these stimuli/defects abnormally high in IVF, and can they be reduced? Or is PCD and embryo wastage a natural predisposition of the human embryo? If IVF is at least partially to blame, it may be possible to determine the triggers of apoptosis by manipulating different culture parameters. This question is difficult to address in humans, although some progress may be made once a suitable animal model is established.

One possible trigger for mammalian embryo fragmentation is the artificial environment for fertilization and cleavage utilized during IVF. For example, hormonal stimulation, inappropriate culture conditions, and exposure to high concentrations of sperm may all contribute to excessive activation of PCD. The influence of various intrinsic and external factors on embryo fragmentation will be examined in detail in the following sections.

Maternal Age

Our preliminary experiments suggest that only 12% of in vivo fertilized murine embryos harvested from young females (8 to 10 weeks of age postpartum) 24 h after mating contain fragments similar to those observed in human embryos. We confirmed that these fragmented embryos were undergoing apoptosis by combined DAPI/TUNEL. The proportion of fragmented embryos increased to 25% when murine oocytes were fertilized in vitro in glucose rich media (12, 25). Of interest, the rate of oocyte fragmentation did not change when 24 to 26 week old females were used. However, when maternal age reached 40 to 44 weeks, the rate of both in vivo and in vitro oocyte fragmentation doubled (up to 42%). These observations support our hypothesis that the artificial environmental conditions used during IVF contributes to the increased rate of embryo fragmentation since in vitro fertilization resulted in a two-fold increase in embryo fragmentation over that observed in embryos obtained by in vivo fertilization. In addition, embryonic factors are important since maternal aging in the model resulted in a significant increase in embryo fragmentation, both in vivo and in vitro. Using mouse embryos as an experimental system, we observed that embryo fragmentation occurs in the time period between activation of the oocyte and the 2-cell stage. This timing is slightly different from that of fragmentation in human embryos, which occurs between the two- to eight-cell stage. The zygotic genome becomes activated in humans at the four- to eight-cell stage and in mice at the 2-cell stage. In both species, therefore, initiation of PCD and fragmentation slightly precedes activation of the embryonic genome.

Culture Conditions

Reactive oxygen species (ROS), such as H_2O_2, have been implicated as mediators of apoptosis in somatic cells (26) and in murine embryonic carcinoma (24). Mouse oocytes and embryos are able to generate small but measurable amounts of endogenous H_2O_2 that increases during the transition from the two- to four-cell stage. This rise in H_2O_2 may be attributed to the in vitro culture conditions where O_2 concentrations are relatively high (27). Whether or not a similar situation exists in vivo, where the O_2 concentration is low, is not known. The unfertilized oocyte contains glutathione, a tripeptide that effectively protects cells against damage from reactive oxygen radicals. Glutathione levels decrease by an order of magnitude in the time between fertilization and cleavage to the blastocyst stage, when glutathione synthesis is renewed. Levels of glutathione are significantly decreased in embryos cultured in vitro compared to those that developed in vivo (28). The exposure to greater concentrations of oxygen free radicals from the culture conditions or from sperm may contribute to the depletion of glutathione and thus may disturb the reduction-oxidation (redox) balance in early embryos. An intracellular decrease of glutathione in somatic cells precedes the onset of apoptosis (29). Compounds that increase intracellular glutathione, such as β-mercaptoethanol, or that stimulate activity of glutathione peroxidase (i.e., N-acetyl-L-cysteine), can prevent the induction of apoptosis (30, 31).

Although glucose seems to interfere with early preimplantation embryo development (32), almost all media used for human embryo culture contain relatively high concentrations of glucose. We recorded an increased rate of murine oocyte fragmentation when maintained for 24 h in HTF (glucose and phosphate rich) medium commonly used for human IVF compared to the rates when a low-glucose medium was used (25). As mentioned above, unfertilized mouse oocytes contain a high concentration of glutathione. Recently, a link between glucose toxicity and glutathione metabolism has been established. Rat postimplantation embryos exposed to a high concentration of glucose were reported to have decreased levels of glutathione, most likely due to decreased activity of glutathione synthetase (33). Glucose inhibition of glutathione synthetase, with the ensuing depletion of intracellular glutathione, may lead to activation of PCD by accumulation of ROS and subsequent cellular damage. Additionally, since mitochondria are readily damaged by oxygen radicals (reviewed in 34), it is likely that increased exposure of the embryo to ROS decreases the number of active mitochondria. Since mitochondrial replication does not occur until the blastocyst stage (35), each blastomere receives a finite number of oocyte-derived mitochondria, a number that decreases by 50% with each cell division. Therefore, a critical reduction in mitochondria could occur after several cell divisions, leading to impaired energy production for the blastomeres and initiation of PCD. Supplementation of media with agents that increase glutathione metabolism, such as β-mercaptoethanol or cysteamine, during bovine oocyte maturation and embryo development have been reported to result in an increased rate of blastocyst formation in vitro (36).

Coculture techniques using a variety of epithelial cells improved embryo morphology, with a decreased amount of fragmentation and an increased rate of blastocyst formation in vitro (37, 38). It is unclear what underlying mechanisms may mediate the positive effect of co-culture. Cells in coculture may secrete growth factors or other embryotrophic products that foster embryonic growth. For example, leukemia inhibitory factor (LIF) plays a crucial role in mouse blastocyst implantation (39) and is able to suppress PCD in mouse primordial germ cells (40). However, our results with human embryos indicate that LIF does not have a significant effect on fragmentation or blastocyst formation in vitro (41). Alternatively, cocultured cells may remove undesirable metabolic toxins and decrease oxygen pressure in the medium, thus protecting embryos from oxidative stress. This is in agreement with the observation that the rate of embryo fragmentation is decreased in coculture when compared to regular culture conditions (37).

Sperm-Associated Factors

The routine IVF practice is to incubate oocytes with 50,000 to 100,000 washed sperm overnight, and to assess fertilization the next morning. In contrast, normal in vivo fertilization probably involves exposure of the oocyte in the fallopian tube to one-thousandth the number of sperm. It is possible that prolonged exposure to a large number of sperm exposes the oocyte to enzymes and other factors released during the acrosome reaction that may be detrimental to embryo development and could trigger PCD. For example, hyaluronidase, a component of the acrosome, will activate oocytes (42) and is known to be toxic with prolonged exposure to embryos cultured in vitro (43). A second potential mediator of PCD during IVF is oxidative stress generated by the sperm suspension as discussed above.

Furthermore, intrinsic sperm factors, such as integrity of the sperm DNA, may also contribute to the high rate of embryo demise. We have used the TUNEL technique with fluorescence-activated cell sorting (FACS) to demonstrate that a low percentage (<5%) of sperm from normal fertile men contains fragmented DNA (44). In contrast, up to 40% of sperm from infertile men has fragmented DNA, and this percentage is negatively correlated with semen analysis parameters, fertilization and cleavage rates in IVF. Interestingly, increased DNA fragmentation has been detected in sperm from smokers versus nonsmokers. It is possible that sperm with damaged DNA could penetrate oocytes in IVF procedures. Perhaps more importantly, it is likely that some sperm with fragmented DNA could be selected for intracytoplasmic sperm injection (ICSI) for assisted fertilization, especially since extremely poor quality semen is the indication for this technique. Although it is not known if sperm with fragmented DNA can fertilize an oocyte, a number of possible outcomes could ensue. The first possibility is that the oocyte may fail to fertilize without sperm decondensation. It has been shown that between 20% and 50% of apparently unfertilized oocytes, or one pronuclear (presumably parthenogenetically-activated) oocytes, in IVF have unrecognized sperm penetration with failure of chromosomal decondensation and pronuclear formation (45, 46). Second, fertilization may occur but the zygote arrests at the two pronuclear stage. Lastly, fertilization may occur with cleavage and subsequent apoptosis and fragmentation. All three outcomes are presently seen in our IVF and ICSI programs.

Developmental Competence of Oocyte

Controlled ovarian stimulation using exogenous gonadotropins to obtain multiple oocytes may lead to an accelerated follicular response, resulting in "biochemical immaturity" of the oocytes. Embryos arising from such oocytes may not be able to proceed through development if, for example, their PCD "machinery" is unbalanced due to abnormal deposition of maternally stored products (45).

It is also possible that specific interactions among oocyte activation, sperm penetration, and zygotic gene activation are required for further embryonic development and survival (46). Interference with these processes may lead to initiation of PCD and elimination of abnormal embryos. Oocytes may contain a built in program of apoptosis which remains quiescent until the oocyte goes through activation. Oocytes arrested at metaphase II that fail to fertilize and do not resume meiosis become necrotic within 2 to 4 days. However, once the killer program is triggered by oocyte activation, it can be suppressed only by proper signal(s), possibly originating from the oocyte or generated by the embryo itself. Thus, it is likely that triggers and effectors are specific to the embryo at this time. This system may ensure that only normal embryos proceed to the implantation stage, and it may also be one of the mechanisms by which development of parthenogenetic embryos is prevented.

Staurosporine (a protein kinase inhibitor) is able to initiate apoptosis in many mammalian somatic cell types. In addition, staurosporine is capable of initiating apoptosis in trophectoderm and ICM cells in blastocyst stage embryos (M. Raff, personal communication). However, mouse blastomeres of the early cleaving embryo are resistant to cell death initiated by staurosporine (47). It has been proposed that blastomeres are unique cells lacking a built in suicide program. An alternative explanation could be that some blastomeres may have distinct effector pathways that are not triggered by staurosporine. It is also possible that blastomeres of normal embryos are very well protected against activation of PCD.

Chromosomal Abnormalities of Human Embryos and Embryo Fragmentation

PCD in human embryos may result from inherent abnormalities of the oocyte or sperm. For example, if chromosomal aneuploidy is present, PCD may be a protective mechanism to prevent implantation of an abnormal embryo. Interestingly, cytogenetic observations of spare human embryos confirmed the presence of a spectrum of nuclear anomalies, for example, multinucleated or anucleated blastomeres, and flocculent and fragmented nuclei (14, 19). A wide range of chromosomal abnormalities, including premature chromosome condensation, aneuploidy and polyploidy, have been reported in spare human embryos (48, 49). Compared to embryos with good morphology, a higher incidence of these cytogenetic abnormalities have been found in embryos with fragmentation (50, 51). However, karyotyping blastomeres from fragmented embryos has not been successful in our experience (52), possibly because of the rapid degradation of DNA by endonucleases during PCD. Only 29% of human embryos produced readable chromosomal spreads (51). In addition, chromosomes can be lost during the metaphase spread and false diagnoses of hypoploidy can occur with karyotyping. An alternative technique is to use labeled centromeric DNA probes (fluorescent in situ hybridization; FISH) for chromosomes that are commonly affected by nondysjunction. Using a multicolor FISH technique, several chromosomes can be simultaneously detected and counted with fluorescence microscopy in interphase cells. However, caution should be exercised in the interpretation of these results. It is possible that chromosomes in which the centromeric DNA has already been cleaved may contain multiple signals. We believe the presence of fragmented DNA may explain the observation of multiploid (up to 20n) FISH signals in human monospermic fertilized embryos with extensive cellular fragmentation (53). A correction factor can be applied, however, by using TUNEL in the same cells to determine which nuclei have initiated DNA fragmentation and which have intact DNA.

Current and Future Research Directions

Could Oxygen Tension, pH in Follicular Fluid, and Granulosa Cell Death From Individual Follicles Reflect Developmental Competence of Oocyte?

Gaulden (54) has hypothesized that oocyte chromosomal aneuploidy may originate as the result of a deteriorating intrafollicular environment secondary to hypoxia. It is known that oxygen tension in follicular fluid decreases as the follicle expands to its normal mature size of approximately 2 cm in diameter (55). Some rapidly enlarging follicles may outpace their blood supply resulting in decreased follicular fluid P_{O_2} and lowered intrafollicular pH, both of which could adversely affect the nuclear spindle in the developing oocyte. Morphologic changes associated with apoptosis appear to be preceded by cellular acidification (56), consistent with the decreased cellular pH detected in some human oocytes (57).

Evidence of apoptosis based on DNA laddering on gel electrophoresis in granulosa cells from some, but not all, aspirated follicles during human IVF has been reported (58). In this small study, no correlation between the extent of cell death in granulosa cells, follicle size, color of follicular fluid, or fertilizability of the oocytes was observed. Unfortunately, no observation of embryo quality or developmental potential was recorded. Morphologic observation of abnormal attachment and spreading of cumulus cells from different oocytes in vitro has also been described (59). Apoptosis of granulosa cells may reflect a premorbid follicular state that will be manifested in poor developmental competence of the oocyte upon fertilization. Thus, the occurrence and extent of granulosa cell death in individual follicles could be used as an indicator for selection of suitable embryos for transfer (60).

Can Aspiration of Cellular Fragments Improve Developmental Potential of Remaining Blastomeres?

We have observed that fragmented human embryos exhibit a certain polarity, with normal appearing blastomeres clustered on one side of the embryo and fragments on the other. This observation suggests the possibility that cells originating from one parental blastomere are preferentially being eliminated. Unequal division of cytoplasmic contents could occur during the first cleavage of the zygote to a 2-cell embryo, resulting in two separate cell lines with different developmental potential. Subsequent cell divisions could result in varying resistance to apoptosis, depending on the number of mitochondria (containing products of the *bcl-2* gene family) in each cell or the amount of ROS scavenger activity (e.g., reduced glutathione) that each daughter cell receives. If unequal division of mitochondria occurs at the first cell division, a critical shortage of these cellular organelles or protective molecules could occur in the subsequent cell line derived from the deficient sibling blastomere. Van Blerkom (59) has suggested that high ATP concentration in fragmented embryos, reflecting mitochondrial activity, indicates increased developmental ability for the intact blastomeres after embryo transfer. Cohen et al (61) has reported aspiration of fragments and transfer of the remaining intact blastomeres from fragmented embryos in an attempt to rescue these embryos. However, our preliminary data, derived from aspiration of cellular fragments from 24 human four- to eight-cell stage embryos in which at least 25% of the volume of the embryo was filled with fragments, suggest that most of the remaining normal appearing blastomeres will continue to fragment. This supports the idea of a preprogrammed set of events within these embryos. However, it should be noted that a small proportion of these embryos continued to blastocyst formation after aspiration of fragments.

Conclusions

We have demonstrated that apoptosis occurs in human embryos prior to the blastocyst stage, and that this cell death is clearly detrimental to the subsequent developmental potential of these embryos. Additional work is needed to determine the nature of both intrinsic and extrinsic triggers of PCD and to develop potential protective measures. Application of this research to clinical IVF may lead to improved implantation and pregnancy rates in the future.

References

1. Society for Assisted Reproductive Technology, The American Fertility Society. Assisted reproductive technology in the United States and Canada; 1991 results from the Society for Assisted Reproductive Technology generated from The American Fertility Society Registry. Fertil Steril 1993;59:956–62.
2. Winston RML, Handyside AH. New challenges in human in vitro fertilization. Science 1993;260:932–6.
3. Leese HJ, Conaghan J, Martin KL, Hardy K. Early human embryo metabolism. Bioassays 1993;15:259–64.
4. Hardy K, Handyside AH, Winston RM. The human blastocyst: cell number, death and allocation during late preimplantation development in vitro. Development 1989; 107:597–604.
5. Schultz GA, Heyner S. Gene expression in preimplantation mammalian embryos. Mutation Res 1992;296;17–31.
6. Braude P, Bolton V, Moore S. Human gene expression first occurs between the four and eight cell stages of preimplantation development. Nature 1988;332: 459–61.
7. Tesarik J, Kopecny V, Plachot M, Mandelbaum J. Early morphological signs of embryonic genome expression in human preimplantation development as revealed by quantitative electron microscopy. Develop Biol 1988;128:15–20.
8. Flach G, Johnson MH, Braude PR, Taylor RAS, Bolton VN. The transition from maternal to embryonic control in the 2-cell mouse embryo. EMBO J 1982; 1:681–6.
9. Bolton V, Braude PR. Development of spare human preimplantation embryos in vitro. Curr Topics Dev Biol 1987;23:93–113.
10. Plachot M, Mandelbaum J. Oocyte maturation, fertilization and embryonic growth in vitro. Br Med Bull 1980;46:675–94.
11. Erenus M, Zoues C, Rajamahendran P, Leung S, Fluker M, Gomel V. The effect of embryo quality on subsequent pregnancy rates after in vitro fertilization. Fertil Steril 1991;56:707–10.
12. Summers MC, Bhatnagar PR, Lawitts JA, Biggers JD. Fertilization in vitro of mouse ova from inbred and outbred strains: complete preimplantation embryo development in glucose-supplemented KSOM. Biol Reprod 1995;53:431–7.
13. Formigli I, Roccio C, Belotti G, Stangalini A, Coglitore MT, Formogli G. Non-surgical flushing of the uterus for the pre-embryo recovery: possible clinical applications. Hum Reprod 1990;5:329–35.
14. Hardy K, Winston RML, Handyside AH. Binucleate blastomeres in preimplanta-

tion human embryos in vitro: failure of cytokinesis during early development. J Reprod Fertil 1993;98:549–58.

15. Jurisicova A, Varmuza S, Casper RF. Involvement of programmed cell death in preimplantation embryo demise. Hum Reprod Update 1995;1:558–66.

16. Jurisicova A, Varmuza S, Casper RF. Programmed cell death and human embryo fragmentation. Mol Hum Reprod 1996;2:101–6.

17. Rotello RJ, Fernandez PA, Yuan J. Anti-apogens and anti-engulfens: monoclonal antibodies reveal specific antigens on apoptotic and engulfment cells during chicken embryonic development. Development 1994;120:1421–31.

18. Drake BL, Rodger JC. Phagocytic properties of cultured murine trophoblast. Placenta 1987;8:129–39.

19. Winston NJ, Braude PR, Pickering SJ, George MA, Cant A, Currie J, Johnson MH. Incidence of abnormal morphology and nucleocytoplasmic ratios in 2-, 3- and 5-day human preembryos. Hum Reprod 1991;6:17–24.

20. El-Shershaby AM, Hinchliffe JR. Cell redundancy in the zona-intact preimplantation mouse blastocyst: a light and electron microscope study of dead cells and their fate. J Embryol Exp Morphol 1974;31:643–54.

21. Mohr LR, Trounson AO. Comparative ultrastructure of hatched human mouse and bovine blastocysts. J Reprod Fertil 1982;66:499–504.

22. Parchment RE. The implications of a unified theory of PCD, polyamines, oxyradicals and histogenesis in the embryo. Int J Dev Biol 1993;37:75–83.

23. Pierce GB, Lewellyn AL, Parchment RE. Mechanism of PCD in the blastocyst. Proc Natl Acad Sci USA 1989;86:3654–8.

24. Pierce GB, Parchment RE, Lewellyn AL. Hydrogen peroxide as a mediator of PCD in the blastocyst. Differentiation 1991;46:181–6.

25. Erbach GT, Lawitts JA, Papaioannou VE, Biggers JD. Differential growth of the mouse preimplantation embryo in chemically defined media. Biol Reprod 1994;50:1027–33.

26. Ratan PR, Murphy TH, Baraban JM. Oxidative stress induces apoptosis in embryonic cortical neurons. J Neurochem 1994;62:376–9.

27. Nasr-Esfahani MH, Aitken JR, Johnson MH. Hydrogen peroxide levels in mouse oocytes and early cleavage stage embryos developed in vitro or in vivo. Development 1990;109:501–7.

28. Gardiner CS, Reed DJ. Status of glutathione during oxidant-induced oxidative stress in the preimplantation mouse embryo. Biol Reprod 1994;51:1307–14.

29. Beaver JP, Waring P. A decrease in intracellular glutathione concentration proceeds the onset of apoptosis in murine thymocytes. Eur J Cell Biol 1995;68:47–54.

30. Ferrari G, Yan CY, Greene LA. N-acetyl cysteine prevents apoptotic death of neuronal cells. J Neurosci 1995;15:2857–66.

31. Tilly JL, Tilly KI. Inhibitors of oxidative stress mimic the ability of follicle-stimulating hormone to suppress apoptosis in cultured rat ovarian follicles. Endocrinology 1995;136:242–52.

32. Bavister B. Culture of preimplantation embryos: facts and artifacts. Hum Reprod Update 1995;1:91–148.

33. Trocino RA, Shoichi A, Ishibashi M, Matsumoto K, Matsuo H, Yamamoto H, Goto S, Urata Y, Kondo T, Nagataki S. Significance of glutathione depletion and oxidative stress in early embryogenesis in glucose-induced rat embryo culture. Diabetes 1995;44:992–7.

34. Barnet D, Bavister B. What is the relationship between the metabolism of preimplantation embryos and their developmental competence. Mol Repr Dev 1996;43: 105–33.
35. Dvorak M, Tesarik J. Differentiation of mitochondria in the human preimplantation embryo grown in vitro. Scr Med 1985;3:161–9.
36. Matos DG, Furnus C, Moses D, Baldassarre H. Effect of cysteamine on glutathione level and developmental capacity of bovine oocyte matured in vitro. Mol Reprod Dev 1995;42:432–6.
37. Wiemer K, Cohen J, Wiker S, Malter H, Wright G, Godke R. Coculture of human zygotes on fetal bovine uterine fibroblasts: embryonic morphology and implantation. Fertil Steril 1989;52:503–8.
38. Bongso A, Ng S, Fong C, Ratnam S. Coculture: a new lead in embryo quality improvement for assisted reproduction. Fertil Steril 1991;56:179–91.
39. Stewart C, Kaspar P, Brunet L, Bhatt H, Gadi I, Konthegn F, Abbondanzo S. Blastocyst implantation depends on maternal expression of leukaemia inhibitory factor. Nature 1992;359:76–9.
40. Pesce M, Farrace M, Piacentini M, Dolci S, De Felici M. Stem cell factor and leukaemia inhibitory factor promote germ cell survival suppressing programmed cell death (apoptosis). Development 1993;118:1089–94.
41. Jurisicova A, Ben-Chetrit A, Varmuza S, Casper RF. Recombinant human leukaemia inhibitory factor (rLIF) does not enhance in-vitro human blastocyst formation. Fertil Steril 1995;64:999–1002.
42. Kaufmann M.H. Early mammalian development. Cambridge University Press, Cambridge 1983;28–32.
43. De Silva M, Stracher K, Sauer S, Horvath P, Butler W. Effect of removal of cumulus cells from 1-cell mouse embryos on in vitro development. J IVF ET 1990;7: 129–33.
44. Guo J, Jurisicova A, Casper RF. Detection of deoxyribonucleic fragmentation in human sperm: correlation with fertilization in vitro. Biol Reprod 1997;56:602–7.
45. Balakier H, Squire J, Casper RF. Characterisation of human abnormal one pronuclear oocytes by morphology, cytogenetics and in situ hybridization. Hum Reprod 1993;8:740–3.
46. Van Blerkom J, Davis PW, Merriam J. A retrospective analysis of unfertilized and presumed parthenogenetically activated human oocytes demonstrates a high frequency of sperm penetration. Hum Reprod 1994;9:2381–8.
47. Raff MC, Barres BA, Burne JF, Coles HSR, Ishizaki Y, Jacobson MD. Programmed cell death and the control of the cell survival Phil Trans R Soc Lond 1994;345: 263–8.
48. Papadopoulos G, Templeton AA, Fisk N, Randall J. The frequency of chromosomal anomalies in human preimplantation embryo after in vitro fertilization. Hum Reprod 1989;4:91–8.
49. Zenzes MT, Casper, RF. Cytogenetics of human oocytes, zygotes, and embryos after in vitro fertilization. Hum Genet 1992;88:367–75.
50. Michaeli G, Fejgin M, Ghetler Y, Ben Nun I, Beyth Y, Amiel A. Chromosomal analysis of unfertilized oocytes and morphologically abnormal preimplantation embryos from an in vitro fertilization program. J IVF ET 1990;7:341–6.
51. Pellestor F, Dufour MC, Arnal F, Humeau C. Direct assessment of the rate of chromosomal abnormalities in grade IV human embryos produced by in vitro fertilization procedure. Hum Reprod 1994;9:293–302.

52. Zenzes MT, Wang P, Casper RF. Chromosome normality of untransferred (spare) embryos correlates with likelihood of pregnancy in the in vitro fertilization procedure. Lancet 1992;340:391–4.
53. Munne S, Cohen J. Monospermic polyploidy and atypical embryo morphology. Hum Reprod 1994;9:506–10.
54. Gaulden M. Maternal age effect: the enigma of Down syndrome and other trisomic conditions. Mutatation Res 1992;296:69–88.
55. Fischer B, Kunzel W, Kleinstein J, Gips H. Oxygen tension in follicular fluid falls with follicle maturation. Eur J Obstet Gynecol 1992;43:39–43.
56. Gotlieb RA, Giesing H, Zhu J, Engler R, Babior B. Cell acidification in apoptosis: granulocyte colony-stimulating factor delays programmed cell death in neutrophils by up-regulating the vascular H-ATPase. Proc Natl Acad Sci USA 1995;92: 5965–8.
57. Van Blerkom J. The influence of intrinsic and extrinsic factors on the developmental potential and chromosomal normality of the human oocyte. J Soc Gynecol Invest 1996;3:3–11.
58. Piquette GN, Tilly JL, Prichard L, Simon C, Polan ML. Detection of apoptosis in human and rat ovarian follicles. J Soc Gynecol Invest 1994;1:297–301.
59. Van Blerkom J, Davis PW, Lee J. ATP content of human oocytes and developmental potential and outcome after in vitro fertilization and embryo transfer. Hum Reprod 1995;10:415–24.
60. Tilly JL. Apoptosis and the ovary: a fashionable trend or food for thought? Fertil Steril 1997;67:226–8.
61. Cohen J, Alikani M, Liu H-C, Rosenwaks Z. Rescue of human embryos by micromanipulation. Bailliere's Clin Obstet Gynecol 1994;8:95–116.

5

Control of Apoptosis in the Uterus During Decidualization

BRUCE C. MOULTON, KAMIL C. AKCALI, THOMAS F. OGLE,
THOMAS L. BROWN, JOAN MOTZ, AND SOHAIB A. KHAN

Apoptosis has a vital function in the uterus in maintaining the homeostasis of cell number during the estrous cycle and in the tissue remodeling inherent in blastocyst implantation and development of the placenta. During the estrous cycle, estrogen and progesterone secretion control not only the proliferation and differentiation of luminal and glandular epithelial cells, but also their death by apoptosis (1). Decreases in the levels of estrogen during the estrous cycle or during later stages following a single estrogen treatment of ovariectomized animals initiate a wave of cell death in the luminal epithelium that is predominantly apoptosis (2–4). Several lines of evidence suggest that during blastocyst implantation the death of uterine epithelial cells adjacent to blastocysts is the result of apoptosis (5, 6).

As the blastocyst initiates apoptosis in luminal epithelial cells, stromal cells of the endometrium are transformed by growth and differentiation into decidual cells. Decidualization begins in the immediate vicinity of the blastocyst in the primary decidual zone at the antimesometrial side of the uterus, expands to form a secondary decidual zone or antimesometrial decidua, and eventually transforms stromal cells in the mesometrial region to form the decidua basalis which persists throughout gestation (5). Development of the hemochorial placenta requires the controlled cell death of decidual cells that occurs in the same order as their differentiation (7). Antimesometrial decidual cells in the immediate vicinity of trophoblast cells die within two days after the initiation of implantation (8). Decidualization in the secondary decidual zone in the antimesometrium begins on day 6, proceeds to a maximum level of complexity and cell number on day 10 or 11 of pregnancy, and then regresses until there is virtually no trace left by day 16 (9). Decidualization of the mesometrial stroma leads to the formation of the decidua basalis that regresses following day 14 (10).

The cells of each of the three regions of the decidualized stroma show pronounced morphological differences and synthesize specific proteins dependent upon not only spatial differentiation but also gestational age (11). Initiation of stromal decidualization initiates an ordered and timely enhanced expression of new genes yielding terminally differentiated cells capable of undergoing apoptosis. In this pattern of gene expression, there are genes necessary for the structural changes associated with decidualization, genes needed for the altered function associated with decidualization and eventual apoptosis, and genes required for the secretion of controlling growth factors or the receptors necessary for responses to these growth factors. A simple scenario would be that decidualization, once initiated, is a fixed or programmed sequence without the requirement for external growth factors for direction. Rat endometrial stromal cells decidualize in vitro (12), particularly well after progestin/estradiol pretreatment in vivo, without the addition of growth factors and perhaps after release of some putative in vivo restraint. However, increased growth factor synthesis is observed in decidual cells in vivo (13) and in vitro (14), and could be a part of this fixed program by directing and/or limiting decidualization and apoptosis by paracrine or autocrine mechanisms. Our discussion here will attempt to tease apart the control of the genes necessary for decidualization from the ones more directly involved with apoptosis while recognizing that stromal cell preparation for apoptosis may be an integral part of decidualization.

Decidual Apoptosis After Loss of Steroid Hormone Support

For many tissues that respond to sex steroids, their regression when these hormones are removed appears to proceed by apoptosis (15). Antiandrogens or androgen withdrawal induce apoptosis in the glandular epithelium of the prostate with increases in internucleosomal DNA fragmentation on days 1 and 2 postcastration (16). After decidualization of the endometrial stroma, decidual cells become exquisitely sensitive to progesterone loss (17, 18). We examined the hypothesis that decidual regression after progesterone withdrawal proceeds by apoptosis. We determined the time course and spatial/cellular location of apoptosis in deciduomal tissues after progesterone withdrawal or treatment with the antiprogestin, RU-486.

Ovariectomized (OVX) rats were pretreated sequentially without (no treatment; 2 days), with estradiol (1 µg/day; 2 days), progesterone pellets (100 mg/pellet) implanted subcutaneously, or medroxyprogesterone acetate (MPA, 3.5 mg; single injection 66 h before intrauterine stimulus) and estradiol (200 ng; single injection 18 h before intrauterine stimulus) (14). Decidualization was initiated by intrauterine administration of 50 µl of mineral oil by transcervical injection. Serum progesterone levels were measured by radioimmunoassay. Genomic DNA was purified by phenol-chloroform-isoamyl alcohol extraction (24:24:1), as described previously (14). DNA samples

(15 μg) were electrophoretically separated on 1.8% agarose gels, and the gel was stained with ethidium bromide (1 μg/mL) to visualize the DNA. The gel image was recorded with an IS-500 Digital Imaging System and the relative amount of the 200 basepair (bp) fragment was determined by image analysis. Cells undergoing apoptosis in paraffin tissue sections were identified using an in situ apoptosis detection kit.

Three days after initiation of decidualization, progesterone pellets were removed and animals killed at several intervals afterwards to determine the time course of cell death. By 8 h after removal of the progesterone pellets, serum progesterone levels decreased to basal levels (Fig. 5.1). DNA was purified from decidual tissues collected at various intervals after progesterone withdrawal, and electrophoresed on agarose gels. As shown in Figure 5.1, progesterone withdrawal increased internucleosomal cleavage in decidual cell DNA. Levels of internucleosomal DNA fragmentation significantly increased to maximal levels by 12 h after progesterone withdrawal (t-test, $p < 0.01$, Fig. 5.1). Uterine sections from rats at 0 and 12 h after progesterone withdrawal were analyzed to determine the tissue location of decidual cells sensitive to progesterone withdrawal. Sections were stained with a proce-

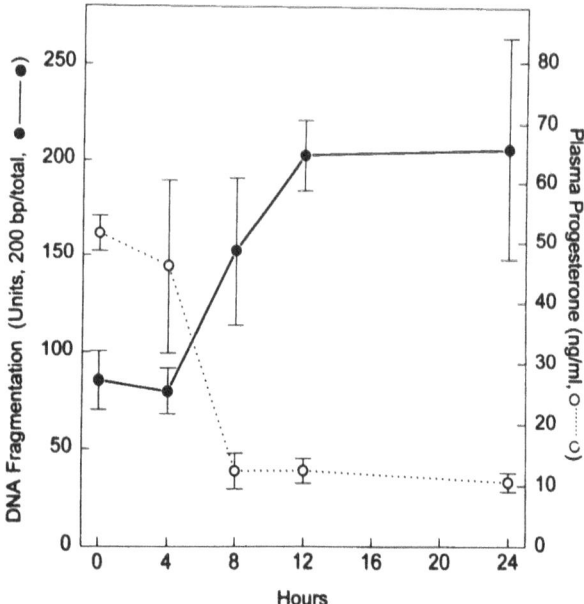

FIGURE 5.1. Effect of progesterone withdrawal on DNA fragmentation in decidual tissue. Three days after initiation of decidualization, progesterone pellets were removed and animals were killed at intervals thereafter. Amounts of the 200-bp fragment were determined after electrophoresis of equal amounts of total DNA (●—●). Serum levels of progesterone were also determined after progesterone withdrawal (o—o). Each point represents the mean ± SEM of 3 rats.

dure designed to identify cellular concentrations of 3'-OH DNA ends generated by the internucleosomal DNA cleavage associated with apoptosis. Concentrations of positively-stained decidual cells were identified on the periphery of the decidual tissue near the myometrium with other isolated positive cells scattered throughout the tissue.

The effect of the antiprogestin, RU-486, on apoptosis in decidual cells was determined in ovariectomized rats pretreated with MPA and estradiol. Three days after initiation of decidualization, RU-486 (1.5 mg/0.1 mL sesame oil, sc) was injected and animals killed at several intervals afterwards to determine the time course of cell death. RU-486 treatment significantly increased internucleosomal cleavage of decidual DNA with the same kinetics as progesterone withdrawal.

Either progesterone withdrawal or anti-progestin treatment during the growth phase of decidualization resulted in rapid increases in apoptosis in specific decidual cells at the periphery of the developing deciduoma. During decidualization, it is these peripheral cells that incorporate labeled thymidine; once stromal cells differentiate they no longer synthesize DNA (19). In several systems, it has been shown that the deregulated progression of cells through the cell cycle results in apoptotic cell death (20). Withdrawal of testosterone causes the apoptotic death of differentiated prostate epithelial cells by driving them from a quiescent G_o state into a cell cycle that they cannot complete (21). For many growth factors that promote passage from G_1 to S phase of the cell cycle, apoptosis resulting from growth factor withdrawal correlates with arrest in G_1 (20). In our studies, apoptosis at the periphery of the deciduoma would appear to result from the loss of a mitogenic factor, progesterone, which coordinates successful cell cycle progression of stromal cells during decidualization. Once stromal cells complete their division and differentiation they are unable to revert to stromal cells, and as differentiated decidual cells they become less sensitive to progesterone withdrawal.

Maximal internucleosomal DNA cleavage at 12 h after progesterone withdrawal occurs within the same time frame as the effects of progesterone withdrawal on protein synthesis observed in the hamster deciduoma (17). After the removal of progesterone implants at three days after initiation of decidualization, methionine incorporation into acid-precipitable deciduoma proteins decreased by as much as 46% within 8 h (17). When labeled decidual proteins were electrophoresed on isoelectric focussing-sodium dodecyl sulfate-polyacrylamide gel electrophoresis 2-dimensional (IEF-SDS-PAGE 2-D) gels, the effects on the synthesis of specific proteins could be identified. As expected, most of the major changes in methionine incorporation into proteins were decreased, but 13 proteins showed increased incorporation with progesterone withdrawal (17). Initiation of apoptosis induces the synthesis of specific proteins in many systems (22). It seems reasonable that several of the 13 proteins showing increased methionine incorporation with progesterone withdrawal could be newly synthesized proteins required for the completion of apoptosis.

Expression of Genes That Regulate Apoptosis During Decidualization

In several systems where apoptosis occurs, the products of several genes control the death or survival of specific cells, and many studies have established the importance of *bcl-2* and related genes in this control (23–25). The product of the *bcl-2* gene, when elevated in cells either in vivo or in vitro, prevents the normal course of cell death induced by trophic factor deprivation or by other stimuli without altering proliferation (23, 25). Recently, several other proteins homologous to BCL-2 have been isolated and their complementary DNAs (cDNAs) have been cloned (26). One of these proteins, BAX, functions to counter the effects of BCL-2 on cellular survival (24). It has also been proposed that the ratio of *bcl-2* to *bax* gene expression determines whether or not a cell undergoes apoptosis (24). Since members of the *bcl-2* gene family are found as final common mediators of survival or death in a wide variety of cell types (22), we hypothesized that the balance between *bax* and *bcl-2* expression would be altered during differentiation of endometrial stromal cells and before their eventual apoptosis (27). To test this hypothesis, we used Northern blot analysis, in situ hybridization, and immunohistochemistry to determine changes in mRNA levels and the cell type-specific localization of mRNA and protein of *bax* and *bcl-2* during decidualization.

To examine the hormonal control of *bax* and *bcl-2* expression, ovariectomized rats were given various progestin and estradiol treatments and levels of *bax* and *bcl-2* mRNAs were determined. Northern blot analysis of uterine RNA revealed two *bax* transcripts, 1.0 kb (*bax-α*) and 1.5 kb (*bax-β*), as shown in Figure 5.2A. Relative expression in treatment groups was quantitated and normalized using levels of ribosomal RNA determined from the agarose gel. Changes in levels of *bax* mRNA in response to hormone treatment and decidualization are shown in Figure 5.2B. The sequence of progestin and estrogen treatments that results in sensitivity to deciduogenic stimulus increased *bax-α* expression by approximately 2.5-fold. Expression of *bax-β* mRNA was unaffected by the progestin/estrogen treatments. Weak basal expression of *bcl-2* mRNA could be quantitated in the control group and decreased after progestin/estradiol treatments (data not shown). During decidualization, the expression of *bax-α* remained unchanged during first 48 h and then increased as decidualization spread throughout the endometrial stroma (Fig. 5.1B). At 96 h, levels of *bax-α* were 1.6-fold greater than levels at 0 h. The expression of *bax-β* remained unchanged during this period. We did not detect *bcl-2* expression during decidualization (data not shown).

Because Northern blot analysis showed that progestin/estradiol treatment and decidualization induced *bax-α* expression, we examined the cell type-specific regulation of expression of the *bcl-2* gene family at the mRNA level by in situ hybridization and at the protein level by immunohistochemistry.

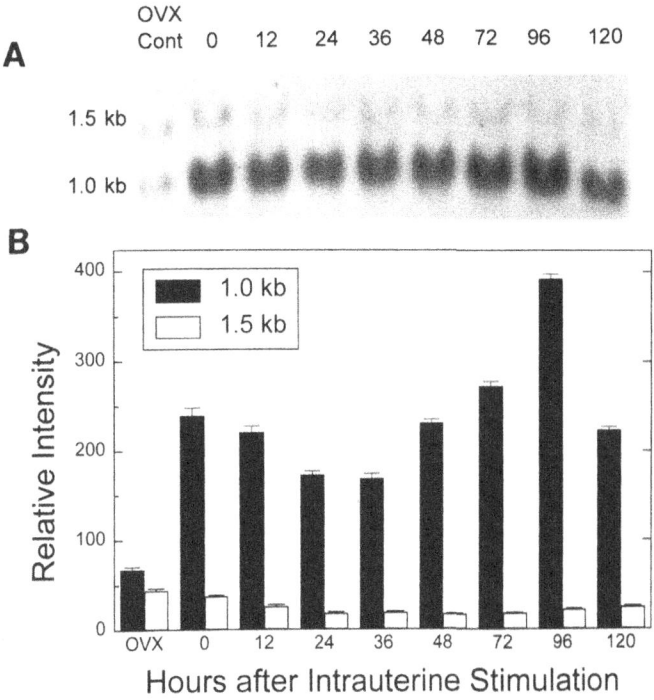

FIGURE 5.2. Time course of the effect of decidualization on expression of *bax* mRNA. Total uterine RNA was prepared from the animals before and after decidualization. Decidualization was initiated by an intrauterine stimulation after estradiol/progestin treatment. (A) Northern blot analysis of total uterine RNA with a ^{32}P-labeled *bax* cDNA. (B) Results were quantitated by the densitometric analysis of autoradiographs. 0 h versus OVX and 96 h versus 0 h, both significantly different $p < 0.001$. Data are the means of three separate experiments ± SEM. (Reproduced with permission from Akcali et al. (27).)

When an antisense *bax-α* probe was used on uterine sections from ovariectomized control rats, low numbers of silver grains were found in the luminal epithelial cells (Fig. 5.3B). Progestin/estradiol treatment to induce uterine sensitivity increased the expression of *bax* mRNA in luminal and glandular epithelia, in stromal cells just beneath the luminal epithelium, and in the circular myometrium (Fig. 5.3C). At 24 h following an intrauterine deciduogenic stimulus, increased *bax* expression was observed in the periluminal stroma (Fig. 5.3D). As decidualization progressively spread throughout the endometrial stroma, the expression of *bax* also spread (Fig. 5.3C). When the differentiation of stromal cells into decidual cells was completed in the antimesometrial region of the uterus, *bax* mRNA expression decreased accordingly (data not shown). Increased expression of *bax* ap-

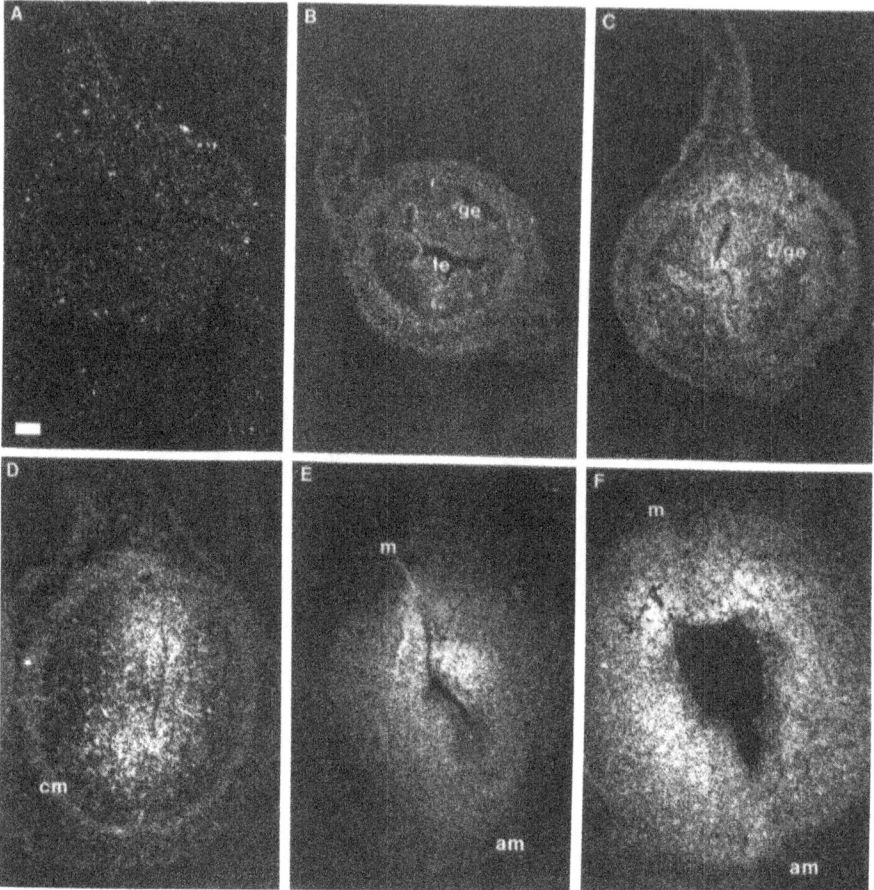

FIGURE 5.3. Effect of decidualization on in situ localization of *bax* mRNA in the rat uterus. Ovariectomized rats were pretreated with progestin and estradiol before initiation of decidualization at 0 h and killed at 24 h intervals thereafter (C–F). Frozen uterine sections were hybridized with a ^{35}S-labeled sense (A) or antisense (B–F) *bax* riboprobe. (A) Section probed with sense *bax* riboprobe. (B) Ovariectomized control, no hormone treatment. Luminal and glandular epithelium are weakly labeled. (C) 0 h, before decidualization. Increased labeling in luminal and glandular epithelia and subepithelial stroma. (D) 24 h, periluminal stroma labeled where decidualization begins. (E) 48 h, increased *bax* mRNA expression spreads through stroma. (F) 72 h, *bax* mRNA expression is seen throughout stroma. Magnification, 40X for all sections. le: luminal epithelium; ge: glandular epithelium; am: antimesometrium; m: mesometrium. (Reproduced with permission from Akcali et al. (27).)

peared to be coordinated with early stages of the decidualization of endometrial stromal cells. Uterine sections that were probed with antisense *bcl-2* probe showed a decrease in *bcl-2* expression following hormonal treatments or decidualization (data not shown). Specific tissue binding was not observed with sense RNA probes for either *bax* (Fig. 5.3A) or *bcl-2* (data not shown), used as negative controls.

Increased expression of *bax* mRNA following progestin/estradiol treatment and during decidualization suggested that increased levels of BAX protein would be an important component of this growth and differentiation. A rabbit polyclonal antibody against BAX protein was used to determine the cell type-specific location of BAX protein following progestin/estradiol treatment and during decidualization (27). In cross-sections of untreated OVX controls, weak BAX staining was observed in luminal epithelial cells but there was strong reactivity in the longitudinal myometrium. After progestin/estradiol treatment, strong staining was detected in the luminal and glandular epithelial cells and in the longitudinal part of the myometrium but not in the circular muscle. By 24 h after the deciduogenic stimulus, BAX was detected in the periluminal stroma where decidualization begins. Strong staining was also observed in the glandular epithelial cells. By 48 h after the stimulation, BAX immunoreactivity was observed in the antimesometrial region of the stroma. Thereafter, BAX immunoreactivity increased in the stroma and spread towards the mesometrial region of the uterus in parallel with the differentiation of stromal cells into the decidual cells. Although the most intense BAX staining was observed at 96 and 120 h after decidualization, there remained an area in the mesometrial region of the uterine section that showed no staining. When tissue sections were treated with an antibody against BCL-2, we observed very weak staining only in the luminal and glandular epithelial cells. This staining was present in the epithelial cells after progestin/estradiol treatment, and decreased and then disappeared as decidualization proceeded. In contrast to BAX staining, BCL-2 was not present in the longitudinal and circular myometrium. These results indicated that the balance between BAX and BCL-2 protein was altered during stromal decidualization.

In our analysis of *bcl-2* and *bax* gene expression in the uterus, progestin/estrogen treatment and decidual differentiation were associated with a marked increase in expression of *bax* and decreased expression of *bcl-2*. Differentiation and apoptosis in the endometrial stroma appear to be linked to the ability of estrogen/progestin treatment and an intrauterine stimulus to increase *bax* expression and to decrease *bcl-2* expression in the uterus, thus increasing the *bax/bcl-2* ratio. Increased *bax* expression was identified in decidualizing stromal cells 4 to 5 days before the development of internucleosomal DNA cleavage in these cells. Increased expression of *bax* during the first 24 h after the initiation of decidualization and the shift of the *bax/bcl-2* ratio towards the death inducer (e.g., BAX) suggests that early stages of stromal cell differentiation involve a commitment to cell death. Effects of increased *bax* expression might be delayed by heterodimenic pairing of BAX protein with the other proteins encoded by the *bcl-2* gene family, most of which are expressed in the rat uterus at different levels. The coordinated expression of the other members of the BCL-2 family may also have important functions in controlling cell death during placental development.

Decidual Expression of Genes Associated With the Degradation Phase of Apoptosis

Cardinal morphological features of apoptosis include the early compaction of nuclear chromatin and condensation of cytoplasm (28). Further condensation is accompanied by convolution of nuclear and cellular outlines and the breakup of the nucleus and cytoplasm into a number of membrane-bounded apoptotic bodies surrounded by plasma membrane. Phagocytosis by neighboring cells or macrophages completes apoptosis without disruption of overall tissue architecture and without triggering inflammation. Several genes necessary for accomplishing these morphologic events, which show increased expression either before or after the initiation of apoptosis have been identified.

Increased transglutaminase expression has been associated with apoptosis in several systems and appears to play a role in achieving structural changes characteristic of apoptotic cells (29). Transglutaminases catalyze calcium dependent cross-linking reactions between proteins establishing stable ε(γg-glutamyl)lysine cross-links and/or of N,N-bis(γ-glutamyl) polyamine cross-bridges leading to protein polymerization. Increased transglutaminase immunoreactivity is observed in endometrial decidual cells at maximal levels in cells in contact with the embryo and with a decreasing gradient outward (30). Transglutaminase expression in human endometrial stromal cells increases with hormone concentration dependence at 6 h after progesterone treatment (31). When transglutaminase activity is blocked by monodansylcadaverine, the morphological transformation of stromal cells into decidual cells is blocked. These data suggest that the commitment of stromal cells to the differentiated phenotype associated with decidualization involves a similar commitment to the eventual apoptosis of these cells.

Progesterone increases the expression of cathepsin D, a lysosomal aspartyl endopeptidase involved in intracellular protein degradation, in uterine luminal epithelial cells during early pregnancy. Cathepsin D functions perhaps as part of the capacitation of these cells for response to the apoptotic stimuli associated with blastocyst implantation (32). Removal of testosterone by castration initiates apoptosis in the ventral prostate and increases the expression of several genes including cathepsin D (33). Extensive expression of cathepsin D is observed in decidual cells only after DNA fragmentation has occurred, suggesting that the increase in cathepsin D is a late manifestation of programmed cell death in the decidua (34).

Another protein closely associated with apoptosis is apolipoprotein J (apoJ, testosterone repressed message-2, sulfated glycoprotein-2, clusterin), an 80 kDa glycoprotein associated with plasma lipids. ApoJ binds extracellular matrix proteins, such as fibronectin and heparin, as well as several membrane-active molecules (35). The increased expression of apoJ at sites of tissue injury or reorganization suggests that apoJ plays a role in cytoprotection, or that it is involved in lipid transport from dying cells or cells undergoing membrane remodelling during differentiation (36). Mouse apoJ mRNA is constitutively expressed in epithelial cells at barrier-fluid interfaces, but can be dramatically induced in epithelial and other cell types under a variety of conditions such as hormone depletion in the regressing mammary gland, prostate (37), and uterus (38). ApoJ mRNA expression was increased in the antimesometrial deciduoma on day 9 before the first detection of internucleosomal DNA cleavage, and on day 10 in both the antimesometrial and mesometrial regions (34). When in situ hybridization was used to identify the location of apoJ expression in mice, apoJ mRNA first appeared in a band of decidualized cells close to the embryo (39) (Fig. 5.4B). As the decidualized region expanded and the border separating decidualized from undecidualized tissue moved away from the embryo, the band of apoJ-expressing cells moved similarly (Fig. 5.4.D). In cells of the fully decidualized region, however, neither apoJ message nor protein was detected (Fig. 5.4E). ApoJ protein appeared to be localized exclusively to the undecidualized stroma. Although the function of apoJ is unknown, these observations and the properties of the protein suggest that apoJ is involved in localized lipid transport and in cytoprotection during apoptosis.

Steroid Hormone and Growth Factor Control of Decidual Apoptosis

Although progesterone withdrawal and antiprogestin treatment initiate decidual apoptosis during decidual growth, decidual cells appear to have a finite life span with eventual apoptosis as a feature of their cellular differentiation. Apoptosis in stromal cells around the lumen on day 7 of pregnancy takes place without loss of progesterone (8). Antimesometrial cells undergo apoptosis on days 10 to 12 of pseudopregnancy, despite the maintenance of plasma progesterone levels (34). Levels of mRNA for progesterone receptor (PR) were measured by RT-PCR during this period. An increase in PR-mRNA levels was observed between days 10 and 15 with no difference in expression between the antimesometrial and mesometrial regions, despite the extensive apoptosis occurring in these tissues at this time. However, localized changes in PR protein expression were observed during decidual growth and regression by Western blot analysis (40).

Three forms of PR detected by Western blotting have been described (A, B, and C) that arise from the use of alternate splicing and are regulated by different promoters (41–43). Gibori et al. (40) showed that antimesometrial cells express principally the A form and little of the B form, whereas mesometrial cells expressed some of the A form but principally the C form. The

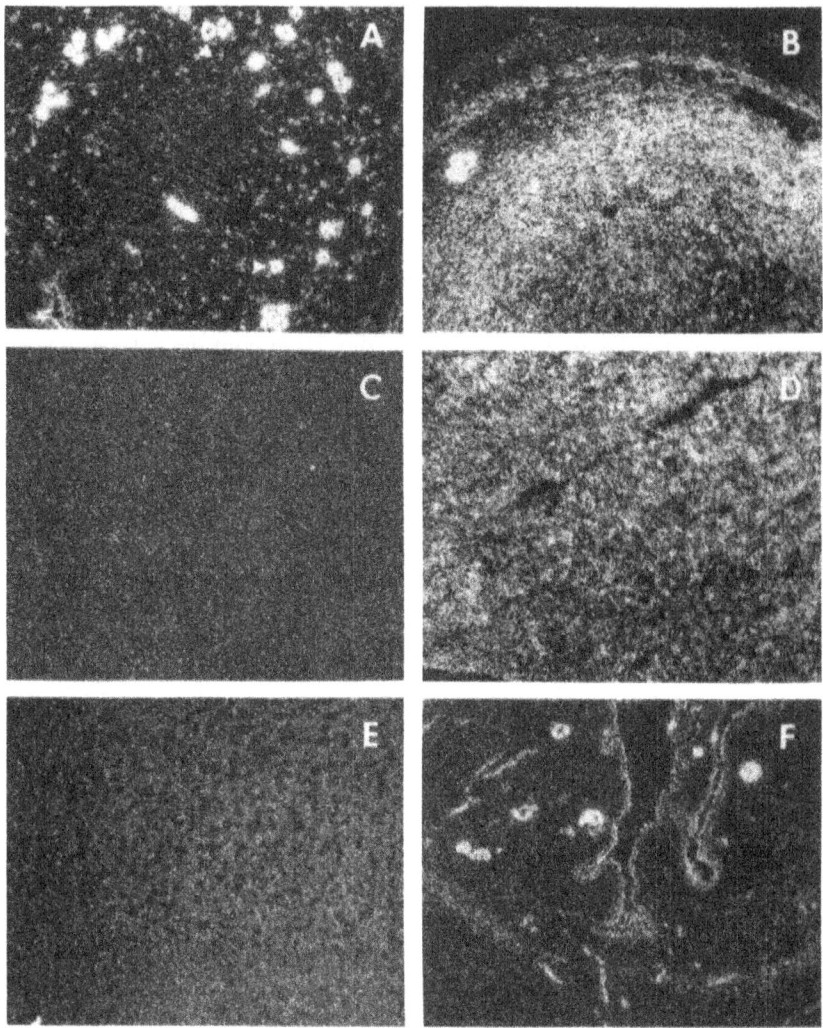

FIGURE 5.4. Expression of apoJ during decidualization. (A) No apoJ mRNA was detected in uterine luminal epithelial cells 3 to 4 days following intrauterine stimulation
(day 3 shown), although abundant apoJ signal was present in glandular epithelial cells
(arrowheads). (B) ApoJ mRNA was strongly expressed in decidual stromal cells and
myocytes of the circular muscle in hormone-treated females on day 5 following stimulation. (C) A control section of the decidua on day 6 probed with an apoJ sense cDNA
indicating background hybridization. (D) A serial section to (C) probed with apoJ
antisense riboprobe, indicating apoJ message in decidual cells at day 6 poststimulation.
(E) ApoJ mRNA was not evident in decidualized cells on days 7 to 8 in hormone-treated
females following stimulation (day 8 shown). (F) In the control unstimulated uterine
horn, apoJ mRNA was highly expressed in both glandular and luminal epithelial, but
not stromal, cells on days 7 to 8 (day 8 shown). Magnification: A–F, 200X. (Reproduced with permission from Brown et al. (39).)

A form and B form disappear from the antimesometrium after day 11 at the time when extensive internucleosomal DNA cleavage takes place in this tissue. The C form continues to be expressed in mesometrial cells but decreases by day 13. All of these changes in PR protein occur without comparable changes in PR mRNA, suggesting regulation of the receptor at the posttranscriptional level. This regulation of PR may provide a mechanism for the control of location and timing of decidual regression. Further evidence for PR regulation of decidual apoptosis has been gained from studies of the decidua basalis, decidual tissue that develops in the mesometrial region of the endometrium.

After day 10 of pregnancy, when antimesometrial decidual cells have begun to regress, the continuing decidualization of the mesometrial region leads to the formation of the decidua basalis and eventual establishment of the definitive chorioallantoic placenta/decidua unit. Levels of progesterone-binding sites were determined in cytosolic and nuclear fractions by exchange assay from day 8 to day 17 (T. Ogle et al., unpublished observations). Internucleosomal DNA cleavage in the decidua basalis remained unchanged until day 12 of pregnancy and then increased by day 14 and day 17 (Fig. 5.5). On the days when nuclear progesterone binding decreased, internucleosomal DNA cleavage increased. Levels of total progesterone-binding (cytosolic + nuclear) remained relatively constant with the exception of an increased level

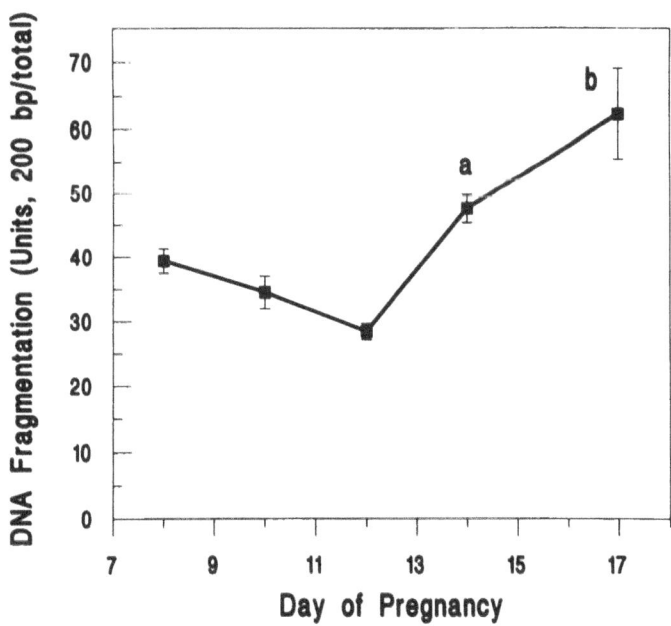

FIGURE 5.5. DNA fragmentation in decidua basalis during pregnancy. Values are means ± SEM, n = 3–4. Statistical analysis by ANOVA followed by Student-Newman-Keuls multiple range test. a, $p < 0.01$ from earlier stages; b, $p < 0.05$ from D14.

on day 11 (Fig. 5.6). Serum progesterone levels remained high during this 17-day period. Despite the maintenance of serum progesterone levels and total progesterone binding, apoptosis increased in the decidua basalis coincident with the loss of nuclear progesterone binding (T. Ogle, unpublished observations). Decreased nuclear PR could involve alterations in receptor phosphorylation. Receptor phosphorylation by various cellular modulators of other signal transduction pathways may alter DNA and steroid binding (44). In addition to its phosphorylation status, the interaction of PR with co-activators and/or co-repressors may lead to a change in the function of PR as a transcription factor (45).

To investigate the interaction between the protein kinase C (PKC) cell signaling pathway and the control of progesterone-binding and apoptosis, active (phorbol 12-myristate 13-acetate, PMA) and inactive (4α-phorbol 12,13-didecanoate, PDD) phorbol esters were given on days 10 or 14 of pregnancy. PMA treatment had no effect on progesterone-binding sites nor internucleosomal DNA cleavage on day 10, but decreased both nuclear and cytosolic progesterone-binding on day 14 and increased internucleosomal DNA cleavage. Apoptotic regression of the decidua basalis during both normal physiologic

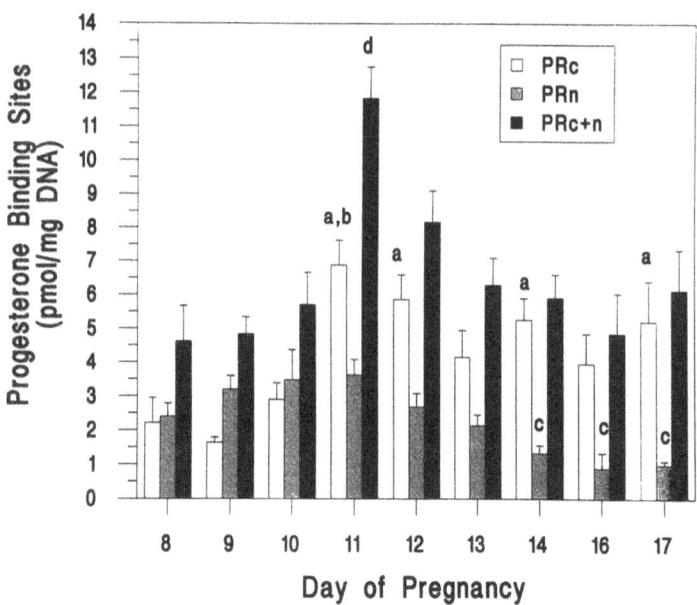

FIGURE 5.6. Changes in the concentration of progesterone binding sites in the decidua basalis during pregnancy. Values are means + SEM, n = 5–8. Statistical analysis by ANOVA followed by Student-Newman-Keuls multiple range test. PRc: progesterone binding in cytosolic fraction; PRn progesterone binding in nuclear fraction; PRc + n: total binding. a, $p < 0.05$ from D 8 and D 9 PRc. b, $p < 0.01$ from D 10 PRc; c, $p < 0.05$ from D 14, D 16, and D 17 PRn; d, $p < 0.01$ from all other PRc + n.

processes and following phorbol ester treatment was preceded by the loss of nuclear progesterone-binding (T. Ogle, unpublished observations).

Although loss of nuclear progesterone binding activity might provide a mechanism for the initiation of decidual apoptosis, the occurrence of apoptosis in specific tissue locations at specific time points requires additional control mechanisms. As suggested earlier, a simple scenario would be that decidualization once initiated is a fixed or programmed sequence without the requirement for external growth factors for direction. However, increased growth factor synthesis is observed in specific decidual regions in vivo at specific times and could be a part of this fixed program by directing and/or limiting decidualization and apoptosis by paracrine or autocrine mechanisms. Members of the transforming growth factor-beta (TGFβ) family have been implicated in these control mechanisms.

TGFβ1 mRNA is expressed in all zones of the decidua with some concentration in the secondary decidua zone during the growth and differentiation of murine decidual cells (days 5 to 8) (13, 46). During this period of decidual growth, which precedes internucleosomal DNA cleavage, intracellular TGFβ1 peptide was detected in the primary decidual zone with extracellular TGFβ1 in the secondary zone. Later in the progression of decidualization, the expression of another member of the TGFβ family, activin, was increased (47). Activin expression was increased at the time and tissue location of internucleosomal DNA cleavage in both the antimesometrial and mesometrial decidua. Additional control mechanisms may involve modulation of the biologic activity of these growth factors in specific regions of the decidua. Two proteins, α2-macroglobulin and follistatin, bind and inactivate TGFβ and activin and are synthesized in specific regions of the decidua during its growth and differentiation (47). It has been proposed that synthesis of α2-macroglobulin by the mesometrial region of the decidua could provide a mechanism of control of apoptosis, preserving these cells to become the decidua basalis (14). It has also been proposed that follistatin and α2-macroglobulin, by binding to activin, prevent activin from stimulating the expression of its own gene (47).

Members of the TGFβ family have been implicated in the regulation of cell proliferation and differentiation, regulation of the formation of extracellular matrix and cell surface molecules, and immunoregulation (48). The addition of TGFβ1 to primary cultures of rabbit uterine epithelial cells both inhibited cell proliferation and increased apoptosis (49). TGFβ1 also induced apoptosis in primary cultures of hepatocytes and in a hepatoma cell line (50, 51). TGFβ1 is one of the genes activated with the initiation of apoptosis in the ventral prostate by castration (52). We have shown that stromal cells isolated and cultured after progestin/estrogen pretreatment will respond to TGFβ1 or β2 with increased internucleosomal DNA cleavage (14) (Fig. 5.7). The stromal cells secreted TGFβ2 rather than TGFβ1, and neutralization of secreted TGFβ inhibited stromal apoptosis. Activin has been shown to be involved in the induction of apoptosis (53, 54).

[TGF-β₁]

Std   Cont   10⁻⁹   10⁻¹⁰   10⁻¹¹   10⁻¹²

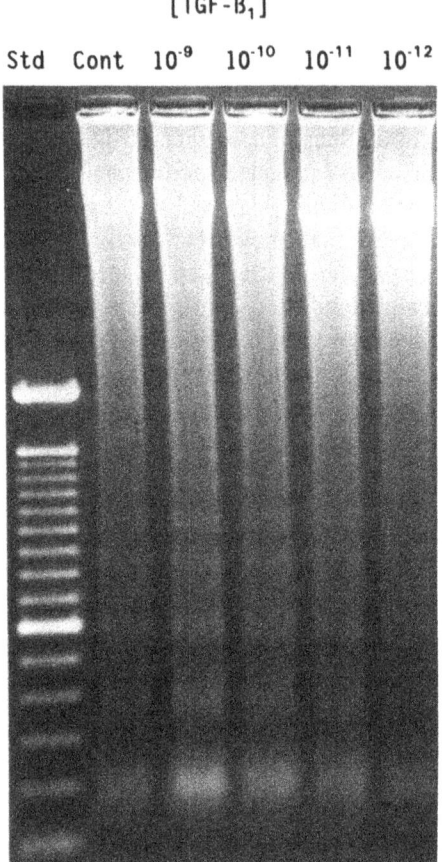

FIGURE 5.7. Effect of TGFβ on internucleosomal DNA cleavage in endometrial stromal cells. After pretreatment of ovariectomized rats with MPA for 72 h, endometrial stromal cells were isolated and incubated for 48 h in DMEM:F-12 containing 10% fetal bovine serum. Incubations were continued for 24 h with various concentrations of TGFβ or control vehicle (Cont), DNA was isolated, and 15 μg were electrophoresed on a 1.8% agarose gel. A 100-bp DNA ladder was used as the size standard (Std).

Control of decidua differentiation and/or apoptosis by TGFβ and activin remains to be elucidated. TGFβ synthesis appears to be increased in the earliest stages of stromal cell differentiation and coincident with the expression of BAX, a gene associated with apoptosis. Stromal activin synthesis increases during internucleosomal DNA cleavage, a terminal stage of both apoptosis and stromal cell differentiation. It is not clear that either TGFβ or activin synthesis can be identified as part of specific control mechanisms in the differentiation events involved in decidualization or apoptosis which is the final stage of this differentiation.

Summary

The growth and differentiation of uterine endometrial cells during decidualization requires continuous support by progesterone. Although apoptosis can be initiated in decidual cells by progesterone withdrawal, apoptosis also appears to be a feature of their terminal differentiation. The earliest stages of decidualization involve localized increases in the expression of BAX and transglutaminase, as well as of TGFβ which might serve to control either decidualization or apoptosis, or both. Increased apoJ expression may serve to control potential inflammatory processes during tissue remodeling. Localized increases in tissues levels of members of the TGFβ family, with modulation of their activity by the localized synthesis of binding proteins, may control the expression of genes necessary for stromal differentiation, including the genes required for apoptosis.

References

1. Sandow BA, West NB, Norman RL, Brenner RM. Hormonal control of apoptosis in hamster uterine luminal epithelium. Am J Anat 1979;156:15–36.
2. Pollard JW, Pacey J, Cheng SVY, Jordan EG. Estrogens and cell death in murine uterine luminal epithelium. Cell Tissue Res 1987;249:533–40.
3. Nawaz S, Lynch MP, Galand P, Gerschenson LE. Hormonal regulation of cell death in rabbit uterine epithelium. Am J Pathol 1987;127:51–9.
4. Rotello RJ, Hocker MB, Gerschenson LE. Biochemical evidence for programmed cell death in rabbit uterine epithelium. Am J Pathol 1989;134:491 5.
5. Parr MB, Parr EL, ed. The implantation reaction. New York and London: Plenum Medical 2nd ed., 1989;233–77.
6. Parr EL, Tung HN, Parr MB. Apoptosis as the mode of uterine epithelial cell death during embryo implantation in mice and rats. Biol Reprod 1987;36:211–25.
7. Welsh AO. Uterine cell death during implantation and early placentation. Microsc Res Tech 1993;25:223–45.
8. Welsh AO, Enders AC. Trophoblast-decidual cell interactions and establishment of maternal blood circulation in the parietal yolk sac placenta of the rat. Anat Rec 1987;217:203–19.
9. Welsh AO, Enders AC. Light and electron microscopic examination of the mature decidual cells of the rat with emphasis on the antimesometrial decidua and its degeneration. Am J Anat 1985;172:1–29.
10. Davies J, Glasser SR. Histological and fine structural observations on the placenta of the rat. Acta Anat 1968;69:542–608.
11. Gu Y, Jayatilak PG, Parmer TG, Gauldie J, Fey GH, Gibori G. Alpha 2-macroglobulin expression in the mesometrial decidua and its regulation by decidual luteotropin and prolactin. Endocrinology 1992;131:1321–8.

12. Vladimirsky F, Chen L, Amsterdam A, Zor U, Lindner HR. Differentiation of decidual cells in cultures of rat endometrium. J Reprod Fertil 1977;49:61–8.
13. Tamada H, McMaster MT, Flanders KC, Andrews GK, Dey SK. Cell type-specific expression of transforming growth factor-beta 1 in the mouse uterus during the periimplantation period. Mol Endocrinol 1990;4:965–72.
14. Moulton BC. Transforming growth factor-β stimulates endometrial stromal apoptosis in vitro. Endocrinology 1994;134:1055–60.
15. Thompson EB. Apoptosis and steroid hormones. Mol Endocrinol 1994;8:665–73.
16. English HF, Kyprianou N, Isaacs JT. Relationship between DNA fragmentation and apoptosis in the programmed cell death in the rat prostate following castration. Prostate 1989;15:233–50.
17. MacDonald RG, Morency KO, Leavitt WW. Progesterone modulation of specific protein synthesis in the decidualized hamster uterus. Biol Reprod 1983;28: 753–66.
18. Finn CA, Pope M. Vascular and cellular changes in the decidualized endometrium of the ovariectomized mouse following cessation of hormone treatment: a possible model for menstruation. J Endocrinol 1984;100:295–300.
19. Lobel BL, Levy E, Shelesnyak MC. Studies on the mechanism of nidation. XXXIV. Dynamics of cellular interactions during progestation and implantation in the rat. Acta Endocrinol (Supplement) 1967;123:7–109.
20. Meikrantz W, Schlegel R. Apoptosis and the cell cycle. J Cell Biochem 1995;58: 160–74.
21. Colombel M, Olsson CA, Ng PY, Buttyan R. Hormone-regulated apoptosis results from reentry of differentiated prostate cells onto a defective cell cycle. Cancer Res 1992;52:4313–9.
22. Steller H. Mechanisms and genes of cellular suicide. Science 1995;267:1445–9.
23. Korsmeyer SJ. Bcl-2 initiates a new category of oncogenes: regulators of cell death. Blood 1992;80:879–86.
24. Oltvai ZN, Milliman CL, Korsmeyer SJ. Bcl-2 heterodimerizes in vivo with a conserved homolog, Bax, that accelerates programmed cell death. Cell 1993; 74:609–19.
25. Reed JC. Bcl-2 and the regulation of programmed cell death. J Cell Biol 1994;124:1–6.
26. Hacker G, Vaux DL. A sticky business. Curr Biol 1995;5:622–4.
27. Akcali KC, Khan SA, Moulton BC. Effect of decidualization on the expression of bax and bcl-2 in the rat uterine endometrium. Endocrinology 1996;137:3123–31.
28. Kerr JFR, Harmon BV. Definition and incidence of apoptosis: an historical perspective. In: Tomei LD, Cope FO, eds. Apoptosis: the molecular basis of cell death. Plainview, NY: Cold Spring Harbor Laboratory Press, 1991:5–29.
29. Fesus L, Thomazy V, Falus A. Induction and activation of tissue transglutaminase during programmed cell death. FEBS Lett 1987;245:150–54.
30. Piacentini M, Autuori F. Immunohistochemical localization of tissue transglutaminase and Bcl-2 in rat uterine tissues during embryo implantation and post-partum involution. Differentiation 1994;57:51–61.
31. Fujimoto M, Kanzaki H, Nakayama H, Higuchi T, Hatayama H, Iwai M, et al. Requirement for transglutaminase in progesterone-induced decidualization of human endometrial stromal cells. Endocrinology 1996;137:1096–101.
32. Moulton BC, Khan SA. Progestin and estrogen control of cathepsin D expression

and processing in rat uterine luminal epithelium and stroma-myometrium. Proc Soc Exp Biol Med 1992;201:98–105.

33. Sensibar JA, Liu X, Patai B, Alger B, Lee C. Characterization of castration-induced cell death in the rat prostate by immunohistochemical localization of cathepsin D. Prostate 1990;16:263–76.

34. Gu Y, Jow GM, Moulton BC, Lee C, Sensibar JA, Park-Sarge OK, et al. Apoptosis in decidual tissue regression and reorganization. Endocrinology 1994;135:1272–9.

35. Fesus L, Davies PJA, Piacentini M. Apoptosis: molecular mechanisms in programmed cell death. Eur J Cell Biol 1991;56:170–7.

36. Ahuja HS, Tenniswood M, Lockshin R, Zakeri ZF. Expression of clusterin in cell differentiation and cell death. Biochem Cell Biol 1994;72:523–30.

37. Sensibar JA, Griswold MD, Sylvester SR, Buttyan R, Bardin CW, Cheng CY, et al. Prostate ductal system in rats: regional variation in localization of an androgen-repressed gene product, sulfated glycoprotein-2. Endocrinology 1991;128: 2091–102.

38. Brown TL, Moulton BC, Baker VV, Mira J, Harmony JAK. Expression of apolipoprotein J in the uterus is associated with tissue remodeling. Biol Reprod 1995;52:1038–49.

39. Brown TL, Moulton BC, Witte DP, Swertfeger DK, Harmony JAK. Apolipoprotein J/clusterin expression defines distinct stages of blastocyst implantation in the mouse uterus. Biol Reprod 1996;55:740–7.

40. Gibori G, Gu Y, Srivastava RK. Differential gene expressions and programmed cell death in the two cell populations forming the rat decidua. In: Dey SK, ed. Molecular and cellular aspects of periimplantation processes. New York: Springer-Verlag, 1995:67–83.

41. Wei LL, Gonzalez-Aller C, Wood WM, Miller LA, Horwitz KB. 5'-Heterogeneity in human progesterone receptor transcripts predicts a new amino-terminal truncated "C" receptor and unique A-receptor messages. Mol Endocrinol 1990;4:1833–40.

42. Schott DR, Shyamala G, Schneider W, Parry G. Molecular cloning, sequence analyses, and expression of complementary DNA encoding murine progesterone receptor. Biochemistry 1991;30:7014–20.

43. Kraus WL, Montano MM, Katzenellenbogen BS. Cloning of the rat progesterone receptor gene 5'-region and identification of two functionally distinct promoters. Mol Endocrinol 1993;7:1603–16.

44. Beck CA, Weigel NL, Edwards DP. Effects of hormone and cellular modulators of protein phosphorylation on transcriptional activity, DNA binding, and phosphorylation of human progesterone receptors. Mol Endocrinol 1992;6:607–20.

45. Onate SA, Tsai SY, Tsai M-J, O'Malley BW. Sequence and characterization of a coactivator for the steroid hormone receptor superfamily. Science 1995;270: 1354–7.

46. Manova K, Paynton BV, Bachvarova RF. Expression of activins and TGF beta 1 and beta 2 RNAs in early postimplantation mouse embryos and uterine decidua. Mech Dev 1992;36:141–52.

47. Gu Y, Srivastava RK, Ou J, Krett NL, Mayo KE, Gibori G. Cell-specific expression of activin and its two binding proteins in the rat decidua: role of alpha-2-macroglobulin and follistatin. Endocrinology 1995;136:3815–22.

48. Sporn MB, Roberts AB. Transforming growth factor-beta: recent progress and new challenges. J Cell Biol 1992;119:1017–21.

49. Rotello RJ, Lieberman RC, Purchio AF, Gerschenson LE. Coordinated regulation of apoptosis and cell proliferation by transforming growth factor beta 1 in cultured uterine epithelial cells. Proc Natl Acad Sci USA 1991;88:3412–5.

50. Oberhammer FA, Pavelka M, Sharma S, Tiefenbacher R, Purchio AF, Bursch W, et al. Induction of apoptosis in cultured hepatocytes and in regressing liver by transforming growth factor-beta 1. Cell Biol 1992;89:5408–12.

51. Lin JK, Chou CK. In vitro apoptosis in the human hepatoma cell line induced by transforming growth factor-beta 1. Cancer Res 1992;52:385–8.

52. Kyprianou N, Isaacs JT. Expression of transforming growth factor-beta in the rat ventral prostate during castration-induced programmed cell death. Mol Endocrinol 1989;3:1515–22.

53. Mundle SD, Sheth NA. Suppression of DNA synthesis and induction of apoptosis in rat prostate by human seminal plasma inhibin (HSPI). Cell Biol Int 1993; 17:587–94.

54. Hully JR, Chang L, Schwall RH, Widmer HR, Terrell TG. Induction of apoptosis in the murine liver with recombinant human activin A. Hepatology 1994;20:854–62.

6

Molecular Mechanisms of Müllerian Inhibiting Substance-Mediated Apoptosis

PATRICIA K. DONAHOE AND TONGWEN WANG

The process of Müllerian duct regression in the male fetus is a remarkable example of apoptosis, triggered by Sertoli cell produced Müllerian inhibiting substance (MIS), a member of the transforming growth factor-β (TGF-β) superfamily. Since the cloning and expression of recombinant MIS (1), and identification of its putative type I (2) and type II (3, 4, 5) serine/threonine kinase receptors, as well as immediate cytosolic interactors (6, 7) has been accomplished, we will attempt in this chapter to correlate the outcome of MIS-mediated apoptosis with these molecular events. In 1965, Hamilton and Teng (8) noted changes in tissue levels of nucleic acids and nucleases and the appearance of proteolytic enzymes during Müllerian duct regression in the chick embryo. We described autophagy of epithelial cells during regression of the rat Müllerian duct in vivo (9) and in vitro (10) by electron microscopy, and in 1982 (11) noted condensation of mesenchyme around the epithelial ducts and breakdown of basement membrane as earlier morphological events in Müllerian duct regression. These early events are accompanied by epithelial to mesenchymal transformation, after which the transformed cells migrated into surrounding mesenchyme (12) where they could be further followed, in experimental chick-quail chimeras, to the mesonephros (13). We determined in vitro that continuous exposure of the rat Müllerian duct to MIS was required for 12 h after which the source of MIS could be removed and the regressive events would proceed autonomously (14). If exposure to MIS was delayed for 12 to 24 h after the ducts were placed in organ culture, then condensation of mesenchyme around the Müllerian duct was the predominant morphologic feature observed (14), indicating that the receptor response to MIS was greatest in the mesenchyme (15). Delayed exposure (e.g., after 24 h) produced few morphologic characteristics of Müllerian duct regression.

Müllerian-inhibiting substance may also initiate programmed cell death in some tumors of Müllerian duct origin. Human ovarian tumors (16), morphologically recapitulate the histology of the fetal Müllerian duct, i.e., serous cystoadenocarcinomas have the appearance of fetal fallopian tube, endometrioid tumors look like endometrium, and mucinous cystoadenocarcinomas have the morphology of the endocervix (17). Thus, analysis of the apoptotic molecular events involved in the regressing Müllerian duct might uncover mechanisms by which programmed cell death could be initiated in such human tumors. A program was therefore undertaken to elucidate the molecular events surrounding Müllerian duct regression with the assumption that ovarian or Müllerian tumors from the fetal celoemic epithelium would maintain receptors and a functional apoptotic response to MIS as observed in the fetus. The specificity of MIS was a factor in its choice as a paradigm for study, with the expectation that broader therapeutic applications for growth control would ensue.

Characterization of MIS

We adapted the organ culture bioassay of Picon (18) using the 14 day rat urogenital ridge as a target organ and scoring Müllerian duct regression in a semiquantitative manner based upon the morphologic criteria of programmed cell death. Using a variety of tissues from various developmental stages, we first found that embryonic and postnatal testis produced high levels of MIS (19), and later discovered that the postnatal ovary throughout adulthood was also capable of inducing regression, though less robust (20, 21). Gonadal production of MIS was confirmed in a number of mammalian and avian species and further delineated to the Sertoli cell of the testis (22) and the granulosa cell of the ovary (21). As a result of these observations, bovine postnatal (23–25) and fetal (26) testes were used as a source to purify MIS to homogeneity. Microsequencing of gel-purified bovine MIS was carried out, and degenerate oligonucleotides covering the N-terminus were used to isolate a bovine cDNA for MIS (1). A genomic clone of MIS was subsequently isolated, which, when transiently expressed in COS cells (1) and stably expressed in CHO cells (27), produced bioactive MIS. A partial bovine cDNA was simultaneously isolated by Picard et al. (28) using expression cloning. Sequence homology at the amino acid level was noted to the then newly cloned TGF-β (29) and inhibin (30). Cloning of decapentaplegia (31) and Vg-1 (32) soon followed, establishing MIS as a member of an enlarging gene family in which the highest homology existed in the C-terminus surrounding conserved cysteine residues. Glycoprotein prohormones of this family are activated by endogenous proteolytic enzymes to yield N- and C-terminal domains (33). Cleavage of MIS occurs at position 427 (33), specifically at a consensus $R^{-4}XXR^{-1}$ Kex 2/furin-like site (34). Although the endogenous proteolytic enzyme responsible for testicular MIS cleavage remains unknown, prohormone cleavage appears to be enhanced by testosterone in vivo (35) and can be accomplished in vitro with high plasmin concentrations (33). The C-terminal fragment so produced maintains biological activity (34) that can be somewhat enhanced by addition of the glycosylated N-terminal domain (35, 36).

Characterization of the MIS Receptor

Although expression cloning successfully yielded the activin and TGF-β type II receptors (37, 38), attempts to use expression cloning techniques to isolate the MIS receptor were unsuccessful since labeling of the ligand by conventional means resulted in a loss of biological activity and presumably binding ability (39). We used a PCR-based cloning strategy with information provided by sequences conserved in the TGF-β and activin type II receptors, and identified four receptors with homology to the known type II receptors (2). All of these receptors are serine/threonine kinase transmembrane receptors, but each has shorter tail regions than those found in the type II receptors, and all contain a juxtamembrane conserved GS box not found in the type II receptors. Thus, they belong to a unique class of type I serine/threonine kinase receptors. Although none of these receptors can bind labeled MIS ligand, expression of one of the receptors was localized to the mesenchyme around the Müllerian duct (R1) (2). We and others continued PCR cloning, using sequences conserved among the type II receptors, but absent in the type I receptors. Three laboratories discovered putative MIS type II receptors based upon their localization by in situ hybridization to the mesenchyme surrounding the Müllerian duct in the rat (3, 5) and the rabbit (4). However, binding of labeled MIS to this type II receptor overexpressed in COS cells showed binding that was either noncompetable (Teixeira et al., unpublished) or nonsaturable (4). The human homolog of this type II receptor was isolated as two overlapping lambda phage clones using the rat type II receptor cDNA to probe a human genomic library, and was localized to chromosome 12 by Visser et al. (40). Simultaneously, a human cDNA was reconstructed from fragments isolated by hybridization and PCR techniques from a human fetal testis library (41). A patient was found to have a mutation at the splice-donor site of intron 2 that leads to a premature stop codon in the human putative type II receptor (41). Furthermore, when the mouse MIS type II receptor was inactivated by nonhomologous recombination, the null mutant mouse displayed retained Müllerian ducts (R. Behringer et al., personal communication). These apparent contradictions between convincing in vivo observations and the inefficiency of receptor binding to labeled ligand in overexpressed COS cells in vitro could be explained by the existence of alternately spliced isoforms. One piece of evidence to support this is the finding that Northern blot analysis failed to detect the predicted 2.5-kb mRNA fragment of the type II receptor in MIS-responsive cell lines. Instead, these cells display a larger molecular species (Teixeira et al., unpublished), the nature of which is now being investigated as a candidate binding isoform of the MIS type II receptor.

Identification of Candidate MIS Type I Receptor Interactors

The yeast two-hybrid system has proved to be extremely powerful in identifying downstream interactors of this serine/threonine kinase receptor family. Since R1 is the candidate MIS type I receptor, the entire cytoplasmic domain of R1 was used as the bait to screen a neonatal rat heart cDNA library in yeast. Two interactors were isolated, the immunophilin, FKBP12 (6), and a novel protein named Myx (Wang et al., unpublished). FKBP12 is an abundant cytoplasmic protein known to be the receptor for two immunosuppressive macrolide molecules, rapamycin and FK506. Interactions were detected between FKBP12 and all known type I receptors of the TGF-β family of ligands, and domains responsible for the interaction, and specifically conserved among the type I receptors, have been mapped. Recently, we demonstrated that FKBP12 functions as a common inhibitor of the type I receptors for TGF-β and MIS (7).

The interaction between R1 and the novel interactor, Myx, has been demonstrated only in vitro (Wang et al., unpublished). Protein sequence analysis indicates that Myx is a new member of the basic-helix-loop-helix-leucine-zipper (bHLH-Zip) family, and is most homologous to the known Max interactors, Mxi1 and Mad. These latter two proteins have been shown to repress Myc/Max heterodimer-mediated transcriptional activation of an E-Box regulated reporter (Min4CAT) in NIH3T3 cells (42, 43). Myx is able to heterodimerize with Max, but not with Myc. Myx/Max heterodimers can also bind to the Myc/Max E-box (CACGTG), and repress transcription from the Min4CAT reporter in NIH3T3 cells (Wang et al., unpublished). Thus, Myx is a new Max interactor. However, different from the other two Max interactors, Myx also forms a homodimer, but this homodimer cannot bind to the CACGTG sequence. Furthermore, it is highly expressed in fetal tissue, but is absent in most adult tissues, except in muscles. These findings suggest that Myx may be an important fetal regulator of Myc. Since Myc is implicated in mediating apoptosis (44), a potential link may exist between Myc-mediated and MIS-mediated apoptosis. Future biochemical characterization of Myx as a signaling molecule for R1, as well as analysis of the regulatory role of Myx on Myc function, are essential for the understanding of the apoptotic events mediated by both Myc and MIS at the molecular level.

References

1. Cate RL, Mattaliano RJ, Hession C, Tizard R, Farber NM, Cheung A, et al. Isolation of the bovine and human genes for Müllerian inhibiting substance and expression of the human gene in animal cells. Cell 1986;45:685–98.
2. He WW, Gustafson ML, Hirobe S, Donahoe PK. The developmental expression of

four novel serine/threonine kinase receptors homologous to the activin/TGF-β II receptor family. Dev Dyn 1993;196:133–42.

3. Baarends WM, van Helmond MJL, Post M, van der Schoot PJCM, Hoogerbrugge JW, de Winter JP, et al. A novel member of the serine/threonine kinase receptor family is expressed in the gonads and in mesenchymal cells adjacent to the Müllerian duct. Development 1994;120:189–97.

4. di Clemente N, Wilson CA, Faure E, Boussin L, Carmillo P, Tizard R, et al. Cloning expression, and alternative splicing of the receptor for anti-Müllerian hormone. Mol Endocrinol 1994;8:1006–20.

5. Teixeira J, He WW, Shah PC, Morikawa N, Lee MM, Catlin EA, et al. Developmental expression of a candidate Müllerian inhibiting substance type II receptor. Endocrinology 1996;137:160–5.

6. Wang TW, Donahoe PK, Zervos AS. Specific interaction of type I receptors of the TGF-β family with the immunophilin FKBP12. Science 1994;265:674–6.

7. Wang TW, Li B-Y, Danielson PD, Shah PC, Rockwell S, Lechleider RJ, Martin J, Manganaro T, Donahoe PK. The immunophilin FKBP12 functions as a common inhibitor of the TGFβ family type I receptors. Cell 1996;86:435–44.

8. Hamilton HH, Teng C-S. Sexual stabilization of Müllerian ducts in the chick embryo. In: DeHaan RL, Ursprung H, eds., Organogenesis. New York: Holt, Rhinehart and Winston, 1965;681–700.

9. Price JM, Donahoe PK, Ito Y, Hendren WH. Programmed cell death in the Müllerian duct induced by Müllerian inhibiting substance. Am J Anat 1977;149:353–76.

10. Price JM, Donahoe PK, Ito Y. Involution of the female Müllerian duct of the fetal rat in the organ culture assay for the detection of Müllerian inhibiting substance. Am J Anat 1979;156:265–84.

11. Trelstad RL, Hayashi A, Hayashi K, Donahoe PK. The epithelial-mesenchymal interface of the male rat Müllerian duct: loss of basement membrane integrity and ductal regression. Dev Biol 1982;92:27–40.

12. Hayashi M, Budzik GP, Trelstad RL. Periductal and matrix glycosamino-glycans in rat Müllerian duct development and regression. Dev Biol 1982;92:16–26.

13. Hutson JM, Fallat ME, Donahoe PK. The fate of the grafted quail Müllerian duct in the chick embryonic coelom. J Pediatr Surg 1984;19:345–52.

14. Donahoe PK, Ito Y, Marfatia S, Hendren WH. A graded organ culture assay for the detection of Müllerian inhibiting substance. J Surg Res 1977;23:141–8.

15. Tsuji M, Shima H, Yonemura CY, Brody J, Donahoe PK, Cunha GR. Effect of human recombinant Müllerian inhibiting substance on isolated epithelial and mesenchymal cells during Müllerian duct regression in the rat. Endocrinology 1992; 131:1481–8.

16. Donahoe PK, Swann DA, Hayashi A, Sullivan MD. Müllerian duct regression in the embryo is correlated with cytotoxic activity against a human ovarian cancer. Science 1979;205:913–5.

17. Scully RE. Recent progress in ovarian cancer. Hum Pathol 1970;1:73.

18. Picon R. Action of the fetal testis on the development in vitro of the Müllerian ducts in the rat. Arch Anat Microsc Morphol Exp 1969;58:1–19.

19. Donahoe PK, Ito Y, Price JM, Hendren WH. Müllerian inhibiting substance activity in bovine fetal, newborn, and prepubertal testes. Biol Reprod 1977;16:238–43.

20. Hutson JM, Ikawa H, Donahoe PK. Estrogen inhibition of Müllerian inhibiting substance in the chick embryo. J Pediatr Surg 1982;17:953–9.

21. Takahashi M, Hayashi M, Manganaro TF, Donahoe PK. The ontogeny of Müllerian inhibiting substance in granulosa cells of the bovine ovarian follicle. Biol Reprod 1986;35:447–53.

22. Josso N. In vitro synthesis of Müllerian-inhibiting hormone by seminiferous tubules isolated from the calf fetal testis. Endocrinology 1973;93:829–34.

23. Swann DA, Donahoe PK, Ito Y, Morikawa Y, Hendren WH. Extraction of Müllerian inhibiting substance from newborn calf testis. Dev Biol 1979;69:73–84.

24. Budzik GP, Swann DA, Hayashi A, Donahoe PK. Enhanced purification of Müllerian inhibiting substance by lectin affinity chromatography. Cell 1980;21:909–15.

25. Budzik GP, Powell SM, Kamagata S, Donahoe PK. Müllerian inhibiting substance fractionation by dye affinity chromatography. Cell 1983;34:307–14.

26. Picard JY, Josso N. Purification of testicular anti-Müllerian hormone allowing direct visualization of the pure glycoprotein and determination of yield and purification factor. Mol Cell Endocrinol 1984;34:23–9.

27. Cate RL, Ninfa DT, Pratt DJ, MacLaughlin DT, Donahoe PK. Development of Müllerian inhibiting substance as an anti-cancer drug. Cold Spring Harbon Symp 1986;51:641–7.

28. Picard JY, Benarous R, Guerrier D, Josso N, Kahn A. Cloning and expression of cDNA for anti-Müllerian hormone. Proc Natl Acad Sci USA 1986;83:5464–8.

29. Derynck R, Jarrett JA, Chen EY, Eaton DH, Bell JR, Assoian RK, et al. Human transforming growth factor-beta complementary DNA sequence and expression in normal and transformed cells. Nature 1985;316:701–5

30. Mason AJ, Hayflick JS, Ling N, Esch F, Ueno N, Ying SY, et al. Complementary DNA sequences of ovarian follicular fluid inhibin show precursor structure and homology with transforming growth factor-beta. Nature 1985;318:659–63.

31. Padgett, RW, St. Johnston RD, Gelbart WM. A transcript from a *Drosophila* pattern gene predicts a protein homologous to the transforming growth factor-beta family. Nature 1987;325:81–4.

32. Weeks DL, Melton DA. A maternal RNA localized to the vegetal hemisphere in *Xenopus* eggs codes for a growth factor related to TGF-β. Cell 1987;51:861–7.

33. Pepinsky RB, Sinclair LK, Chow EP, Mattaliano RJ, Manganaro TF, Donahoe PK, et al. Proteolytic processing of Müllerian inhibiting substance produces a transforming growth factor-β-like fragment. J Biol Chem 1988;263:18961–4.

34. MacLaughlin DT, Hudson PL, Graciano AL, Kenneally MK, Ragin RC, Manganaro TF, et al. Müllerian duct regression and anti-proliferative bioactivities of Müllerian inhibiting substance resides in its carboxy-terminal domain. Endocrinology 1992; 131:291–6.

35. Kuroda T, Lee MM, Ragin RC, Donahoe PK. MIS production and cleavage is modulated by gonadotropins and steroids. Endocrinology 1991;129:2985–93.

36. Wilson CA, di Clemente N, Ehrenfels, Pepinsky RB, Josso N, Vigier B, Cate RL. Müllerian inhibiting substance requires its N-terminal domain for maintenance of biological activity, a novel finding within the transforming growth factor-beta superfamily. Mol Endocrinol 1993;7:247–57.

37. Mathews LS, Vale WW. Expression cloning of an activin receptor, a predicted transmembrane serine kinase. Cell 1991;65:973–82.

38. Lin HY, Wang XF, Ng-Eaton E, Weinberg RA, Lodish HF. Expression cloning of the TGF-beta type II receptor, a functional transmembrane serine/threonine kinase. Cell 1992;68:775–85.

39. Catlin EA, Ezzell RM, Donahoe PK, Gustafson ML, Son EV, MacLaughlin DT. Identification of a receptor for human Müllerian inhibiting substance. Endocrinology 1993;133:3007–13.
40. Visser JA, McLuskey A, van Beers T, Weguis DO, van Kessel AG, Grootegoed JA. Structure and chromosomal localization of the human anti-Müllerian hormone type II receptor gene. Biochem Biophys Res Comm 1995;215:1029–36.
41. Imbeaud S, Faure E, Lamarre I, Mattei MG, di Clemente N, Tizard R, et al. Insensitivity to anti-Müllerian hormone due to a mutation in the human anti-Müllerian hormone receptor. Nat Genet 1995;11:382–8.
42. Ayer DE, Kretzner RN, Eisenman RN. Mad: a heterodimeric partner for Max that antagonizes myc transcriptional activity. Cell 1993;72:211–22.
43. Schreiber-Agus N, Chin L, Chen K, Torres R, Rar G, Guida P, et al. An amino-terminal domain of Mxi1 mediates anti-myc oncogenic activity and interacts with a homolog of the yeast transcriptional repressor SIN3. Cell 1995;80:777–86.
44. Evan GI, Littlewood TD. The role of c-myc in cell growth. Curr Biol 1993;3:44–9.

7

A Program of Cell Death and Extracellular Matrix Degradation in Fetal Membranes Prior to Labor

HANQIN LEI, VIOLETTA DELGADO, EMMA E. FURTH, LAURIE G. PAAVOLA, FELIPE VADILLO-ORTEGA, AND JEROME F. STRAUSS, III

Membranes surround the fetus to retain amniotic fluid and prevent infectious agents ascending the reproductive tract from entering the amniotic compartment. The fetal membranes also have important transport and endocrine functions during pregnancy in addition to their structural roles (1). The fetal membranes usually rupture during labor and it has generally been held that the breaking of the membranes is the consequence of physical forces associated with uterine contraction. However, recent observations suggest that structural changes occur in the membranes before labor. Thus, biochemical changes, as well as physical forces, are involved in membrane rupture.

The fetal membranes consist of an outer chorion in the human or visceral yolk sac placenta in rodents. Beneath the chorion or visceral yolk sac placenta lies the amnion, an avascular membrane consisting of fibroblasts, a matrix of fibrillar collagen and proteoglycans, and a basement membrane upon which amnion epithelial cells sit. In the human, the chorion and amnion become adherent as pregnancy progresses. In rodents, the visceral yolk sac placenta and amnion remain as separate structures.

We have characterized the structural changes in the rat fetal membranes at the end of pregnancy with the goal of using this animal model as a system to define the biochemical and molecular basis of the modifications of the fetal membranes in preparation for parturition (2, 3). Rodent models offer advantages of being amenable to experimental manipulation. Moreover, transgenic and gene "knock-out" technologies, as well as the power of mouse genetics can be applied in these systems.

Changes in the Amnion During Parturition in the Rat

The amnion undergoes striking changes in the rat as parturition approaches. Prior to day 20 of pregnancy, it is a translucent sheet. Sometime on day 20, the integrity of this membrane begins to change, becoming an amorphous gel by the morning of day 21, the day on which the animals deliver. In our colony, animals complete delivery in the early afternoon of day 21. Histologically, the dramatic change in the consistency of the amnion is associated with a widening of the membrane, the delamination of amnion epithelial cells and the loss of the collagen extracellular matrix (2). Ultrastructural analysis revealed that the basement membrane upon which the epithelial cells rest is lost by the morning of day 21, causing the cells to detach from the underlying matrix. The collagen fibrils in the amnion matrix are also depleted along with the associated proteoglycans. The amnion epithelial cells show internalized desmosomes, cytoplasmic vacuolization, and loss of nuclear euchromatin and heterochromatin. These histologic and ultrastructural alterations suggest a process of cell death in association with degradation of the extracellular matrix. Notably, these changes are evident 12 to 18 h before the animals begin active labor.

The mechanism of extracellular matrix degradation associated with the structural dissolution of the amnion appears to involve the induction of specific matrix degrading enzymes. The type IV collagenase, gelatinase B, also called matrix metalloproteinase-9 (MMP-9), is not observed by zymography of rat amnion extracts until the evening of day 20 of pregnancy (3). At this time activity is detected and by day 21, MMP-9 antigen is readily demonstrable by Western blotting. Zymography and Western blotting document that the activated form of MMP-9 (83 kDa) is present on day 21. The appearance of MMP-9 mRNA is correlated with the appearance of MMP-9 activity and protein in the amnion, suggesting induction of this gene in preparation for labor. The appearance of MMP-9 also correlates with the degradation of the amnion basement membrane. Unlike MMP-9, the 72-kDa gelatinase A (MMP-2) is present in amnion extracts on days 18 to 21 (3). There is a modest increase in MMP-2 activity as assessed by zymography on day 21, but this change is nowhere near as dramatic as the increase in MMP-9.

Our additional studies suggest that another matrix degrading enzyme, interstitial collagenase, which catalyzes the first step in the catabolism of fibrillar collagens (type I and type III), is similarly induced in the amnion prior to the onset of labor (4). The appearance of this enzyme correlates with the loss of type I collagen from the amnion and the appearance of type I collagen degradation products in extraembryonic fluid. The denatured collagen fragments resulting from interstitial collagenase action are substrates for further catabolism by MMP-9. These observations indicate that there is a concerted up-regulation of matrix degrading enzymes in the amnion at term.

The detection of nuclear DNA fragmentation in amnion epithelial cells by in situ 3'-end-labeling of DNA and the presence of nucleosomal DNA cleavage products in amnion by day 21 indicate that amnion cells undergo apoptotic cell death coincident with degradation of amnion extracellular matrix (4). Selective catabolism of 28S rRNA and 28S ribosomal subunit proteins has also been found at this time, which are known to be features of apoptotic cell death. The degradation of 28S rRNA and nuclear DNA degradation appear to be late events that follow the induction of matrix degrading enzymes. Thus, there is an orchestrated program of dissolution of both the matrix and cellular components of the rat amnion in preparation for delivery. Since the extracellular matrix, particularly the basement membrane, is known to maintain cell viability, the loss of type IV collagen and the subsequent detachment of the amnion cells may be the initiating event in apoptosis.

Do the structural and biochemical changes found in the rat fetal membranes also occur in human fetal membranes? It is much more difficult to establish the temporal changes in the structure of human fetal membranes because of limited opportunities to sample the tissues in relationship to the time of parturition. However, existing data suggest that the changes in the rat amnion described above are in many respects similar to changes reported to occur in the human amnion in the peripartal period. Malek and Bell (5) reported a zone of "extreme" morphological modification that includes detachment of amnion epithelial cells and loss of collagen in a restricted area surrounding the rupture site of human membranes. The histological features of this zone are reminiscent of the structural changes in the rat amnion at term. We have also reported that MMP-9 expression is markedly increased with labor in human fetal membranes and that the enzyme is produced by the amnion epithelial cells as well as the chorion trophoblast cells (6). These findings are consistent with the concept of the induction of MMP-9 before or with active labor as has been seen in the rat. If the matrix degrading enzymes like MMP-9 are induced in the fetal membranes prior to the onset of labor or membrane rupture, the detection of these enzymes may be used as a biochemical marker for impending labor or membrane rupture.

Conclusions

An important issue that remains to be explored is whether or not the program of cell death described above in the rat amnion is also played out in other tissues of the reproductive tract, in particular the cervix which is known to undergo marked extracellular matrix remodelling in preparation for labor (7). A second question that arises is the identity of the factors that initiate the program of cell death and extracellular matrix catabolism in the amnion. At present, we can answer this question only with speculations. These changes could reflect a withdrawal of factors essential for the maintenance of pregnancy and integrity of the reproductive tract (e.g., progesterone) or the appearance of a "death-inducing" signal that is coordinated with the initiation of labor. However, with the knowledge that specific genes are induced in the amnion epithelium as parturition approaches, it should be possible to map promoter elements in these genes that are required for their induction, and to then use this information to identify the transcription factors that activate the key *cis*-elements required for gene induction. This information, in turn, may suggest a signalling cascade that leads to transcriptional activation and, ultimately, to the factor(s) that triggers it.

Acknowledgments. We thank Mrs. Marianne Winberg for her help in preparation of this manuscript. Work described in this chapter was supported in part by grants from the March of Dimes National Foundation and the Rockefeller Foundation.

References

1. Schmidt W. The amniotic fluid compartment: the fetal habitat. Anat Embryol Cell Biol 1992;124:1–100.
2. Paavola LG, Furth EE, Delgado V, Boyd CO, Jacobs CC, Lei H, Strauss JF III. Striking changes in the structure and organization of rat fetal membranes precede parturition. Biol Reprod 1995;53:321–38.
3. Lei H, Vadillo-Ortega F, Paavola LG, Strauss JF III. 92-KDa gelatinase (matrix metalloproteinase-9) is induced in rat amnion immediately prior to parturition. Biol Reprod 1995;53:339–44.
4. Lei H, Furth EE, Kalluri R, Chiou T, Tilly KI, Tilly JL, Elkon KB, Jeffrey JJ, Strauss JF III. A program of cell death and extracellular matrix degradation is activated in the amnion prior to the onset of labor. J Clin Invest 1996;98:1971–8.
5. Malak TM, Bell SC. Structural characteristics of term human fetal membranes: a novel zone of extreme morphological alteration within the rupture site. Br J Obstet Gynecol 1994;101:375–86.
6. Vadillo-Ortega F, Gonzalez-Avila G, Furth EE, Lei H, Muschel RJ, Stetler-Stevenson WG, Strauss JF III. 92-Kd type IV collagenase (matrix metalloproteinase-9) activity in human amniochorion increases with labor. Am J Pathol 1995;146:148–56.
7. Rajabi JR, Dodge GR, Solomon S, Poole AR. Immunochemical and immunohistochemical evidence of estrogen-mediated collagenolysis as a mechanism of cervical dilatation in the guinea pig at parturition. Endocrinology 1991;128:371–8.

8

Apoptosis-Susceptible Versus -Resistant Granulosa Cells From Hen Ovarian Follicles

ALAN L. JOHNSON

The Avian Ovary as a Model System

Before addressing the process of apoptosis and follicle atresia, it is useful to briefly describe the hen ovarian model system with regard to follicle growth and granulosa cell differentiation. Follicles selected into the initial growth phase from the resting stage are subsequently organized in an orderly and morphologically distinguishable fashion consisting of many (50 to 100) slowly growing follicles measuring 1 to 8 mm in diameter (prehierarchal follicles) and five to eight yellow, yolk-filled follicles ranging from 9 mm to greater than 35 mm in diameter (preovulatory follicles). It is of interest to note that one can accurately predict the day and approximate time of ovulation for the five to six largest preovulatory follicles. Preovulatory follicles are committed to ovulation, and under normal physiological conditions are not subject to becoming atretic. By contrast, most slowly growing follicles less than 9 mm in diameter (prehierarchal follicles) will undergo follicle atresia, and it is estimated that fewer than 1 in 20 of these follicles express the potential to enter the preovulatory hierarchy. Follicle selection into the preovulatory hierarchy occurs from a small group of eight to ten 6 to 8 mm diameter, prehierarchal follicles at the rate of one follicle per day. Although the mechanism(s) responsible for such a selection process are as yet unknown, it is generally assumed that follicles not selected subsequently undergo atresia.

The slow-growth phase of follicle development is associated with the highest rate of granulosa cell proliferation, and granulosa cells from 1 to 8 mm diameter follicles exhibit a five- to six-fold greater rate of ^{3}H-thymidine incorporation compared to granulosa from the third largest (F3) and largest (F1) preovulatory follicles (2). This higher rate of granulosa cell proliferation in prehierarchal follicles is associated with elevated mRNA and protein levels for the nuclear transcription factor, c-myc (3; Williams and Johnson, unpublished data), and these latter findings are consistent with a role for c-myc as a mediator of normal cell proliferation and an inhibitor of cell differentiation (4, 5). At the prehierarchal follicle stage of development, granulosa cells are under the regulatory control of the gonadotropin, follicle-stimulating hormone (FSH), but not luteinizing hormone (LH), as evidenced by the comparatively high levels of FSH receptor mRNA and FSH-induced cAMP formation, and the lack of detectable LH receptor mRNA and of granulosa cell responsiveness to exogenous LH (Fig. 8.1) (6–8).

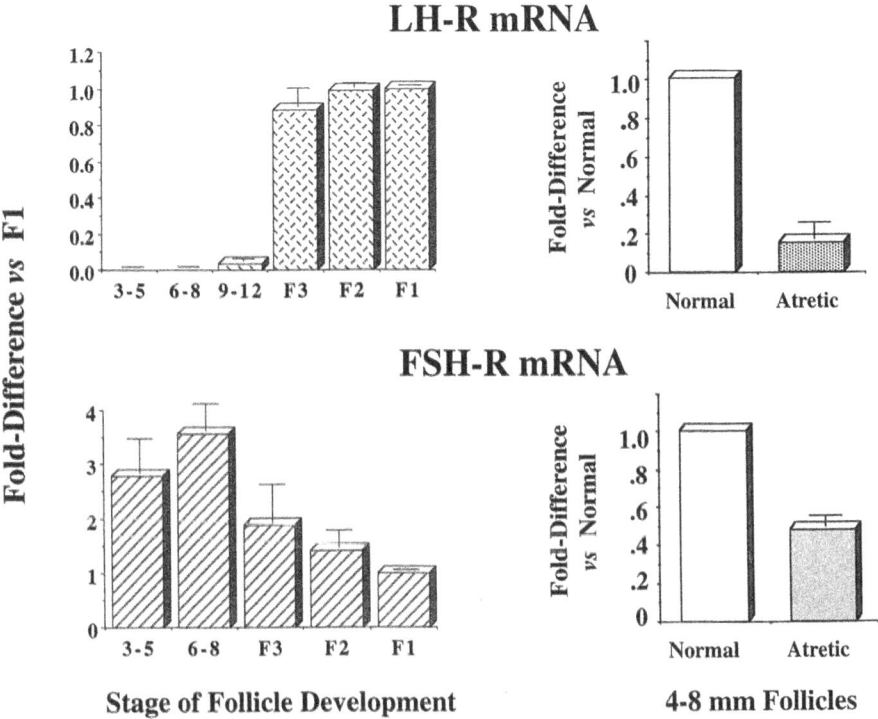

FIGURE 8.1. Luteinizing hormone receptor (LH-R) and follicle-stimulating hormone receptor (FSH-R) mRNA in granulosa cells during follicle differentiation (left panels) and in whole, morphologically normal and atretic follicles (right panel). Note that the increased expression of LH-R in the granulosa layer occurs only subsequent to follicle selection into the preovulatory hierarchy and that atretic follicles express significantly less LH receptor (of theca origin) and FSH receptor than the comparably sized (3–5 mm) normal follicles. At this time it is not possible to conclude whether the loss of gonadotropin receptor mRNA is a cause, or alternatively an effect, of the atretic process. (LH-R data redrawn from Johnson et al. (7); FSH-R data from You et al. (8).)

Following selection into the preovulatory hierarchy, granulosa cell proliferation declines dramatically, and cells begin the process of terminal differentiation. Granulosa cell differentiation is likely initiated by the actions of
FSH and possibly vasoactive intestinal peptide (VIP). This process includes
the induction of progesterone production resulting from increased expression of cytochrome P_{450} cholesterol side-chain cleavage enzyme mRNA, protein and activity (9, 10). On the other hand, premature granulosa cell
differentiation and the selection of more than a single follicle into the preovulatory hierarchy per day are presumably prevented, at least in part, by
several growth factors, including transforming growth factor-α (TGFα) and
epidermal growth factor (EGF), which can completely suppress FSH- and VIP-
induced differentiation (9, 10). Consistent with the inhibitory effects of EGF
and TGFα on cellular differentiation is the related finding that these growth
factors actively promote granulosa cell proliferation (11, 12).

As noted above, prehierarchal follicles are highly susceptible to atresia,
and follicles undergoing atresia are easily identified based on the presence of
infiltrated blood (follicle hemorrhagia), and a collapsed and opaque appearance (1). For several years now it has been recognized that vertebrate ovarian
follicle atresia is mediated via the process of apoptosis, and that it is specifically the granulosa cell layer that succumbs to this form of programmed cell
death (13). Morever, it has recently been established that granulosa cells from
atresia-prone, prehierarchal follicles are inherently more susceptible to undergoing apoptosis, in vitro, when compared to granulosa cells from atresia-
resistant, preovulatory follicles (Fig. 8.2) (1). Although limited, but detectable,
oligonucleosome formation is evident by 3 h of incubation, both morphological and biochemical indications of apoptosis are prominent at 6 h of
incubation. For example, following a 24 h culture of granulosa cells in serum-
free medium (M199 containing HEPES), less than 30% of the cells from
prehierarchal follicles survive and successfully attach to culture plates whereas
greater than 80% of F1 follicle granulosa cells survive to form a primary cell
line (Witty and Johnson, unpublished data). In addition, isolated granulosa
cells from 4 to 8 mm follicles incubated for 6 h in defined medium exhibit a
20- to 23-fold higher incidence of pyknotic nuclei (a hallmark of apoptosis)
when compared to unincubated control cells (1). It is important to note that
the cessation of granulosa cell division within preovulatory follicles is associated with an apoptosis-resistant cell type whereas the high rate of cell division in prehierarchal follicles is correlated with a propensity for apoptotic
cell death. A similar relationship between granulosa cell proliferation and
cell death has been noted in the rat ovary (14). It is also of interest that the
highest levels of Myc protein in hen granulosa cells are associated with an
increased susceptibility of apoptotic cell death, consistent with previous reports that increased levels of Myc in the absence of a supportive growth
factor environment facilitate programmed cell death (15, 16). As a final point,
it should be emphasized that the hen follicle model system permits the col-
lection of large numbers of granulosa cells (e.g., $> 5 \times 10^7$ per F1 follicle)

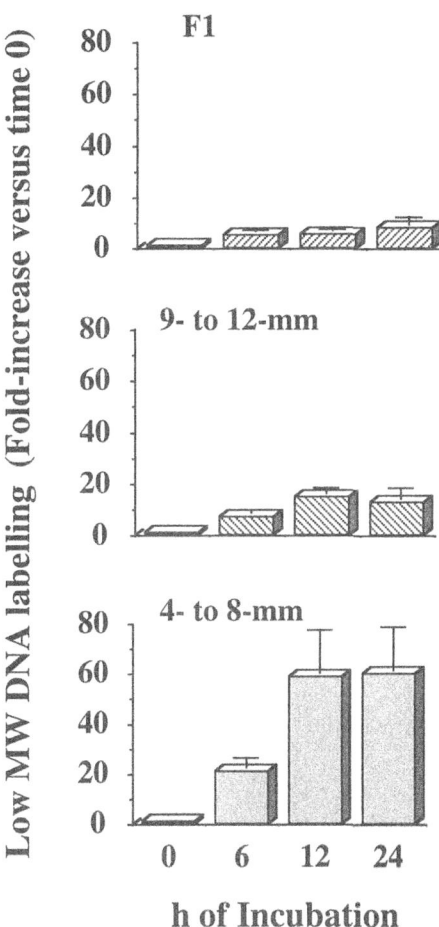

FIGURE 8.2. Demonstration of apoptosis-susceptible and -resistant populations of granulosa cells relative to stage of follicle development. Cells were incubated in serum-free
medium (M199/HEPES) for 0 to 24 h, then collected and processed for biochemical
analysis of oligonucleosome formation following 3'-end-labeling with [^{32}P]dideoxy-
ATP. Data represent the fold-increase in oligonucleosome formation compared to
unincubated (0 h) granulosa cells from prehierarchal (4–8 mm diameter) follicles, follicles recently selected into the preoovulatory hierarchy (9–12mm diameter), or the
largest preovulatory (F1) follicle. Note that only granulosa cells from prehierarchal
follicles are susceptible to undergoing apoptosis. (Reproduced with permission from
Johnson et al. (1).)

without contamination by cells from the theca layer. This inherent difference
in susceptibility to apoptosis at two well-defined stages of normal follicle
development (e.g., in the absence of hormonal priming) using a pure population of granulosa cells thus provides a unique model for studying physiological cell death in the ovary.

Attenuation of Apoptosis by Gonadotropins and Growth Factors

As previously documented for rat preovulatory and preantral follicles (17, 18), FSH attenuates the progression of apoptotic cell death in isolated hen granulosa cells (1). This effect is likely mediated, in part, via FSH-induced cyclic adenosine monophosphate (cAMP) production as treatment of cells with the cell-permeable analog of cAMP, 8-bromo-cAMP (8-br-cAMP), suppresses the progression of oligonucleosome formation to a similar extent (e.g., both FSH and 8-br-cAMP attenuate oligonucleosome formation to approximately 40% to 50% of the levels detected in incubated, untreated control cells). As alluded to above, it is unlikely that LH has a physiological role in the process of maintaining cell viability at this stage of development due to the presence of low- to nondetectable levels of LH receptor expression in prehierarchal follicle granulosa cells (Fig. 8.1). Both FSH receptor and LH receptor mRNA levels are greatly reduced in atretic follicles compared to morphologically normal follicles (Fig. 8.1), and this is consistent with findings from the pig (19). Nevertheless, in neither model has it been established whether the decrease in gonadotropin receptor message is a cause or an effect of the atretic process. Vasoactive intestinal peptide has similarly been determined to attenuate the process of apoptosis in rat antral follicles and hen granulosa cells, presumably via its ability to stimulate the formation of cAMP (20). As VIP has previously been localized within nerve terminals of the hen theca layer (10), it is likely that this factor can act via a paracrine mode of action, in vivo, to supplement the viability supporting effects of FSH.

In addition to antiatretogenic factors acting via the adenylyl cyclase/cAMP second messenger pathway, several growth factors, including TGFα (1) EGF and FGF (Johnson, unpublished data), have been demonstrated to attenuate the progression of apoptotic cell death in hen granulosa cells. For instance, TGFα, at levels of 0.3 and 3.3 nM, suppresses apoptosis to 56.8% and 43.2% of control (untreated cells incubated for 6 h) levels. Such results are not unlike those from studies with rat antral follicles and isolated granulosa cells following treatment with growth factors (21). Futhermore, Tilly et al. (21) determined that the actions of these growth factors are mediated via activation of the tyrosine kinase signaling pathway. Taken collectively, it is concluded that the inability of any single hormone or growth factor described above to completely suppress granulosa cell death to levels normally found in healthy follicles is due to the requirement for multiple factors acting simultaneously via similar and/or different signaling pathways, in vivo.

Expression of *bcl-2*-Related Genes

The BCL-2-related family of proteins is homologous to the death-repressing CED-9 protein from *C. elegans,* primarily in two highly conserved domains, BH1 and BH2 (22). There is considerable information indicating that several members of the BCL-2 family of proteins serve to block apoptosis in a variety of mammalian cell types. For instance, in Chinese hamster ovary cells BCL-2 inhibits apoptotic cell death induced by overexpression of the c-*myc* protooncogene (15), and overexpression of BCL-X_{LONG} recues B lymphocytes from oxidative stress-induced cell death (23). By contrast, elevated expression of an alternatively-spliced variant of BCL-X, BCL-X_{SHORT}, is proposed to promote cell death by its ability to counter the death-repressing effects of BCL-2 and BCL-X_{LONG} (24, 25).

Both *bcl-2* and *bcl-x* mRNA transcripts are now known to be expressed in hen granulosa cells at all stages of follicle development, but constitutively expressed levels of *bcl-2* are considerably lower than those for *bcl-x* (Fig. 8.3A). In light of the considerable data from mammals demonstrating the expression of death-suppressing and -inducing forms of mRNA encoded by the *bcl-x* gene, initial efforts were directed towards establishing whether or not alternatively spliced forms of *bcl-x* mRNA are expressed in the hen ovary. Results from both polymerase chain reaction and ribonuclease protection assays indicate the presence of only the putative death-suppressing form of *bcl-x*, *bcl-x*$_{LONG}$ (1). Although levels of *bcl-x*$_{LONG}$ mRNA do not vary in theca tissue throughout follicle development, there is a five- to nine-fold increase in *bcl-x*$_{LONG}$ mRNA levels in granulosa cells associated with the transition from an apoptosis-susceptible to apoptosis-resistant state. Moreover, there is a significant decline in *bcl-x*$_{LONG}$ mRNA in atretic, as compared to morphologically normal, 4 to 8 mm follicles (Fig. 8.3B). Such data provide a strong correlation between elevated expression of the death-suppressing form of *bcl-x* mRNA and granulosa cell survival. It has previously been proposed that follicle survival in the rat may be linked to some ratio of a death-suppressor protein (e.g., BCL-2 or BCL-X_{LONG}) compared to a death inducing protein (e.g., BAX, or alternatively, BAK or BAD) (25–27). The ability to test this hypothesis in avian granulosa cells awaits the cloning and characterization of *bcl-2* related death-inducing genes in the chicken.

As noted above, we have previously observed that a minor proportion of granulosa cells from 4 to 8 mm follicles is capable to forming a primary culture in Medium 199/HEPES containing 2.5% fetal bovine serum, whereas the remaining cells fail to attach to the plates and subsequently die via programmed cell death. Studies were therefore conducted to test the hypothesis that elevated expression of *bcl-x*$_{LONG}$ mRNA may be associated with prehierarchal follicle granulosa cell survival following culture.

Dispersed granulosa cells from F1 follicles or from pools of fifteen to twenty 4 to 8 mm follicles were prepared and frozen at −70°C prior to culture or were plated in six-well culture plates for 24 h, then washed with fresh medium and subsequently collected by trypsinization. Following the preparation of total cellular RNA, Northern blot analysis was utilized to quantify levels of *bcl-x*$_{LONG}$ mRNA. Although there was no change in *bcl-x*$_{LONG}$ mRNA levels in F1 follicle granulosa cells following 24 h of culture, *bcl-x*$_{LONG}$ mRNA levels were increased by 2.5-fold in viable cells from 4 to 8 mm follicles that had

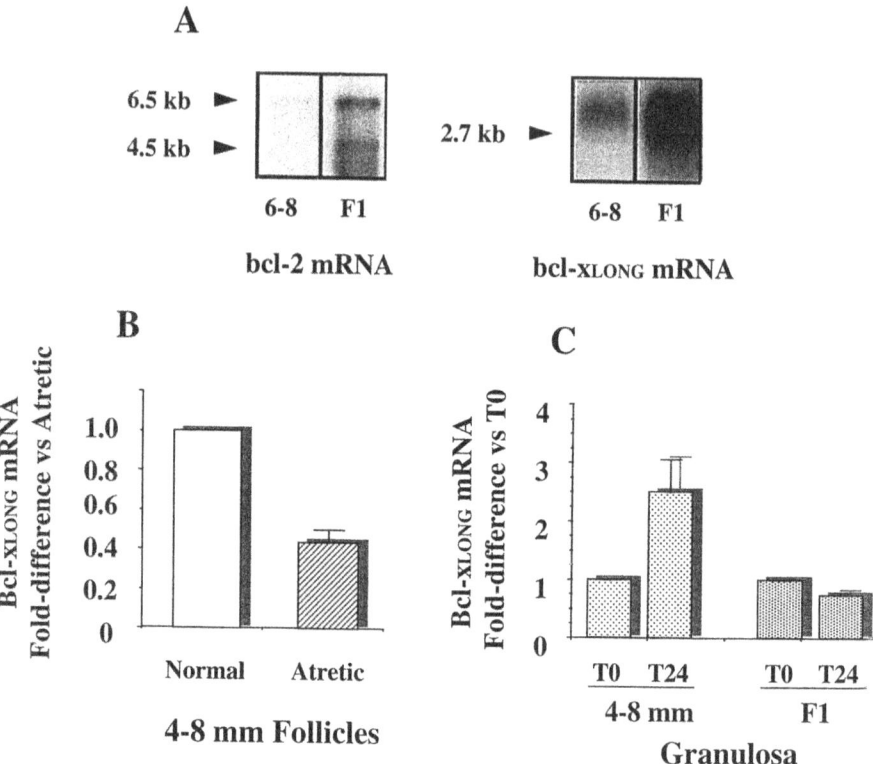

FIGURE 8.3. (A) Relative expression of *bcl-2,* compared to *bcl-x*$_{LONG}$, mRNA in 6–8 mm diameter and F1 follicle granulosa cells. Note the comparatively greater levels of *bcl-x*$_{LONG}$ expression (Northern blot derived from 15 μg total cellular RNA) compared to *bcl-2* (1 μg poly A+– enriched RNA), and the higher mRNA levels for both genes in F1 versus 6–8 mm follicles. Levels of *bcl-x*$_{LONG}$ mRNA in morphologically normal and atretic 3–5 mm follicles ((B) with permission from ref. 1), and in granulosa cells from 4–8 mm and F1 follicles incubated for 0 (T0) or 24 (T24) h in serum-free medium (C). Note that *bcl-x*$_{LONG}$ mRNA levels are decreased in atretic versus normal prehierarchal follicles, whereas levels are increased in granulosa cells from prehierarchal follicles that successfully culture following a 24 h incubation. Such data, albeit correlative, indicate that there may be selective survival of those granulosa cells constitutively expressing higher levels of *bcl-x*$_{LONG}$. (Reproduced with permission from Johnson et al. (1).)

formed a primary culture (Fig. 8.3C). It is concluded that survival of prehierarchal follicle granulosa cells in culture is associated with elevated *bcl-x*$_{LONG}$ mRNA. Furthermore, it is tempting to speculate that surviving granulosa cells may have originated primarily from follicles that were destined for selection into the preovulatory hierarchy. Nevertheless, it is important to note that all of our *bcl-2* and *bcl-x* data generated to date provide only a strong correlation to suggest a role for the proteins encoded by each transcript in protecting granulosa cells from apoptotic cell death. Currently, studies are underway to monitor levels of BCL-2 and BCL-X proteins at similar stages of follicle development and under similar experimental conditions, and to evaluate the effects of antisense bcl-x_{LONG} oligodeoxynucleotides and bcl-x overexpression on granulosa cell survival in the face of apoptosis-inducing or -attenuating experimental conditions.

Finally, a novel BCL-2-related protein originally described in quail but not yet identified in mammals, NR-13 (28), shares amino acid homology with BCL-2-related proteins, particularly within the highly conserved BH1 and BH2 domains. Increased quail NR-13 expression is associated with Rous sarcoma virus-induced cell transformation, and has been linked to anitapoptotic activities. A partial cDNA for chicken NR-13 has recently been cloned, and NR-13 mRNA is expressed at relatively high levels in the brain, kidney, and adrenal, and in particular, ovarian stromal, theca, and granulosa tissues (Johnson and Wagner, unpublished data). Unlike bcl-2 and bcl-x_{LONG}, however, constitutive expression of NR-13 mRNA is not significantly altered within either granulosa or theca tissues during follicle development. Thus, an association between NR-13 expression and follicle viability is yet to be established, and studies are ongoing to evaluate the relationship, if any, of NR-13 to hen follicle atresia.

Proteases Related to the Progression of Apoptosis

Interleukin-Converting Enyzme (ICE)-Related Protease Activity

The family of ICE-related proteases is homologous to CED-3 from *C. elegans,* a protein that in this nematode is required for programmed cell death. This ever-expanding gene family now numbers at least eleven, and includes ICE, CPP32 (also called apopain of Yama), ICH-1 (Nedd2), ICH-2 (TX, ICE-rel II), ICE-rel III, MCH2, and MCH3 (ICE-LAP3) (for original references, see ref. 29). Each of these cysteine proteases is composed of two subunits of approximately 20 kDa and 10 kDa in size that are derived from processing of a proenzyme, and all ICE-related family members share conserved amino acid residues for binding and catalysis of the active site sequence, QACXG. Over-expression of all ICE-related proteases thus far described results in apoptotic death in many different cell types following transfection (30). Nevertheless, although such studies clearly support a role for one or more of these porteases in mediating apoptosis, their mechanism of action is not yet understood.

Evidence for the existence of ICE-related proteases in hen ovarian follicles is derived, in part, from the identification of *Ich-1* mRNA in granulosa and theca tissues (31). Ribonuclease protection assays have determined that only the Ich-1_{LONG} (death-inducing form), but not Ich-1_{SHORT} (death-suppressing form), of the mRNA is expressed in the hen ovary, and that levels of Ich-1_{LONG} mRNA do not change significantly during follicle growth and differentiation. On the other hand, such results are not necessarily unexpected, as post-translational modifications and protease activation are more likely sites for the regulation of enzyme activity during the process of cell death.

Much of the initial information regarding the physiological relevance of ICE-related proteases has been derived from the use of pharmacological inhibitors of ICE-related proteases. To evaluate whether or not ICE-related protease activity is associated with apoptosis in hen follicles, granulosa cells were incubated for 6 h in the absence or presence of either of two putative and non-competitive inhibitors of ICE-related proteases, sodium aurothiomalate or iodoacetic acid (32). Both inhibitors markedly attenuate oligonucleosome formation, with the highest dose of sodium aurothiomalate (1 mM) reducing levels to less than those found by treatment with 8-br-cAMP and not different from the unincubated control (Fig. 8.4A). On the other hand, sodium aurothiomalate fails to prevent the morphological signs of programmed cell death in incubated 3 to 5 mm follicles as indicated by the formation of pyknotic nuclei, the loss of cell-to-cell contacts within the granulosa layer, and the disorganization of the granulosa cell layer (Fig. 8.4B). It should be noted that sodium aurothiomalate is known to have additional effects on cultured cells, including an inhibition of cell adhesion molecules in endothelial cells (33) and a potential direct inhibitory action on nuclear endonuclease activity (34). Nevertheless, such actions cannot account for the appearance of pyknotic nuclei, a hallmark of apoptosis, indicating the programmed cell death in granulosa cells can progress in the absence of oligonucleosome formation. The results presented for these pharmacologic inhibitors are similar to those recently reported from studies of the rat ovary (35).

Endogenous inhibitors of ICE-related protease activity in vertebrates have yet to be identified, and one of the only natural inhibitors of ICE-related proteases described to date is the cowpox virus serpin, CrmA (36). Recent work has centered on synthesizing specific peptide inhibitors that

---

FIGURE 8.4. (A) Indirect evidence for ICE-related protease (IRP) activity in mediating oligonucleosome formation in prehierarchal (4–8 mm follicle) granulosa cells. Sodium aurothiomalate (SAM) and iodoacetic acid, inhibitors of IRP activity, attenuate oligonucleosome formation in granulosa cells from 4–8 mm follicles incubated for 6 h (T6) in defined medium; the level of attenuation is equal to or greater than that previously reported for the cAMP analog, 8-bromo-cAMP (8-br-cAMP). (B) On the other hand, SAM fails to block the progression of apoptotic cell death as indicated by the presence of pyknotic nuclei (arrows), the loss of cell-to-cell contacts within the granulosa layer, and the disorganization of the granulosa cell layer (compare normal follicle on left to a follicle incubated for 6 h with 1 mM SAM in serum-free medium on right. Gr, granulosa layer; TI, theca interna; TE, theca externa; magnification, 130X. Sections are stained with Lee's methylene blue basic fuchsin, and are provided courtesy of Dr. A. Hirshfield and A. DeSanti, University of Maryland, School of Medicine). (C) The tetrapeptide aldehyde inhibitor of ICE-like protease activity (YVAD-CHO) is ineffective in attenuating the progression of oligonucleosome formation, suggesting that the IRP is not ICE itself.

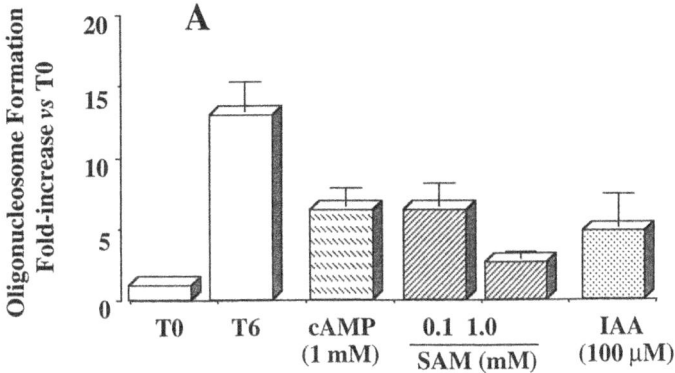

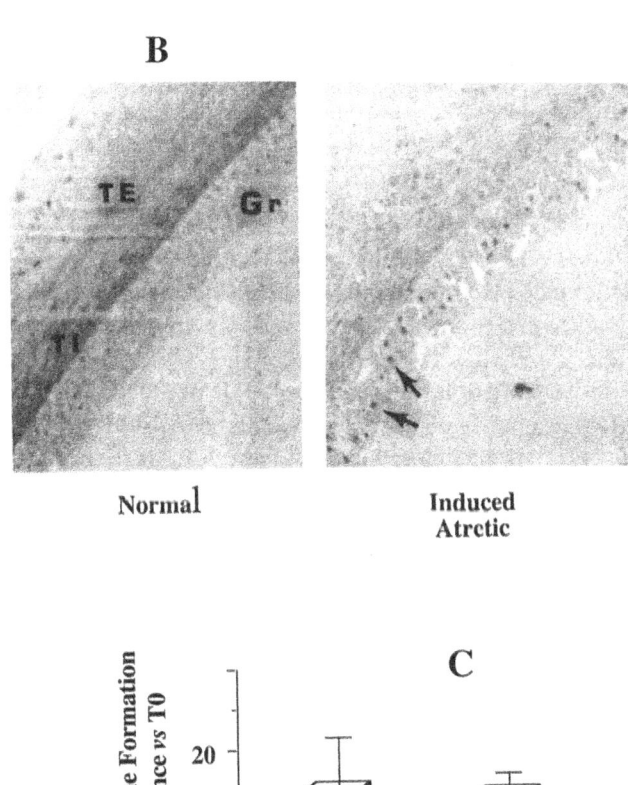

demonstrate family member specificity. For instance, one of the most specific and potent reversible inhibitors of ICE is the tetrapeptide aldehyde, Ac-Tyr-Val-Ala-Asp-CHO (YVAD-CHO; 37). This ICE-specific inhibitor has been shown to block cell death in several specific cell types, including chick neurons deprived of neurotrophic factors (38) and primary cultures of rat hepatocytes (39). By comparison, YVAD-CHO fails to attenuate the progression of apoptotic cell death in 4 to 8 mm follicle granulosa cells (Fig. 8.4C). These results, together with additional data that IL-1β, the primary product of ICE activity, does not induce apoptotic cell death in hen granulosa cells (Johnson et al., unpublished data), fail to support a role for ICE activity in mediating the progression of apoptotic cell death in hen granulosa cells. Nevertheless, it is becoming increasingly evident that ICE-related proteases other than ICE itself, potentially including CPP32 and MCH3 (29), are important participators in mediating programmed cell death. To this end, screening of a chicken cDNA library is currently underway to identify additional ICE-related protease genes expressed in hen granulosa cells, and granulosa cell extracts are being evaluated for ICE-related protease activity(ies).

Urokinase Plasminogen Activator

Additional proteases related to cellular reorganization and extracellular matrix degradation induced during apoptotic cell death include collagenases, stromelysin-1, cathepsins B and D, and the plasminogen activators (tissue-type and urokinase). Although the activation of such proteases has not been directly linked to the initiation of apoptotic cell death, they are clearly involved in the progression of apoptosis by causing the dissolution of cell-to-cell and cell-to-basement membrane contacts that lead to tissue involution (40). In addition, urokinase plasminogen activator (uPA) is proposed to play an essential role in cancerous cell invasion through the extracellular matrix and basement membrane during tumor metastasis (41), and thus this protease represents a focal point of attention in the diagnosis and therapeutic treatment of ovarian cancer (42).

We have previously reported that the highest levels of cell-associated and secreted PA activity in hen granulosa cells are found in cells from prehierarchal follicles, and suggested that such elevated levels of protease activity may be related to cellular reorganization at the time of follicle selection (2). More recent studies, however, have determined that in direct contrast to most other mRNA transcripts evaluated to date (e.g., LH and FSH receptor, Fig. 8.1; $bcl\text{-}x_{LONG}$, Fig. 8.3B), urokinase PA mRNA levels are 10.9-fold higher in atretic, as compared to normal, 4 to 8 mm follicles (Fig. 8.5A). These findings suggest an active induction of urokinase PA during the early stages of atresia. Moreover, urokinase PA mRNA levels are increased by 7.5- to 9.1-fold during the early stages of apoptotic cell death induced in 4 to 8 mm follicle granulosa cells following incubation in serum-free medium (Fig. 8.5B). These data demonstrate that urokinase PA is produced directly by the granulosa layer at a

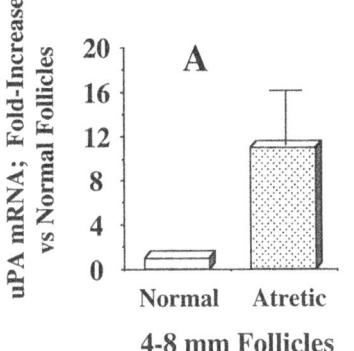
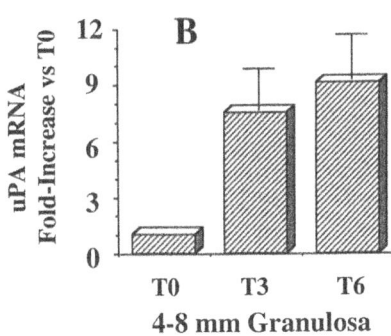

FIGURE 8.5. Increased expression of urokinase plasminogen activator (uPA) mRNA during follicle atresia (A), and in 4–8 mm follicle granulosa cells induced to undergo apoptotic cell death by incubation in serum-free medium (B). The rapid increase in uPA mRNA levels in granulosa cells at 3 h precedes the ability to detect significant oligonucleosome formation, and indicates the importance of this serine protease in extracellular matrix and basement membrane dissolution leading to follicle involution.

time prior to the detection of significant oligonucleosome formation, and generally indicate the importance of protease activity to the progression of follicle atresia as depicted in Figure 8.4B. Finally, a broader implication of these findings is that upon commitment to programmed cell death, specific genes may be targeted for inactivation, while other genes are apparently required to be activated, to facilitate the progression of apoptotic cell death. Whether such activation is the result of increased gene transcription, or alternatively reduced mRNA degradation, remains to be determined.

Summary

This brief overview has provided evidence for a variety of mechanisms related to the maintenance of ovarian follicle granulosa cell viability, or alternatively the progression of apoptotic cell death, that are well conserved between the hen and several mammalian models. In general, much of the data related to programmed cell death in vertebrate species, including that presented herein, is either correlative or derived from nonphysiological systems (e.g., cell incubations or cultures, cell-free systems, overexpression studies, single gene knock-out animals). The challenge for the immediate future will be to establish novel model systems that are well-defined and easily manipulated, yet will yield results that can be interpreted within a physiological context. It is becoming increasingly clear that extracellular, as well as many intracellular, effectors of apoptotic cell death are tissue specific. On the other hand, within a given tissue, including the ovary, many of the intracellular signaling pathways appear to be evolutionarily conserved across animal species and even classes. At the very least, the study of cellular events leading to programmed cell death in hen granulosa cells will provide a basis for understanding this complex process from an evolutionary context. Ideally, this model system will provide novel information from which new hypotheses can be directly tested in the mammalian ovary.

Acknowledgments. I thank Drs. J.L. Tilly and J.P. Witty, and J.T. Bridgham, A. Ott, K.I. Tilly, M. Williams, and B. Wagner, for their assistance and helpful comments throughout the course of these studies. Supported by grants from the USDA (92-37203-8337 and 95-37203-1998), the NSF (IBN94-19613) and a Postdoctoral Fellowship to Dr. J.P. Witty from the Walther Cancer Institute (Indianapolis, IN).

References

1. Johnson AL, Bridgham JT, Witty JP, Tilly JL. Susceptibility of avian granulosa cells to apoptosis is dependent upon stage of follicle development and is related to endogenous levels of bcl-x_{LONG} gene expression. Endocrinology 1996;37:2059–66.
2. Tilly JL, Kowalski KI, Li Z, Levorse JM, Johnson AL. Plasminogen activator and DNA synthesis in avian granulosa cells during follicular development and the periovulatory period. Biol Reprod 1992;46:195–200.
3. Johnson AL, Li Z, Tilly JL. Expression of c-*myc* and bcl-2 mRNA during ovarian follicular development. Biol Reprod 1993;48(Suppl. 1):152.
4. Marcu KB, Bossone SA, Patel AJ. *myc* function and regulation. Annu Rev Biochem 1992; 61:809–60.
5. Freytag SO, Dang CV, Lee WMF. Definition of the activities and properties of c-*myc* required to inhibit cell differentiation. Cell Growth Differ 1990;1:339–43.
6. Tilly JL, Kowalski KI, Johnson AL. Stage of ovarian follicular development associated with the initiation of steroidogenic competence in avian granulosa cells. Biol Reprod 1991;44:305–14.
7. Johnson AL, Bridgham JT, Wagner B. Characterization of a chicken luteinizing hormone receptor (cLH-R) cDNA, and expression of cLH-R mRNA in the ovary. Biol Reprod 1996;55:304–9.
8. You S, Bridgham JT, Foster DN, Johnson AL. Characterization of a chicken follicle-stimulating hormone receptor (cFSH-R) cDNA, and expression of cFSH-R mRNA in the ovary. Biol Reprod 1996;55:1055–62.
9. Li Z, Johnson AL. Regulation of P450 cholesterol side-chain cleavage mRNA expression and progesterone production in hen granulosa cells. Biol Reprod 1993;49:463–9.
10. Johnson AL, Li Z, Gibney JA, Malamed S. Vasoactive intestinal peptide-induced expression of cytochrome P450 cholesterol side-chain cleavage and 17α-hydroxylase enzyme activity in hen granulosa cells. Biol Reprod 1994;51:327–33.
11. Peddie MJ, Onagbesan OM, Williams J. Epidermal growth factor and thecal secretions stimulate proliferation and inhibit progesterone production by cultured chicken granulosa cells. Gen Comp Endocrinol 1994;94:341–56.
12. Lafrance M, Croze F, Tsang BK. Influence of growth factors on the plasminogen activator activity of avian granulosa cells from follicles at different maturational stages of preovulatory development. J Mol Endocrinol 1993;11:291–304.

13. Tilly JL, Kowalski KI, Johnson AL, Hsueh AJW. Involvement of apoptosis in ovarian follicular atresia and postovulatory regression. Endocrinology 1991;129:2799–801.
14. Richards JS. Hormonal control of gene expression in the ovary. Endocr Rev 1994; 15:725–51.
15. Bissonnette RP, Echeverri F, Mahboubi A, Green DR. Apoptotic cell death induced by c-*myc* is inhibited by *bcl*-2. Nature 1992;359:552–4.
16. Evan GI, Wyllie AH, Gilbert CS, Littlewood TD, Land H, Brooks M, Waters CM, Penn LZ, Hancock DC. Induction of apoptosis in fibroblasts by c-myc protein. Cell 1992;69:119–29.
17. Chun, S-Y, Billig H, Tilly JL, Furata I, Tsafriri A, Hsueh AJW. Gonadotropin suppression of apoptosis in cultured preovulatory follicles: mediatory role of endogenous insulin-like growth factor I. Endocrinology 1994;135:1845–53.
18. Chun S-Y, Eisenhauer KM, Minami S, Billig H, Perlas E, Hsueh AJW. Hormonal regulation of apoptosis in early antral follicles: follicle-stimulating hormone as a major survival factor. Endocrinology 1996;137:1447–56.
19. Tilly JL, Kowalski KI, Schomberg DW, Hsueh AJW. Apoptosis in atretic ovarian follicles is associated with selective decreases in messenger ribonucleic acid transcripts for gonadotropin receptors and cytochrome P450 aromatase. Endocrinology 1992; 131:1670–6.
20. Flaws JA, DeSanti A, Tilly KI, Javid RO, Kugu K, Johnson AL, Hirshfield AN, Tilly JL. Vasoactive intestinal peptide-mediated suppression of apoptosis in the ovary: potential mechanisms of action and evidence of a conserved antiatretogenic role through evolution. Endocrinology 1995;136:4351–9.
21. Tilly JL, Billig H, Kowalski KI, Hsueh AJW. Epidermal growth factor and basic fibroblast growth factor suppress the spontaneous onset of apoptosis in cultured rat granulosa cells and follicles by a tyrosine kinase-dependent mechanism. Mol Endocrinol 1992;6:1942–50.
22. Yin X-M, Oltvai, Korsmeyer SJ. BH1 and BH2 domains of Bcl-2 are required for inhibition of apoptosis and heterodimerization with Bax. Nature 1994;369:321–3.
23. Fang W, Rivard JJ, Ganser JA, LeBien TW, Nath KA, Mueller DL, Behrens TW. Bcl-x_L rescues WEHI B lymphocytes from oxidant-mediated death following diverse apoptotic stimuli. J Immunol 1995;155:66–75.
24. Boise LH, Gonzalez-Garcia M, Postema CE, Ding L, Lindsten T, Turka LA, Mao X, Nunez G, Thompson CB. Bcl-x, a bcl-2-related gene that functions as a dominant regulator of apoptotic cell death. Cell 1993;74:597–608.
25. Minn AJ, Boise LH, Thompson CB. Bcl-x_S antagonizes the protective effects of Bcl-x_L. J Biol Chem 1996;271:6306–12.
26. Tilly JL, Tilly KI, Kenton ML, Johnson AL. Expression of members of the *bcl*-2 gene family in the immature rat ovary: equine chorionic gonadotropin-mediated inhibition of granulosa cell apoptosis is associated with decreased *bax* and constitutive *bcl*-2 and *bcl*-x_{LONG} messenger ribonucleic acid levels. Endocrinology 1995;136:232–41.
27. Chittenden T, Flemington C, Houghton AB, Ebb RG, Gallo GJ, Elangovan B, Chinnadurai G, Lutz RJ. A conserved domain in Bak, distinct from BH1 and BH2, mediates cell death and protein binding functions. EMBO J 1995;14:5589–96.
28. Gillet G, Guerin M, Trembleau A, Brun G. A Bcl-2-related gene is activated in avian cells transformed by the Rous sarcoma virus. EMBO J 1995;14:1372–81.
29. Chinnaiyan AM, Orth K, O'Rourke K, Duan H, Poirier GG, Dixit VM. Molecular ordering of the cell death pathway. J Biol Chem 1996;271:4473–6.
30. Kumar S, Harvey NL. Role of multiple cellular proteases in the execution of programmed cell death. FEBS Lett 1995;375:169–73.

31. Johnson AL, Bridgham JT, Bergeron L, Yuan J. Characterization of the avian *Ich-1* cDNA and expression of *Ich-1*$_{LONG}$ mRNA in the hen ovary. Gene 1997;192:227–33.
32. Wilson KP, Black JF, Thompson JA, Kim EE, Griffith JP, Navia MA, Murcko MA et al. Structure and mechanism of interleukin-1β converting enzyme. Nature 1994;370:270–5.
33. Newman PM, To SS, Robinson BG, Hyland VJ, Schrieber L. Effect of gold thiomalate and its thiomalate component on the in vitro expression of endothelial cell adhesion molecules. J Clin Invest 1994;94:1864–71.
34. Trbovich AM, Hughes FM, Perez GI, Tilly KI, Cidlowski JA, Tilly JL. Requirement for ICE-related protease (IRP) activity in oligonucleosomal endonuclease activation during apoptosis resides in the cytoplasmic compartment of rat ovarian granulosa cells. Serono Symposium on Cell Death in Reproductive Physiology. April 11–14, 1996, Chicago, IL, p 36.
35. Flaws JA, Kugu K, Trbovich AM, DeSanti A, Tilly KI, Hirshfield AN, Tilly JL. Interleukin-1β-converting enzyme-related proteases (IRPs) and mammalian cell death: dissociation of IRP-induced oligonucleosomeal endonuclease activity from morphological apoptosis in granulosa cells of the ovarian follicle. Endocrinology 1995;136: 5042–53.
36. Komiyama T, Ray CA, Pickup DJ, Howard AD, Thornberry NA, Peterson EP, Salvesen G. Inhibition of interleukin-1β converting enzyme by the cowpox virus serpin crmA. J Biol Chem 1994;269:19331–7.
37. Thornberry NA, Molineaux SM. Interleukin-1β converting enzyme: a novel cyteine protease required for IL-1β production and implicated in programmed cell death. Protein Sci 1995;4:3–12.
38. Milligan CE, Prevette D, Yaginuma H, Homma S, Cardwell C, Fritz LC, Tomaselli KJ, Oppenheim RW, Schwartz LM. Peptide inhibitors of the ICE protease family arrest programmed cell death of motoneurons in vivo and in vitro. Neuron 1995;15:385–93.
39. Cain K, Inayat-Hussain SH, Couet C, Cohen GM. A cleavage-site-directed inhibitor of interleukin-1β-converting enzyme-like proteases inhibits apoptosis in primary cultures of rat hepatocytes. Biochem J 1996;314:27–32.
40. Lund LR, Romer J, Thomasset N, Solberg H, Pyke C, Bissell MJ, Dano K, Werb Z. Two distinct phases of apoptosis in mammary gland involution: proteinase-independent and -dependent pathways. Development 1996;122:181–93.
41. Schmitt M, Janicke F, Moniwa N, Chucholowski N, Pache L, Graeff H. Tumor-associated urokinase-type plasminogen activator: biological and clinical significance. Biol Chem Hoppe Seyler 1992;373:611–22.
42. Schmalfeldt B, Kuhn W, Reuning U, Pache L, Dettmar P, Schmitt M, Janicke F, Hofler H, Graeff H. Primary tumor and metastasis in ovarian cancer differ in their content of urokinase-type plasminogen activator, its receptor, and inhibitors types 1 and 2. Cancer Res 1995;55:3958–63.

9

Ultrastructural Aspects of cAMP and p53-Mediated Apoptosis in Normal and *ras*-Transformed Granulosa Cells

ABRAHAM AMSTERDAM, MAREN BRECKWOLDT, ADA DANTES,
SELVARAJ NATARAJAGOUNDER, AND DORIT AHARONI

Granulosa cells, the main constituents of the ovarian follicle, nurse the eggs and serve as the main source for progesterone and estradiol biosynthesis (1–3). Of the approximately four-hundred-thousand follicle-enclosed oocytes present at puberty in the human female, only very few will reach full maturation and ovulation whereas the rest will degenerate via a process termed atresia. In each reproductive cycle the newly formed corpora lutea will develop following ovulation, while the corpora lutea of the previous cycle will regress and will be eliminated by a process termed luteolysis (1–3). In contrast, the life span of the corpus luteum will be markedly prolonged during pregnancy.

We and others have recently found that cells dying by both atresia and luteolysis exhibit features characteristic of programmed cell death (PCD) or apoptosis, including condensation and fragmentation of DNA (4–8). Therefore, it has become evident that the control of granulosa PCD plays an important role in the maintenance of the normal ovarian function during the estrous cycle as well as during pregnancy. Moreover, it appears that initial steps of apoptosis in ovarian cells are associated with enhanced steroidogenesis (6–7). Since follicular and corpora luteal PCD is not a synchronized event among the total cell population, it is not clear if the same cells that undergo apoptosis maintain their steroidogenic activity.

In primary cultures of granulosa cells, complete synchrony cannot be achieved since the cells derived from different layers of granulosa cells demonstrate different degrees of differentiation (2, 9). Therefore, in order to obtain synchronization in the induction of apoptosis, we have established granulosa cell lines in which PCD can be induced in a synchronous manner. These cells were transfected with a temperature-sensitive mutant of the tumor suppressor gene, p53, and retain their steroidogenic response to cyclic aden-osine monophosphate (cAMP) stimulation. However, upon temperature shift from 37 °C to 32 °C, that leads to the manifestation of the wild-type activity of p53, the cells undergo massive and rapid apoptosis (7).

Apoptosis Is Controlled by Multiple Extracellular and Intracellular Signals

In primary cultures of granulosa cells obtained from preantral or preovulatory follicles of the rest ovary, follicle-stimulating hormone (FSH), as well as basic fibroblast growth factor (bFGF), serve as survival factors that protect the cells from undergoing apoptosis in serum-free medium (6, 10). Both FSH and sub-stances elevating intracellular cAMP, such as 8-Bromo-cAMP (8-Br-cAMP), forskolin, and cholera toxin, were found to induce differentiation of granu-losa cells derived from preantral follicles in serum-free medium in vitro. This differentiation involves de novo synthesis of luteinizing hormone (LH) re-ceptors (11, 12). However, primary cultures of granulosa cells obtained from preovulatory follicles were found to be more sensitive to serum deprivation in terms of apoptosis induction (10, 13, and Fig. 9.1). Moreover, gonadotro-pic hormones could not prevent apoptosis induced by serum deprivation, although forskolin and 8-Br-cAMP enhanced apoptosis in both human and rat preovulatory granulosa cells (6, 13); however, forskolin has been reported to inhibit apoptosis in intact preovulatory follicles of the rat (14). Interest-ingly, bFGF, which stimulates progesterone production in these cultures and synergizes with cAMP-generated signals in the stimulation of progesterone, reduced the incidence of apoptosis induced by either serum deprivation or by prolonged cAMP stimulation (6, 10). Thus, we can conclude that activation of tyrosine phosphorylation serves as a critical step in the survival of granu-losa cells; granulosa cells acquire different sensitivity to cAMP depending on the degree of their maturation and both the amplitude and the duration of cAMP stimulation; and, cross-talk between these two major signalling path-ways can control the developmental decision of whether the steroidogenic cell will undergo luteinization followed by programmed cell death, or will survive in its differentiated state during pregnancy.

Tumor Suppressor and Survival Genes as Modulators of Apoptosis

bcl-2 and dad-1 has been shown to act as survival genes that may prevent apoptosis in granulosa cells, as well as in other cell types (15–17). On the other hand, Fas/APO-1, p53 and its target genes, waf-1 and bax, have been found to induce apoptosis in numerous cell types, including cells of the

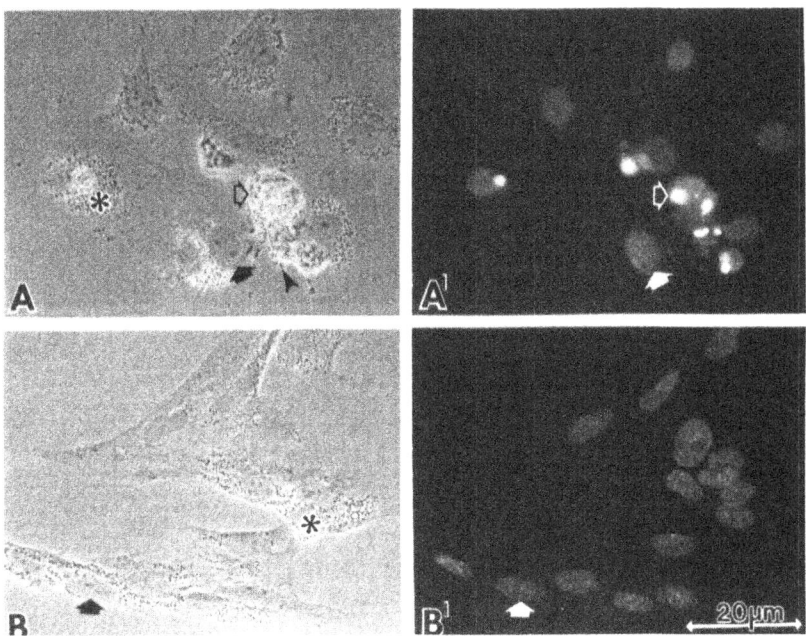

FIGURE 9.1. Morphology of human granulosa cells during apoptosis in vitro. Phase (A, B) and fluorescent (A', B') images of 4'6-diamido-2-phenylindole hydrochloride (DAPI)-stained human granulosa cell cultures in the absence or presence of serum. Cells were incubated for 48 h at 37°C in serum-free Dulbecco's modified Eagles' medium (DMEM)/F12 (A, A') or in the presence of 5% fetal calf serum (FCS) (B, B'). Empty arrows indicate DNA fragmentation in apoptotic nuclei, whereas full arrows mark healthy nuclei with prominent nucleoli. Black arrowheads point to membrane-enclosed blebs. The asterisk marks an accumulation of lipid droplets. All phase and fluorescent images (A and A' or B and B') were taken through a 25 × magnifying objective lens of the same fields. (Reproduced with permission from Breckwoldt et al. (1).)

ovary (15–18). By immunocytochemistry it was demonstrated that the level of p53 increases in apoptotic granulosa cells both in the intact follicle (19) as well as in primary cultures (Amsterdam et al., in preparation). Blocking the activity of wild-type p53 by transfection of granulosa cells with SV40 DNA leading to the expression of T antigen, blocks the cAMP induction of progesterone production as well as the induction of apoptosis by 8-Br-cAMP and forskolin (6).

In order to examine the direct effect of p53 on steroidogenesis and apoptosis, we transfected primary granulosa cells with a temperature-sensitive mutant of p53 (Val 135 p53) (7). This mutant at 37 °C is not capable of binding DNA, and thus cannot function as a tumor suppressor. In contrast, at 32 °C, it binds DNA, exhibiting tumor suppressor activity by the induction of the *waf-1/cip-1* gene (7). Interestingly, these cells are not steroidogenic during their proliferative stage, but become highly steroidogenic, expressing the P450 side-chain cleavage enzyme system upon cAMP stimulation. A temperature shift from 37 °C to 32 °C was sufficient to suppress cell growth, but not to induce apoptosis. However, at 32 °C in the presence of cAMP, the transformed granulosa cells underwent massive and rapid apoptosis with a temporary, but significant, elevation of progesterone production (7). Thus, in the immortalized cell lines, as well as in primary cells, the induction of apoptosis correlates with a temporary elevation of progesterone production (6, 7).

Intracellular Compartmentalization of Steroidogenic Organelles Permits Progesterone Production and Apoptotic Processes to Coexist Within the Same Cell

We have demonstrated that the induction of steroidogenesis in granulosa cells is associated with reorganization of the cytoskeleton. This involves centripetal movement of actin filaments and clustering of steroidogenic organelles, such as lipid droplets and mitochondria (20, 21). This process is accompanied by down-regulation of expression of actin and actin-binding proteins, such as vinculin, α-actinin and the five isoforms of tropomyosin (22, 23). Ultrastructural analysis of primary granulosa cells (Fig. 9.2), as well as immortalized cells (Fig. 9.3), showed that the mitochondria, lipid droplets, lysosomes, and smooth endoplasmic reticulum (organelles associated with steroidogenesis) remained intact and highly clustered during the first 24 h of the induction of apoptosis, concomitant with secretion of high levels of progesterone (6, 7). Interestingly, the apoptotic blebs were almost completely devoid of steroidogenic organelles while condensation of chromatin and initial breakdown of the nuclear membrane was evident in the vast majority of the cells (7). These data strongly suggest that compartmentalization of the steroidogenic organelles during initial steps of apoptosis plays an important role in the preservation of steroidogenesis, and that progesterone production can be continued in the same cells that show progressive signs of apoptosis, such as chromatin condensation and bleb formation.

Actin Cytoskeleton Rearrangement During Apoptosis

Apoptosis involves activation of both endonucleases as well as proteases. Recent investigations have revealed that fodrin, one of the main actin binding proteins localized at the cortex of the actin cytoskeleton, is associated with the cell membrane and is cleaved during apoptosis in lymphoid cells (24). Moreover, it has been recently demonstrated that actin can undergo limited proteolysis (25) by interleukin-1-converting enzyme (ICE). Ubiquitinated

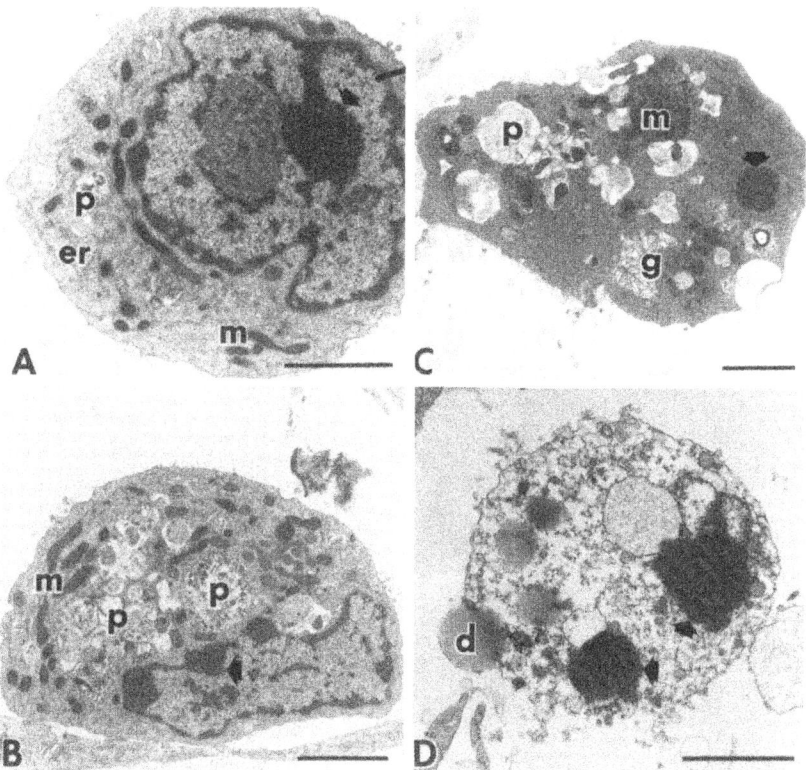

FIGURE 9.2. Various degrees of disorganization in apoptotic preovulatory granulosa cells treated with 50 μm forskolin for 40 h at 37 °C. (A) Large inclusions of chromatin are visible within the nuclear matrix (darkened arrowhead). Electrondense mitochondria (m) and endoplasmic reticulum (er) are well preserved and a few autophagic vesicles (p) are visible. (B) Cell with numerous large vacuoles. Organelles involved in steroidogenesis (i.e., mitochondria and endoplasmic reticulum) are well preserved. (C) Cytoplasmic organization is heavily deteriorated, including disrupted Golgi complexes (g) and clustered mitochondria. Condensed chromatin inclusion is visible (darkened arrowhead). (D) Cell in progressive lysis with remnants of membrane structures, lipid droplets (d) and two large irregular aggregates of condensed chromatin (darkened arrowheads). Bar, 2 μm. (Adapted with permission from Aharoni et al. (6).)

proteins, which can serve as substrates for the multicatalytic proteinase termed the proteasome, have been reported to be modulated during apoptosis in human lymphocytes (26). Therefore, we have examined the possible rearrangement of the actin cytoskeleton and spatial distribution of proteasomes during apoptosis. For this purpose, we have utilized immunocytochemical methods and laser confocal microscopy, to reveal the spatial organization and three-dimensional reconstruction of the steroidogenic organelles, the

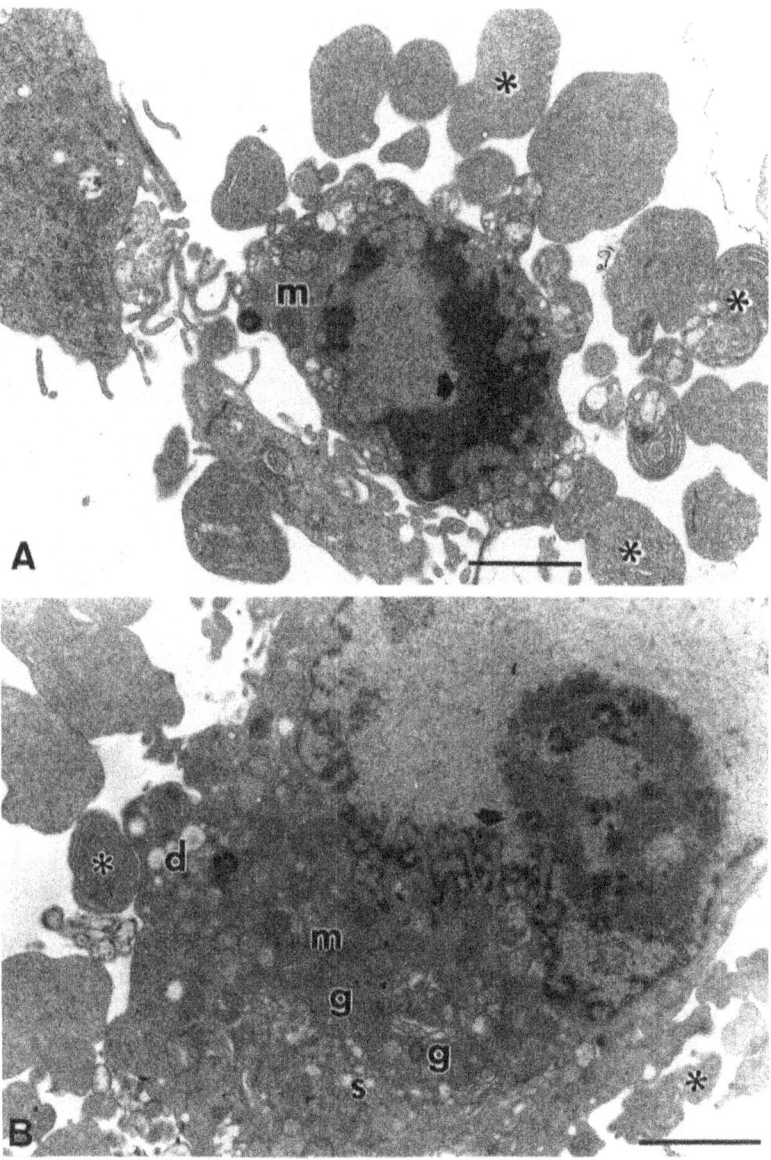

FIGURE 9.3. Apoptosis in immortalized granulosa cells (GTS-5) expressing the temperature sensitive mutant of p53. GTS-5 cells were stimulated for 24 h with 50 μM forskolin
at 32 °C (A). Note condensation of chromatin in the nucleus (darkened arrowhead),
concentration of mitochondria (m) in the perinuclear region and formation of numerous
apoptotic blebs at the periphery of the cells containing mainly polyribosomes but essentially devoid of mitochondria. (B) Intensive clustering of steroidogenic organelles.
Note the initial condensation of chromatin in the nuclear periphery (darkened

actin cytoskeleton, and the proteasomes. Using phalloidin-rhodamine, we have found that the actin filaments in the apoptotic cells are reorganized in a sphere composed of actin networks which serve as a barrier between the apoptotic blebs and the main bulk of the cell (27). Interestingly, the proteasomes, located both in the cytoplasm and the nucleus of the non-apoptotic cells (28, 29), are only found in the apoptotic blebs during cel death. By staining the mitochondria with anti-adrenodoxin antibodies we followed their distribution in the cell. We verified (Fig. 9.3) that the mito-chondria were highly clustered in the perinuclear region while the apoptotic blebs were essentially devoid of mitochrondria (27). Thus, reorganization of the cytoskeleton may provide an efficient barrier between the proteasomes in the apoptotic blebs and the mitochondria in the perinuclear region.

Presence of Components Controlling Steroidogenesis in Apoptotic Blebs

It has recently been demonstrated that the "orphan" nuclear transcription factor Ad4BP, also called steroidogenic factor-1 (SF-1), and the steroidogenic acute regulatory (StAR) protein are essential for steroidogenesis (30–33). We have found that the immortalized granulosa cell lines established in our labo-ratory express both the Ad4BP and StAR protein only in steroidogenic cell lines transfected with SV40 DNA and the *Ha-ras* oncogene, whereas cells transfected with SV40 DNA alone lose their steroidogenic ability and do not express the Ad4BP (34) or StAR protein (our unpublished observations). This suggests that *ras* may play an important role in signal transduction pathways leading to the expression of these proteins during gonadal development. Our preliminary experiments indicates that levels of Ad4BP, StAR protein, and adrenotoxin expression are not down-regulated during the initial phase of apoptosis, thus providing the cell with the necessary machinery to allow steroidogenesis to continue during this process.

Prospective Research

It is still not clear whether or not the cellular and molecular events character-istic of granulosa cell apoptosis are common to other steroidogenic cells,

◄————————————————————————————————————

arrowhead) and a massive aggregation of mitochondria (m), lipid droplets (d), Golgi complexes (g) and small vesicles of smooth membranes (s) characteristic of a highly steroidogenic cell. Apoptotic cytoplasmic blebs (asterisk) are devoid of steroidogenic organelles. Bar, 2 µm. (Reproduced with permission from Keren-Tal et al. (7).)

such as those of the adrenal medulla and the testicular Leydig cells. If these mechanisms are common, it becomes important to identify cell-specific signals that induce or prevent apoptosis. Another important issue concerns the physiological significance of continued, and possibly enhanced, steroidogenic activity during apoptosis. For the granulosa-luteal cells, it would appear that the corpus luteum, during its peak of activity in progesterone production, is already committed to the apoptotic process. Therefore, experiments are neded to identify the signals that extend the lifespan of the corpus luteum during pregnancy. As for the cellular level, the mechanism by which the cytoskeleton acquires its new conformation during programmed cell death is a challenge for future research.

Conclusion

Elucidation of the steroidogenic pathway at the molecular level has dramatically progressed in the last few years, mainly due to the cloning of two essential proteins regulating steroidogenesis, Ad4BP and StAR protein (29–32). High resolution microscopy, immunocytochemistry, and three-dimensional reconstruction of the steroidogenic cell provides detailed resolution of the organization of the steroidogenic elements in the intact cell and the modulation of intracellular organization. Such an approach will provide us with insight into the in situ regulation of steroidogenesis in the intact cell as well as in the whole organism.

Acknowledgments. We thank Dr. M. Walker for helpful discussions and Ms. Vivienne Laufer for excellent secretarial assistance. This work was supported in part by grants from the Leo and Julia Forchheimer Center of Molecular Genetics and from the Dr. Joseph Cohn Minerva Center for Biomembrane Research at the Weizmann Institute of Science. A.A. is the Joyce and B. Eisenberg Professorial Chair of Molecular Endocrinology and Cancer Research at the Weizmann Institute of Science.

References

1. Hsueh AJW, Adashi EY, Jones PBC, Welsh JTH. Hormonal regulation of the differentiation of cultured ovarian granulosa cells. Endocr Rev 1984;5,76–127.
2. Amsterdam A, Rotmensch S. Structure-function relationships during granulosa cell differentiation. Endocr Rev 1987;8:309–37.
3. Richards JS, Hedin L. Molecular aspects of hormone action in ovarian follicular development, ovulation and luteinization. Ann Rev Physiol 1988;50:441–63.
4. Tilly JL, Kowalski KI, Johnson AL, Hsueh AJW. Involvement of apoptosis in ovarian follicular atresia and postovulatory regression. Endocrinology 1991;129:2799–801.
5. Tilly JL. Apoptosis and ovarian function. Rev Reprod 1996;1:162–72.
6. Aharoni D, Dantes A, Oren M, Amsterdam A. cAMP-mediated signals as determinants for apoptosis in primary granulosa cells. Exp Cell Res 1995;218:271–82.

7. Keren-Tal I, Suh B-S, Dantes A, Lindner S, Oren M, Amsterdam A. Involvement of p53 expression in cAMP-mediated apoptosis in immortalized granulosa cells. Exp Cell Res 1995;218:283–95.
8. Hsueh AJW, Billig H, Tsafriri A. Ovarian follicle atresia: a hormonally controlled apoptotic process. Endocr Rev 1994;15:707–24.
9. Amsterdam A, Koch Y, Lieberman ME, Lindner HR. Distribution of binding sites for human chorionic gonadotropin in the preovulatory follicle of the rat. J Cell Biol 1975;67:894–900.
10. Tilly JL, Billig H, Kowalski KI, Hsueh AJW. Epidermal growth factor and basic fibroblast growth factor suppress the spontaneous onset of apoptosis in cultured rat ovarian granulosa cells and follicles by a tyrosine kinase-dependent mechanism. Mol Endocrinol 1992;6:1942–50.
11. Amsterdam A, Knecht M, Catt KJ. Hormonal regulation of cytodifferentiation and intercellular communication in cultured granulosa cells. Proc Natl Acad Sci (USA) 1981;78:30000–4.
12. Knecht M, Amsterdam A, Catt KJ. The regulatory role of cyclic AMP in hormone-induced granulosa cell differentiation. J Biol Chem 1981;256:10628–33.
13. Breckwoldt M, Selvaraj N, Aharoni D, Barash A, Segal L, Insler V, Amsterdam A. Expression of Ad4-BP/cytochrome P450 side-chain cleavage enzyme and induction of cell death in long-term cultures of human granulosa cells. Mol Hum Repr 1996; 2:391–400.
14. Flaws JA, DeSanti A, Tilly KI, Javid RO, Kugu K, Johnson AL, Hirshfield AN, Tilly JL. Vasoactive intestinal peptide-mediated suppression of apoptosis in the ovary: potential mechanisms of action and evidence of a conserved anti-atretogenic role through evolution. Endocrinology 1995;136:4351–9.
15. Tilly JL, Tilly KI, Kenton ML, Johnson AL. Expression of members of the *bcl-2* gene family in the immature rat ovary: equine chorionic gonadotropin-mediated inhibition of granulosa cell apoptosis is associated with decreased *bax* and constitutive *bcl-2* and *bcl-x*$_{LONG}$ messenger ribonucleic acid levels. Endocrinology 1995;136:232–41.
16. Kroemer G, Petit P, Zamzami N, Vayssiere JL, Mignotte B. The biochemistry of programmed cell death. FASEB J 1995;9:1277–87.
17. Martimbeau S, Tao XJ, Tilly KI, Tilly JL. Enhanced expression of the *dad-1* death-repressor gene in rat ovarian granulosa cells during gonadotropin-promoted follicular survival. Proceedings of the 10th International Congress of Endocrinology, San Francisco, CA 1996;p 751 (Abstract).
18. Haffner R, Oren M. Biochemical properties and biological effects of p53. Curr Opin Genet Dev 1995;5:84–90.
19. Tilly KI, Banerjee S, Banerjee PP, Tilly JL. Expression of the p53 and Wilms' tumor suppressor genes in the rat ovary: gonadotropin repression in vivo and immunohistochemical localization of nuclear p53 protein to apoptotic granulosa cells of atretic follicles. Endocrinology 1995;136:1394–402.
20. Amsterdam A, Rotmensch S, Ben-Zeev A. Coordinated regulation of morphological and biochemical differentiation in a steroidogenic cell: the granulosa cell model. TIBS 1989;14:377–82.
21. Amsterdam A, Aharoni D. Plasticity of cell organization during differentiation of normal and oncogene transformed granulosa cells. Microsc Res Tech 1994;27: 108–24.
22. Ben-Ze'ev A, Amsterdam A. In vitro regulation of granulosa cell differentiation. Involvement of cytoskeletal protein expression. J Biol Chem 1987;262:5366–76.

23. Ben-Ze'ev A, Baum G, Amsterdam A. Regulation of tropomyosin expression in the maturing ovary and in primary granulosa cell cultures. Dev Biol 1989;135:191–201.
24. Martin SJ, O'Brien GA, Nishioka EK, McGahon AJ, Mahboubi A, Saido A, et al. Proteolysis of fodrin (non-erythroid spectrin) during apoptosis. J Biol Chem 1995; 270:6425–8.
25. Mashima T, Naito M, Fujita N, Noguchi K, Tsuruo T. Identification of actin as a substrate of ICE and an ICE-like protease and involvement of an ICE-like protease but not ICE in VP-16 induced U937 apoptosis. Biochem Biophy Res Comm 1995;217: 1185–92.
26. Delic J, Morange M, Magdelenat H. Ubiquitin pathway involvement in human lymphocyte gamma-irradiation-induced apoptosis. Mol Cell Biol 1993;13:4875–83.
27. Amsterdam A, Dantes A, Fuchs T, Baumeister W, Pitzer F. Maintenance of granulosa cell steroidogenesis during early stages of apoptosis. Proceedings from the 10th International Congress of Endocrinology, San Francisco, CA 1996;p 598 (Abstract).
28. Amsterdam A, Pitzer F, Baumeister W. Changes in intracellular localization of proteasomes in immortalized ovarian granulosa cells during mitosis associated with a role in cell cycle control. Proc Natl Acad Sci USA 1993;90:99–103.
29. Amsterdam A, Pitzer F, Santarius U, Dantes A, Baumeister W. Possible role of the multi-catalytic proteinase (proteasome) in regulation of the cell cycle. In: Hu VW, ed. The cell cycle '93: regulators, targets and clinical applications. New York: Plenum Press, 1994;203–9.
30. Lin D, Sugawara T, Strauss JF, Clark BJ, Stocco DM, Saenger P, Rogol A, Miller WL. Role of steroidogenenic acute regulatory protein in adrenal and gonadal steroidogenesis. Science 1995;267:1828–31.
31. Sugawara T, Holt JA, Driscoll D, Strauss JF, Lin D, Miller WL, Patterson D, Clancy KP, Hart IM, Clark BJ, Stocco DM. Human steroidogenic acute regulatory protein: functional activity in COS1 cells, tissue-specific expression and mapping of the structural gene to 8p11.2 and a pseudogene to chromosome 13. Proc Natl Acad Sci USA 1995;92:4778–82.
32. Morohashi K, Honda S, Inomata Y, Handa H, Omura T. A common trans-acting factor, Ad4-binding protein, to the promoters of steroidogenic P-450s. J Biol Chem 1992; 267:17913–19.
33. Lynch JP, Lala DS, Peluso JJ, Luo W, Parker KL, White BA. Steroidogenic factor 1, an orphan nuclear receptor, regulates the expression of the rat aromatase gene in gonadal tissues. Mol Endocrinol 1993;7:776–86.
34. Keren-Tal I, Dantes A, Plehn-Dujowich D, Amsterdam A. Expression of Ad4BP/SF-1 transcription factor determines steroidogenic activity in oncogene-transformed granulosa cells. Proceedings from the 10th International Congress of Endocrinology, San Francisco, CA 1996;p 579 (Abstract).

10

Regulation of Apoptosis Versus Mitosis in Immature Granulosa Cells

DAVID W. SCHOMBERG AND JONATHAN L. TILLY

Recent advances in the study of growth regulation at the cellular and molecular level have made it clear that "growth" of a tissue is the result of a tightly monitored and regulated balance between the processes of mitosis and apoptosis. Prior to this, studies of cell growth regulation in the ovarian follicle, as in other systems, emphasized the regulation of mitosis. Now, however, with the realization that apotosis is the fundamental cellular mechanism underlying the process of follicular atresia (1, 2), it is readily apparent that in order to understand the process of follicle selection we need to understand more about the regulation of both apoptosis and mitosis. In this chapter, studies will be reviewed that address both issues in the context of ovarian follicular growth. The focus will be to review the information relevant to the premise that regulatory hormones and growth factors act more effectively in combination to stimulate mitosis and to attenuate apoptosis.

Regulation of Mitosis in Granulosa Cells

The growth-modulating effects of various hormones, growth factors, cytokines, and their combinations upon ovarian follicular cells have been investigated by many laboratories over the past several years. For convenience, these studies have generally focused upon in vitro culture systems utilizing granulosa cells of various species.

Hormonal, Cytokine, or Growth Factor Control

The following studies are representative examples of growth stimulation in cultured granulosa cell (GC) models of several species. Almost without exception, they show that the mitogenic response was superior when the individual growth factors selected for study were used in combination with other growth factors, cytokines, or serum. The study of defined growth factors and their effects upon GC growth began in earnest with the studies of Gospodarowicz and coworkers who showed that epidermal growth factor (EGF) in serum-containing medium was stimulatory to cultured bovine, porcine, rabbit, and human GCs (3, 4). Of these, the porcine GC system has been the most widely used model for the study of growth control. It has since been refined, however, to demonstrate cellular responsiveness to growth modulators under defined conditions in the absence of serum (5–7). In systems developed in our laboratories, combinations of insulin (INS)/EGF/ transforming-growth factor-beta (TGF-β), or follicle-stimulating hormone (FSH)/INS or FSH/insulin-like growth factor-I (IGF-I) were particularly and equally effective (7). In one exception to the above, INS alone markedly stimulated tritiated-thymidine (^{3}H-T) incorporation whereas human chorionic gonadotropin (hCG) or 8-bromo-cyclic adenosine monophosphate (8-br-cAMP) significantly attenuated the INS response; FSH was neutral (9). Differences in cell type (GC from immature versus mature animals) and culture conditions (length of culture and/or cell density) may account for these discrepancies. This study also showed an excellent correlation between ^{3}H-T uptake and cell number, thus validating the use of ^{3}H-T incorporation as an index of GC proliferation (9).

In contrast to the above studies with porcine GC, growth responses in cultured rat GC have been difficult to demonstrate. For example, EGF is not mitogenic to GC obtained from estrogen-primed, immature, hypophysectomized or adult rats (10). Although EGF alone does not induce proliferation under these conditions, EGF in combination with IGF-I and TGF-β, the same combination that was effective in the porcine system (7), is able to stimulate DNA synthesis in the absence of FSH (10). In an example of the growth stimulatory action of TGF-β, the growth factor alone stimulated ^{3}H-T incorporation; again, however, a combination of TGF-β and FSH is superior (11). In rat GC purified by gradient centrifugation prior to culture, 8-br-cAMP alone was ineffective, but INS plus 8-br-cAMP or FSH stimulates ^{3}H-T incorporation (12). GC division, stimulated by 8 Br-cAMP, FSH, or phorbol ester (TPA), has also been noted in follicle culture where several factors probably interact to enhance mitosis (13).

Other effectors, such as the cytokines, tumor necrosis factor-alpha (TNF-α), interleukin-1 (IL-1), and interferon-γ, and members of the TGF-β family, specifically inhibin and activin, have been investigated for their growth-promoting effects in GC culture systems. The general conclusions from a number of studies indicate the absence of a pronounced and consistent mitotic effect (14–18). With respect to inhibin, this may indicate that the in vitro culture systems are not adequate models for the study of autocrine/ paracrine growth regulation, since inhibin gene "knock-out" experiments indicate an ovarian growth regulatory role for inhibin in vivo (19). Collectively, these studies indicate that GC growth in a number of species is stimulated most effectively by a combination of effectors involving the "traditional" growth factors or hormones, setting the stage for further inquiry into combinatorial mechanisms of action, both in terms of mitosis and apoptosis.

Combinatorial Paradigms of Endocrine and Paracine/Autocrine Signaling

The fact that growth factor combinations and FSH/growth factor combinations are the most effective mitotic stimuli suggests various alternative growth regulatory paradigms. One example would be a general model in which gonadotropins stimulate the production of all of the necessary growth factors that in turn signal via their cognate receptor systems. Information is sketchy on the issue of whether or not production of all of the above-mentioned growth factors is induced and/or stimulated by gonadotropins. The available information on this topic, as well as constitutive production of growth factos by ovarian cells, has been summarized in several recent reviews (20–24). Virtually no information on growth factor regulation in the ovary in vivo is available, although developmental and species differences have been described (25–29). The second example would be the situation in which gonadotropins and growth factors each initiated their specific receptor-activated signal transduction pathways. The evidence available to date supports this alternative. In both of the above examples, it is likely that the intracellular signalling pathways activated by the different ligand-receptor interactions converge downstream at specific intracellular points. Although these intracellular pathways have not been defined, most reports show that growth in GC systems is stimulated most effectively by a combination of effectors.

Combinatorial Paradigms of Intracellular Signaling

The fact that various growth factor or growth factor/gonadotropin combinations are the most effective mitotic stimuli also suggests that the signalling pathways used by these effectors are not independent. It is becoming increasingly clear that the actions of FSH and LH involve the protein kinase C (PKC) and tyrosine kinase pathways in addition to the protein kinase A (PKA) pathway (30). Keel and coworkers have shown in porcine GCs that EGF stimulates the mitogen-activated protein kinase (MAP kinase) pathway (31). Other novel pathways may also be involved. For example, recent evidence indicates that FSH can also induce the expression of immediate-early response genes, such as serum-inducible kinase (snk), glucocorticoid-regulated kinase (sgk), and pole kinases that encode novel kinases or kinase-like proteins (32). The regulation of cyclin D2 may also be unique in the granulosa cell since cyclin D2 "knock-out" animals do not develop normal preovulatory-size follicles; in a recent report, the number of GC formed in response to pregnant mare's serum gonadotropin (PMSG) was found to be insufficient (33).

Although the effects of steriods on GC growth in vitro are not uniformly demonstrable, presumably because of the rapid disappearance of functional steriod receptors in primary cultured cells, their regulatory role in vivo is well established and obviously should not be overlooked.

How various signaling pathways interact in terms of mitotic regulation presents one of the great challenges for the future in the area of growth control in mammalian cells. Many general aspects will be applicable to a wide variety of cells. The challenge to ovarian physiologists will be to determine the specific points of regulation that are unique to the cells of the ovary.

Regulation of Apoptosis in Granulosa Cells

Apoptosis in ovarian cells, and the hormonal regulation of apoptosis in the ovary, have been the subject of recent in-depth reviews (1, 2). As is the case for mitotic regulation, the hormonal and growth factor control of apoptosis in GC is complex, may be species-specific with respect to individual effectors, and is combinatorial in certain instances.

General Paradigms of Apoptotic Regulation

A very large amount of work in other systems now points to two general avenues by which many cells are induced to undergo apoptosis: first, the removal of an essential survival factor(s), or second, the action of a dominant cell death-inducing effector acting essentially like a ligand via receptor interaction to inititate the apoptotic signaling pathway. These conditions lead to the induction and/or activation of several cell death-associated genes and/ or proteases that can initiate apoptosis. Several of these intracellular effectors have been identified in GCs, including Fas, Bcl-2, Bcl-x, Bax, p53, Ich-1, and CPP32 (34–40). Although the proteins encoded by most of the genes cloned have been isolated, it is not yet clear how the cell death-associated genes are linked mechanistically or functionally to the apoptotic pathway in the ovary. It is known that in GC, removal of growth-promoting effectors (steroids, gonadotropins, or growth factors) increases the degree of apoptosis and alters the pattern of cell death gene expression in vivo and in vitro (1, 2). Moreover, the death-dominant action of the Fas-Fas ligand system has been demonstrated in cultured human and rat GC (34, 40). These general mechanisms may not be mutually exclusive, and their relative predominance in a given cell type is likely to vary among species, change with development, and be modulated by paracrine and endocrine controls.

Combinatorial Signaling in Mitotic Regulation Contrasts With That in Apoptotic Regulation

The combinatorial control paradigm discussed previously as the most effective for the stimulation of mitosis in cultured porcine GC is also effective, but not required, to attenuate apoptosis in these same cells. In porcine GC cultured under serum-free conditions. EGF/INS/TGF-β and FSH/INS or FSH/IGF-I are clearly the most effective stimulators of ^{3}H-T incorporation (6, 7). These treatments are also among the most effective in attenuating DNA fragmentation; however, EGF/INS and INS alone are as effective as EGF/INS/TGF-β and FSH/INS in this regard (8).

In cultured rat GC or follicles, the above general paradigm appears to be similar, but different growth factors seem to be "coupled" to the attenuation of apoptosis. In cultured rat GC, for example, EGF/TGF-α or bFGF alone (rather than IGF-I or INS in the porcine GC system) attenuates apoptosis under defined conditions (41), but each factor alone is unable to stimulate mitosis (10, 11). In the rat follicle culture system, FSH or LH-hCG alone (42, 43) and TGF-α alone (41) have been reported to attenuate apoptosis, but in this multicellular system other factors could also be present to help mediate the effect. More information is needed, but these initial studies suggest that in GCs certain effectors used "alone" in vitro are necessary and sufficient to attenuate apoptosis. On the other hand, they are necessary but not sufficient to stimulate mitosis. As a working hypothesis, this concept should initiate very interesting experimentation at the molecular level to discover how the various signaling pathways interrelate under specific gonadotropin and growth factor controls to commit cells to apoptosis versus mitosis.

Apoptosis Versus Mitosis: Concurrent Intracellular Signaling

Using cultured rat PC-12 pheochromocytoma cells, Xia and coworkers showed very recently that a dynamic balance between ERK (growth factor-activated extracellular signal-regulated kinase) and stress-activated JNK (c-JUN NH_2-terminal protein kinase)-p38 pathways may be important in determining whether a cell survives or undergoes apoptosis (44). Using several molecular approaches they demonstrated that nerve growth factor (NGF) withdrawal leads to a sustained activation of the JNK and p38 enzymes involved in apoptotic signalling, and to a concurrent inhibiton of the ERK enzymes involved in mitotic signalling. Both activation (JNK, p38) and inactivation (ERK) seem critical for apoptosis. Conversely, direct activation of the ERK pathway and suppression of the JNK and p38 pathways prevents apoptosis and promotes the survival of PC-12 cells. The ERKs are activated in response to growth factor stimulation, whereas the JNK and p38 MAP kinases are activated by various forms of environmental stress. It is therefore possible, if not likely, that cross-talk between these signal transduction pathways exists. In preliminary studies, JNK, p38, and ERK have been reported to be regulated by stress (growth factor deprivation) and by growth factor stimulation (TGF-α addition) in the rat follicle culture system (45). This, therefore, appears to be an exciting area for future research into cellular mechanisms underlying growth versus apoptosis in ovarian follicular cells, and makes the concept that these processes depend upon combinatorial regulation and the integration of multiple signals even more compelling.

References

1. Tilly JL. Apoptosis and ovarian function. Rev Reprod 1996;1:162–72.
2. Hsueh AJW, Billig H, Tsafriri A. Ovarian follicle atresia: a hormonally controlled apoptotic process. Endocr Rev 1994;15:707–24.
3. Gospodarowicz D, Birdwell CR III. Effects of fibroblast and epidermal growth factors on ovarian cell proliferation in vitro. I. Characterization of the response of granulosa cells to FGF and EGF. Endocrinology 1977;100:1108–21.
4. Gospodarowicz D, Bialecki H. Fibroblast and epidermal growth factors are mitogenic agents for cultured granulosa cells of rodent, porcine and human origin. Endocrinology 1979;104:757–64.
5. Baranao JL, Hammond JM. Serum-free medium enhances growth and differentiation of cultured pig granulosa cells. Endocrinology1985;116:51–8.
6. Buck PA, Schomberg DW. A serum-free defined culture system which maintains follicle-stimulating hormone responsiveness and differentiation of porcine granulosa cells. Biol Reprod 1987;36:167–74.
7. May JV, Frost JP, Schomberg DW. Differential effects of epidermal growth factor, somatomedin-C/insulin-like growth factor-I, and transforming growth factor-β on porcine granulosa cell deoxyribonucleic acid synthesis and cell proliferation. Endocrinology 1988;123:168–79.
8. Schomberg DW, Tilly KI, Tilly JL. Growth factors suppress porcine granulosa cell apoptosis via a pathway independent of *bax* gene expression. Biol Reprod 1995;52(Suppl 1):90.
9. Hammond JM, English HF. Regulation of deoxyribonucleic acid synthesis in cultured porcine granulosa cells by growth factors and hormones. Endocrinology 1987;120:1039–46.
10. Bendell JJ, Dorrington JH. Epidermal growth factor influences growth and differentiation of rat granulosa cells. Endocrinology 1990;533–40.
11. Dorrington J, Chuma AV, Bendell JJ. Transforming growth factor-beta and follicle-stimulating hormone promote rat granulosa cell proliferation. Endocrinology 1988;123:353–9.
12. Bley MA, Simon JC, Estevez AG, Deasua L, Baranao JL. Effect of follicle-stimulating hormone on insulin-like growth factor-I-stimulated rat granulosa cell deoxyribonucleic acid synthesis. Endocrinology 1992;131:1223–9.
13. Peluso JJ, Pappalardo A, White BA. Control of rat granulosa cell mitosis by phorbol ester-, cyclic AMP-, and estradiol-17β-dependent pathways. Biol Reprod 1993;49:416–22.
14. Woodruff TK, Lyon RJ, Hansen SE, Rice GC, Mather JP. Inhibin and activin locally regulate rat ovarian folliculogenesis. Endocrinology 1990;127:3196–205.
15. Findlay, JK. An update on the roles of inhibin, activin, and follistatin as local regulators of folliculogenesis. Biol Reprod 1993;48:15–23.
16. Findlay JK, Xiao S, Shukovski L, Michel U. Novel peptides in ovarian physiology. Inhibin, activin, and follistatin. In: Adashi EY, Leung PCK, eds. The Ovary. New York: Raven Press, 1993:413–32.
17. Eokia E, Adashi EY. Potential role of cytokines in ovarian physiology. The case for interleukin-1. In: Adashi EY, Leung PCK, eds. The Ovary. New York: Raven Press, 1993:383–94.
18. Terranova PF, Sancho-Tello M, Hunter VJ. Tumor necrosis factor-α and ovarian function. In: Adashi EY, Leung PCK, eds. The Ovary. New York: Raven Press, 1993:395–411.

19. Matzuk MM, Finegold MJ, Su J-G, Hsueh AJW, Bradely A. Alpha-inhibin is a tu-mour-suppressor gene with gonadal specificity in mice. Nature 1992;360:313–9.
20. Giudice LC. Insulin-like growth factors and ovarian follicular development. Endocr Rev 1992;13:641–69.
21. Hammond JM, Samara SE, Grimes R, Leighton J, Barber J, Canning SF, Guthrie HD. The role of insulin-like growth factors and epidermal growth factor-related peptides in intraovarian regulation in the pig ovary. J Reprod Fertil Suppl 1993;48:117–25.
22. Adashi EY, The intraovarian insulin-like growth factor system. In: Adashi EY, Leung PCK, eds. The Ovary. New York: Raven Press, 1993:319–35.
23. Mulheron GW, Schomberg DW. The intraovarian transforming growth factor system. In: Adashi EY, Leung PCK, eds. The Ovary. New York: Raven Press, 1993:337–61.
24. Ojeda SR, Dissen GA, Malamed S, Hirshfield AN. A role for neurotrophic factors in ovarian development. In: Hsueh AJW, Schomberg DW, eds. Ovarian Cell Interactions, Genes to Physiology. Springer-Verlag, New York. 1993:181–202.
25. Roy SK, Greenwald GS. Mediation of follicle-stimulating hormone action on follicular deoxyribonucleic acid synthesis by epidermal growth factor. Endocrinology 1991;129:1903–8.
26. Lobb DK, Korbin MS, Kudlow JE, Dorrington JH. Transforming growth factor-alpha in the adult bovine ovary: identification in growing ovarian follicles. Biol Reprod 1989;40:1087–93.
27. Oliver JE, Aitman TJ, Powell JF, Wilson CA, Clayton RN. Insulin-like growth factor I gene expression in the rate ovary is confined to the granulosa cells of developing follicles. Endocrinology 1989;124:2671–9.
28. Singh B, Armstrong DT. Transforming growth factor α gene expression and peptide localization in porcine ovarian follicles. Biol Reprod 1995;53:1429–35.
29. Roy SK, Ogren C, Roy C, Lu B. Cell-type-specific localization of transforming growth factor-beta 2 and transforming growth factor-beta 1 in the hamster ovary: differential regulation by follicle-stimulating hormone and luteinizing hormone. Biol Reprod 1992;46:595–606.
30. Richards JS. Hormonal control of gene experssion in the ovary. Endocr Rev 1994;15.525–51.
31. Keel BA, Hildebrandt JM, May JV, Davis JS. Effects of epidermal growth factor on the tyrosine phosphorylation of mitogen-activated protein kinases in monolayer cul-tures of porcine granulosa cells. Endocrinology 1995;136:1197–204.
32. Richards JS, Fitzpatrick SL, Clemens JW, Morris JK, Alliston T, Sirois J. Ovarian cell differentiation: a cascade of multiple hormones, cellular signals, and regulated genes. Rec Prog Horm Res 1995;50:223–54.
33. Sicinski P, Donaher JL, Geng Y, Parker SB, Gardner H, Park MY, Robker RL, Richards JS, McGinnis LK, Biggers JD, Eppig JJ, Bronson RT, Elledge SJ, Weinberg RA. Cyclin D2 in an FSH-responsive gene involved in gonadal cell proliferation and onco-genesis. Nature 1996;384:470–4.
34. Quirk SM, Cowan RG, Joshi SG, Henrikson KP. Fas antigen-mediated apoptosis in human granulosa/luteal cells. Biol Reprod 1995;52:279–87.
35. Tilly JL, Tilly KI, Kenton ML, Johnson AL. Expression of members of the *bcl-2* gene family in the immature rat ovary: equine chorionic gonadotropin-mediated inhibition of granulosa cell apoptosis is associated with decreased *bax* and constitutive *bcl-2* and *bcl-x*$_{LONG}$ messenger ribonucleic acid levels. Endocrinology 1995;136:232–41.
36. Ratts VS, Flaws JA, Kolp R, Sorenson CM, Tilly JL. Ablation of *bcl-2* gene expression decreases the numbers of oocytes and primordial follicles established in the post-natal female mouse gonad. Endocrinology 1995;136:3665–8.

37. Knudson CM, Tung KSK, Tourtellotte WG, Brown GAJ, Korsmeyer SJ. Bax-deficient mice with lymphoid hyperplasia and male germ cell death. Science 1995;270:96–9.
38. Flaws JA, Kugu K, Trbovich AM, DeSanti A, Tilly KI, Hirshfield AN, Tilly JL. Interleukin-1β-converting enzyme-related proteases (IRPs) and mammalian cell death: dissociation of IRP-induced oligonucleosomal endonuclease activity from morphological apoptosis in granulosa cells of the ovarian follice. Endocrinology 1995;136:5042–53.
39. Tilly KI, Banerjee S, Banerjee PP, Tilly JL. Expression of the p53 and Wilms' tumor suppressor genes in the rat ovary: gonadotropin repression in vivo and immunological localization of nuclear p53 protein to apoptotic granulosa cells of atretic follicles. Endocrinology 1995;136:1394–402.
40. Hakuno N, Koji T, Yano T, Kobayashi N, Tsutsumi O, Taketani Y, Nakane PK. Fas/APO-1/CD95 system as a mediator of granulosa cell apoptosis in ovarian follicle atresia. Endocrinology 1996;137:1938–48.
41. Tilly JL, Billig H, Kowalski KI, Hsueh AJW. Epidermal growth factor and basic fibrolast growth factor suppress the spontaneous onset of apoptosis in cultured rat granulosa cells and follicles by a tyrosine kinase-dependent mechanism. Mol Endocrinol 1992;6:1942–50.
42. Chun SY, Billig H, Tilly JL, Furuta I, Tsafriri A, Hsueh AJW. Gonadotropin suppression of apoptosis in cultured preovulatory follices: mediatory role of endogenous insulin-like growth factor-I. Endocrinology 1994;135:1845–53.
43. Tilly JL, Tilly KI. Inhibitors of oxidative stress mimic the ability of follicle-stimulating hormone to suppress apoptosis in cultured rat ovarian follicles. Endocrinology 1995;136:242–52.
44. Xia Z, Dickens M, Raingeaud J, Davis RJ, Greenberg ME. Opposing effects of ERK and JNK-p38 MAP kinases on apoptosis. Science 1995;270:1326–31.
45. Flaws JA, Hiershfield AN, Tilly IL, DeSanti AM, Davis MA. Activation of mitogen activated protein kinases during follicular atresia and survival. Biol Reprod 1996;54(Suppl 1):86.

11

Cell-to-Cell Contact and the Role of Cadherins in Cell Survival

JOHN J. PELUSO

There are two epithelial cell types within the ovary: granulosa cells and ovarian surface epithelial cells. Both cell types die via an apoptotic mechanism (1, 2). Apoptotic granulosa cells are observed within atretic follicles that do not ovulate, while the surface epithelial cells undergo apoptosis as part of the process that allows the oocyte to be released from the ovulatory follicle. The hormonal factors that inhibit granulosa cell apoptosis have been the subject of numerous studies. Many of these studies will be reviewed at this symposium. Briefly, several hormones and growth factors, including epidermal growth factor (EGF), progesterone and basic fibroblast growth factor (bFGF) inhibit granulosa cell apoptosis in vitro. In addition to hormonal factors, cell-to-cell contact also plays an important role in maintaining the viability of both rat granulosa cells and rat ovarian surface epithelial cells. This chapter will discuss the relationship that exists between hormonal/growth factor-mediated and cell contact-mediated cell survival. The primary focus will be on cell contact as a survival signal since other chapters will discuss in detail the mechanisms through which hormones and growth factors promote cell viability.

Granulosa Cell Apoptosis

Hormonal Factors Regulating Granulosa Cell Apoptosis

Over the course of the reproductive life span, only a select few follicles mature and ovulate with the majority of follicles ultimately undergoing atresia (3). Thus, atresia is a dominant physiological event within the ovary. In spite of this fact, the mechanism responsible for follicular atresia has not been clearly defined. Atresia seems to be initiated with the death of the granulosa cells (3) and recent studies have shown that granulosa cell death is characterized by DNA fragmentation. This suggests that granulosa cells die through an active process referred to as programmed cell death or apoptosis (2, 4). In several nonovarian cell types, DNA fragmentation is triggered by an increase in intracellular calcium ($[Ca^{2+}]_i$) that activates an endonuclease that eventually results in the cleavage of the DNA into multimers of about 180 basepair (bp) nucleosomal units with free 3'OH ends (5, 6). Since granulosa cells possess a calcium/magnesium-dependent endonuclease that is activated by the addition of calcium (7), it has been proposed that an increase in $[Ca^{2+}]_i$ initiates granulosa cell apoptosis.

Recent work has focused on the identification of hormonal factors that control granulosa cell DNA fragmentation. These studies have shown that hypophysectomy induces and estrogen prevents rat granulosa cell DNA fragmentation (8). In addition, decreased levels of aromatase mRNA (9) and intrafollicular estrogen are associated with granulosa cell degeneration (3). These studies support the concept that gonadotropins maintain estrogen biosynthesis within the follicle and that estrogen acts to prevent granulosa cell apoptosis. Furthermore, initial degenerative changes occur within the nuclei of many dying granulosa cells (10). These nuclear changes are characteristic of cells undergoing apoptosis (11) and are induced by antiestrogens (10), thus supporting the concept that estrogen prevents the granulosa cells from undergoing apoptosis (8).

Granulosa cells also exhibit DNA fragmentation in vitro and this fragmentation is prevented by either EGF, bFGF (12) or insulin-like growth factor-I (13). It is possible that estrogen, acting either directly on granulosa cells or indirectly by stimulating the production of intraovarian growth factors, controls $[Ca^{2+}]_i$ levels and thus apoptotic DNA fragmentation. However, in rat granulosa cells isolated from gonadotropin-primed immature rats, estrogen neither stimulates nor suppresses basal $[Ca^{2+}]_i$ levels (14). This suggests that estrogen prevents apoptosis indirectly by stimulating the synthesis of intraovarian factors. Recent studies have shown that several different growth factors influence granulosa cell viability. Since granulosa cell viability is essential for ovarian function, it is not surprising that there are redundant systems that inhibit granulosa cell apoptosis.

Although the previously cited studies implicate apoptosis as the mechanism of granulosa cell death, most of these investigations assessed apoptotic changes within populations of granulosa cells. As a result, two important aspects were not considered. First, ovarian follicles are composed of two different sizes of granulosa cells, with only large granulosa cells capable of secreting steroids (15). Second, these different-sized granulosa cells are connected by gap and adhesion-type junctions. These junctional complexes appear to coordinate functional responses between granulosa cells (16). The remainder of this chapter will focus on studies that develop the hypothesis that cell contact acts to maintain the viability of large granulosa cells in a hormone/growth factor-independent manner.

Apoptotic Cell Death in Large Granulosa Cells

Because different-sized granulosa cells perform different functions, it is important to determine if all granulosa cells undergo apoptosis in vitro. To answer this question, small and large granulosa cells were cultured in serum-free medium for 24 h. After culture, fragmented DNA with the associated 180 bp ladder is observed in the large granulosa cell fractions (Fig. 11.1). Since DNA within cells undergoing apoptosis is cleaved and generates numerous DNA fragments, in situ methods were used to identify granulosa cells with fragmented DNA. To further insure proper identification of apoptotic cells, nuclei of granulosa cells were also stained with specific DNA stains. This assessment is classic and possibly the best measure of apoptosis (19). Thus, in situ 3'OH end-labeling and DNA staining procedures were used to simultaneously assess both cell size and apoptosis. As

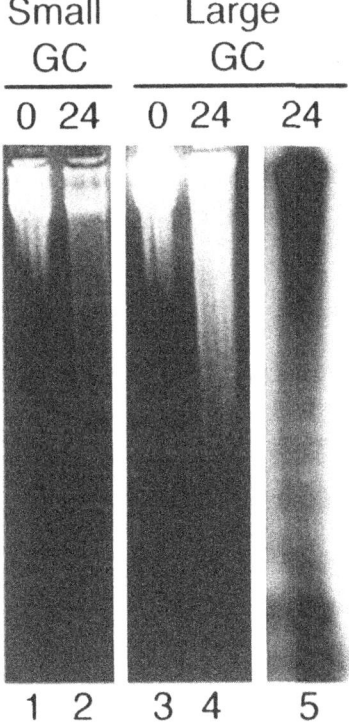

FIGURE 11.1. The integrity of DNA isolated from small and large granulosa cells (GC) of immature rat ovaries. Granulosa cells were separated on a Percoll gradient and cultured for 0 or 24 h in serum-free medium. Cells were prepared for analysis of DNA fragmentation as described by Barry and Eastman (35) (lanes 1–4) or Tilly et al. (9) (lane 5).

expected based on the absence of the DNA ladders, apoptotic cells are rarely observed prior to culture. After 24 h of culture, large granulosa cells with fragmented DNA are frequently observed. These degenerative changes in DNA are not observed within small granulosa cells. These studies demonstrate that in serum-free medium, most large, but not small, granulosa cells die via an apoptotic mechanism within 24 h.

Interestingly, EGF (20), bFGF (unpublished observation), and progesterone (20) inhibit nonaggregated large granulosa cells from undergoing apoptosis. However, aggregated granulosa cells are less likely to be apoptotic after culture (Fig. 11.2), and various hormones that suppress single granulosa cell apoptosis do not enhance the viability of aggregated granulosa cells (21). This latter observation suggests that cell-to-cell contact regulates granulosa cell viability in a hormone/growth factor-independent manner (21, 22).

Hormonal Versus Cell Contact-Mediated Cell Survival Mechanisms

As indicated, a single granulosa cell is two-times more likely to be apoptotic than a granulosa cell that has formed a cell contact (Fig. 11.2). This observation could be explained simply by assuming that granulosa cells that do not form cell associations are nonviable prior to culture and, therefore, are unlikely to form a cell contact in vitro. However, virtually all granulosa cells are viable prior to culture (15). Without exposure to EGF, progesterone or bFGF, most of the single granulosa cells undergo apoptosis. The ability of these hormones and growth factors to prevent single granulosa cells from undergoing apoptosis in vitro attests to the initial viability of the cells.

A second explanation to account for the relationship between cell contact and granulosa cell viability is that aggregated granulosa cells could produce more progesterone than single granulosa cells. This could result in a higher progesterone concentration within granulosa cell aggregates. This concept is supported by studies which showed that an interaction of granulosa cells with extracellular matrix proteins enhances progesterone secretion (16). Furthermore, estrogen secretion is greater in aggregated granulosa cells than dispersed granulosa cells (23). These observations suggest that aggregated granulosa cells are more steroidogenically active.

To eliminate steroidogenesis as a variable, granulosa cells were cultured for 24 h in the presence or absence of aminoglutethimide, an inhibitor of progesterone synthesis. The data from this study demonstrate that in the presence of aminoglutethimide, the relationship between granulosa cell apoptosis and cell contact is still observed (Fig. 11.2A). In this study both control and aminoglutethimide treatment groups had progesterone levels of <3 ng/mL. In addition, a minimum of 100 ng/mL of progesterone is required to inhibit apoptosis of single granulosa cells. This dose-response curve is not affected by aminoglutethimide (21). This suggests that in granulosa cell cultures en-

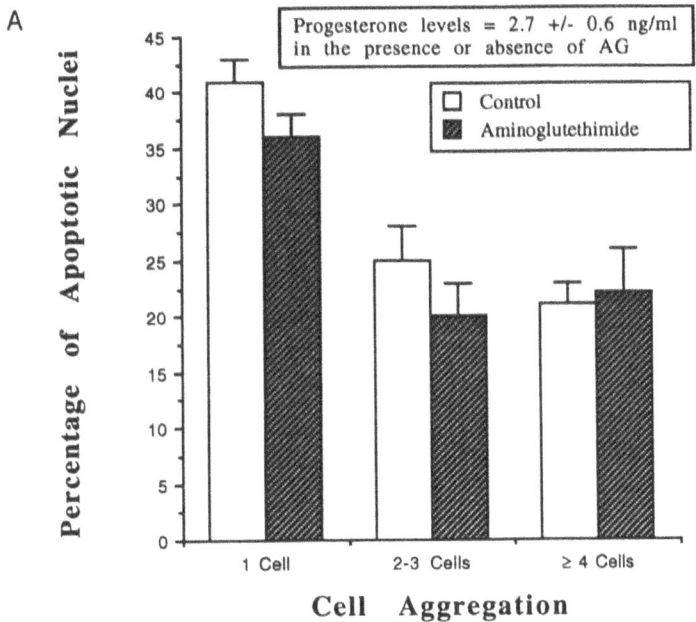

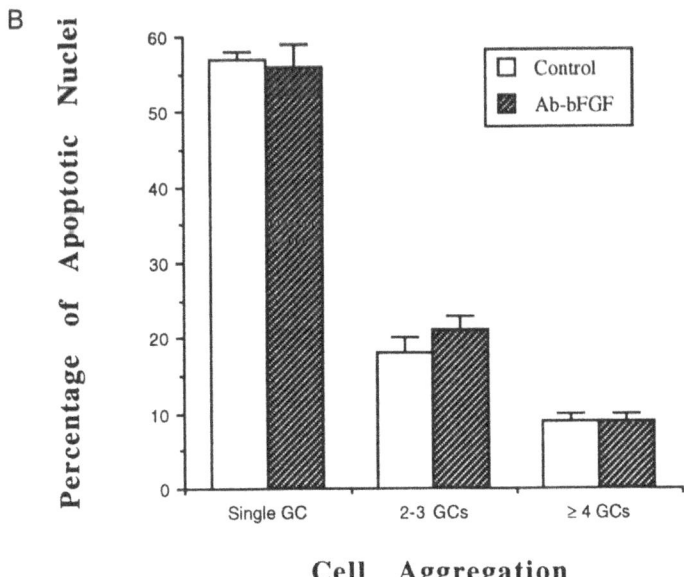

FIGURE 11.2. The effect of cell aggregation and either aminoglutethimide (A) or an antibody to basic fibroblast growth factor (B) on the percentage of granulosa cells undergoing apoptosis. The mean progesterone levels ± standard error is shown above the upper panel. (Data shown in (A) are reproduced with permission from Peluso et al. (21).)

dogenous progesterone levels are too low to maintain granulosa cell viability. Since EGF may mediate its antiapoptotic action by stimulating progesterone secretion (20), these data likely rule out the possibility that endogenous EGF accounts for the enhanced viability of aggregated granulosa cells. Finally, the viability of aggregated granulosa cells is not reduced in the presence of a neutralizing antibody to bFGF (Fig. 11.2B). Collectively, these data support the concept that cell contact, and not endogenous hormones or growth factors, accounts for the enhanced viability of aggregated granulosa cells.

The data from these studies also clearly demonstrate that granulosa cells within 2- to 3-cell clusters have the same likelihood of being apoptotic as those granulosa cells in ≥ 4-cell aggregates (Fig. 11.2). Based on this observation, studies were then conducted to determine if just a single granulosa cell contact was sufficient to maintain granulosa cell viability. The results of this study confirm this hypothesis (Fig. 11.3A). Furthermore, cell contact with either another large steroidogenic granulosa cell or with a small nonsteroidogenic granulosa cell reduces the rate of apoptosis (Fig. 11.3A). Although a large:large cell doublet may be associated with a higher progesterone concentration than a single large granulosa cell, a small:large cell doublet would likely have a steroid milieu comparable to that of a single large granulosa cell, since small granulosa cells are not very steroidogenically active. Taken together, these data provide additional evidence that cell contact, and not a local elevation in progesterone, acts to keep large granulosa cells alive.

Gap Junctions and the Maintenance of Granulosa Cell Viability

Previous ultrastructural studies demonstrate that isolated granulosa cells form aggregates in vitro that are connected by both gap and adhesion-type junctions (21, 22). Both types of junctions convey information between cells and, therefore, could be involved in the mechanism through which cell contact promotes granulosa cell viability. To gain insight into whether or not either of these junctions play an important role, a co-culture study was conducted. In this study, granulosa cells were cultured with R2C cells for 24 h, and the granulosa cells were then assessed for apoptosis and/or prepared for electron microscopic evaluation. The R2C cells are derived from a rat Leydig cell tumor and are easily identified at the light microscope level in that they are several times larger than granulosa cells and possess numerous large lipid droplets (Fig. 11.4D). This study reveals that a single contact with an R2C cell is as effective in maintaining granulosa cell viability as another granulosa cell (Fig. 11.3B). However, normal gap junctions are not observed between granulosa cells and R2C cells. Rather, "junctional complexes" are observed that have a large amount of protein within the intracellular space (21). Although unlikely, it is possible that

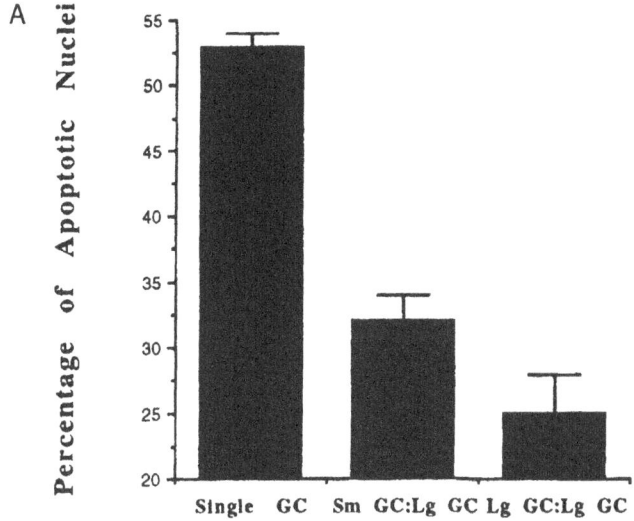

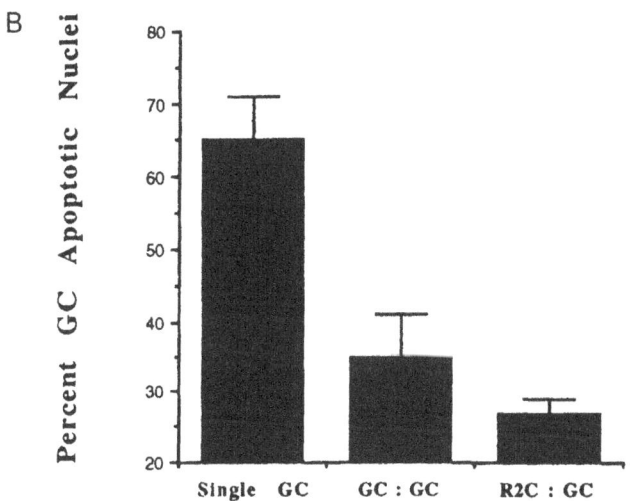

FIGURE 11.3. The effect of a single cell contact on large granulosa cell (GC) apoptosis when contact is with either a single small (Sm) or large (Lg) granulosa cell (A) or an R2C cell (B). For the R2C cell studies, R2C cells were cocultured with GCs. (Reproduced with permission from Peluso et al. (21).)

these gap junction-like structures are functional. In addition, functional gap junctions could form between granulosa cells and R2C cells that are not observed at the ultrastructural level.

To evaluate these possibilities, the functionality of these gap junction-like contacts was assessed by monitoring the transfer of a fluorescent dye from one cell to another (24). Briefly, R2C and granulosa cells were preloaded with two fluorescent dyes: calcein (green fluorescence) and DiI (red fluorescence). Preloading with both DiI and calcein results in 100% of the cells incorporating both dyes, regardless of cell type. These preloaded cells were then plated with non-labeled granulosa cells. The co-cultural were maintained for 3 h and then assessed for the establishment of functional gap junctions. Most granulosa cells attach to the culture dish by 3 h of culture (Fig. 11.4A) and the preloaded granulosa cells are easily identified by the presence of DiI (Fig. 11.4B). As indicated by the presence of calcein and absence of DiI, many of the nonloaded granulosa cells form functional gap junctions with the preloaded granulosa cells (Fig. 11.4C). This confirms published reports (25), in addition to validating this dye-transfer procedure. In contrast, granulosa cells form contacts with R2C cells but fail to establish functional gap junctions (Fig. 11.4D and E). Since R2C cell contact with granulosa cells maintains cell viability in the absence of functional gap junctions, it is con-

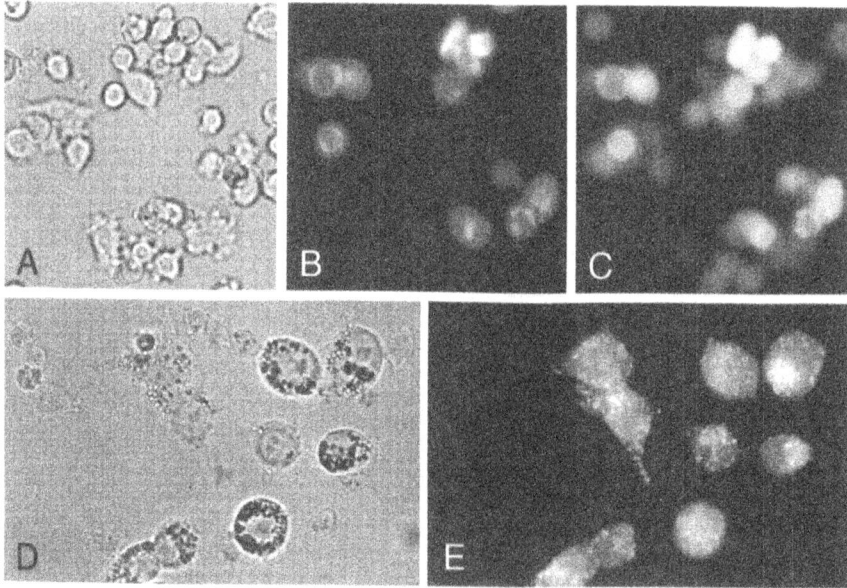

FIGURE 11.4. Assessment of functional gap junctions. In separate incubations, R2C cells and granulosa cells were pre-loaded with DiI and the gap-junction permeable dye, calcein AM. After loading, the cells were washed and both R2C and granulosa cells were co-cultured with nonloaded granulosa cells. In (A) a co-culture between loaded and nonloaded granulosa cells is shown. For this culture, DiI and calcein staining is shown in (B) and (C), respectively. Note that calcein is detected in many granulosa cells that do not stain for DiI, indicating that functional gap junctions have been established. In (D) a R2C-granulosa cell co-culture is shown. Note that the R2C cells are larger than the granulosa cells and that calcein is not transferred between the R2C-granulosa cell aggregates (E). (Reproduced with permission from Peluso et al. (21).)

cluded that gap junctions are not part of the mechanism by which cell contact prevents granulosa cell apoptosis.

N-Cadherin as a Mediator of Granulosa Cell Contact

Since gap junctions do not seem to be involved, it is possible that adhesion junctions convey the protective effects of cell contact. Adhesion-type junctions are composed of several different proteins, one of which is N-cadherin. N-cadherin is a transmembrane protein that forms an adhesion-type junction with N-cadherin expressed by adjacent cells. N-cadherin may also be involved in bFGF-mediated signal transduction pathways (27). N-cadherin is expressed by mouse (26) and rat (21) granulosa cells. As expected, immunocytochemical studies localize N-cadherin to the cell surface with N-cadherin being concentrated at the junctional interface between granulosa cells (21).

Previous work by Farooki and associates has shown that antibodies to cadherin inhibit granulosa cells from aggregating (28). Furthermore, N-cadherin antibody interferes with the ability of progesterone (21) and bFGF to prevent single granulosa cells from undergoing apoptosis (Fig. 11.5A). Recent reports also indicate that N-cadherin interacts with the FGF receptor and this interaction is required for signal transduction via this receptor (27). If this relationship exists for granulosa cells, then this could explain why N-cadherin antibody interferes with the ability of bFGF to maintain the viability of single granulosa cells.

These studies also illustrate that exposure to N-cadherin antibody or to a synthetic N-cadherin peptide reduces the number and size of aggregates, but does not completely prevent all granulosa cells from aggregating. In fact, the N-cadherin antibody is detected between some of the aggregated granulosa cells. Importantly, the rate of apoptosis in aggregated granulosa cells that form in the presence of N-cadherin antibody is greater than that of aggregates exposed to ascites fluid (i.e. monoclonal antibody control) (Fig. 11.5A). Similarly, synthetic N-cadherin peptide allows for some granulosa cells to aggregate but increases the rate of apoptosis of these aggregated cells (21). These experiments are important in that they demonstrate that cell contact per se is not essential to maintain cell viability. Rather, it is a homophilic binding of N-cadherin between adjacent cells that may transduce a signal cascade involved in maintaining cell viability.

Cell Contact, N-Cadherin, and Protein Tyrosine Phosphorylation

Very little is known about the signal transduction mechanism that maintains granulosa cell viability. It is known that genistein, a tyrosine kinase inhibitor, blocks the ability of bFGF to prevent granulosa cell apoptosis (12). Since

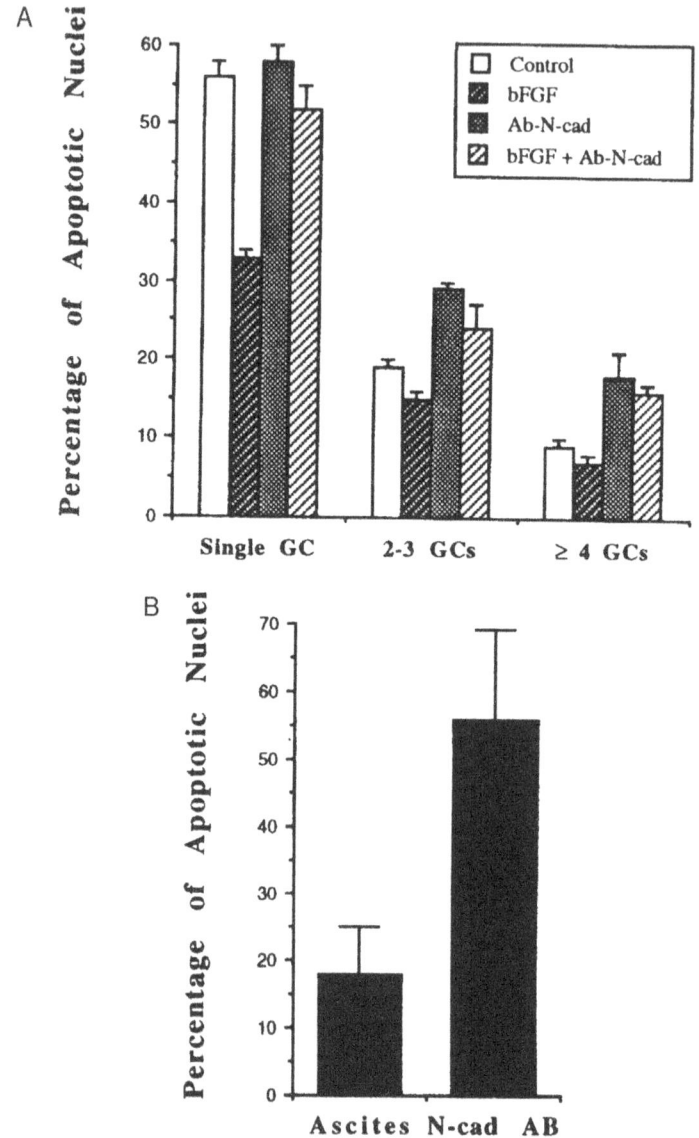

FIGURE 11.5. The effect of an N-cadherin antibody on the ability of either (A) bFGF to inhibit granulosa cell apoptosis or (B) serum to prevent ROSE cell apoptosis.

bFGF mediates its action through a receptor tyrosine kinase (29), the relationship between N-cadherin-mediated cell survival and tyrosine phosphorylation was investigated by culturing granulosa cells with ascites fluid (control) or N-cadherin antibody-supplemented media. After 24 h, the profile of tyrosine-phosphorylated proteins was examined. It is important to note that due to slight variations in experimental trials, minor changes in the intensity of various tyrosine-phosphorylated protein bands are often observed. To avoid focusing on these minor changes, only changes that were detected in at least three experiments are pointed out. With this conservative approach, these studies reveal that proteins with a molecular weight between 150 and 180 kDa are present after ascites treatment but are absent or reduced after N-cadherin antibody exposure (Fig. 11.6A). These bands are the predicted size of the FGF receptors and could indicate that the N-cadherin antibody interferes with the tyrosine phosphorylation of these receptors. In addition, a new tyrosine phosphorylated protein(s) with a molecular weight of about 50 kDa is detected after N-cadherin antibody treatment. The identity of this protein is unknown. The percentage of apoptotic single cells is the same in both ascites and N-cadherin antibody treatment groups but the N-cadherin antibody treatment increases the percentage of aggregated apoptotic cells (Fig. 11.5A). It is therefore likely that the changes in tyrosine-phosphorylated proteins occur within the cell aggregates and are due to a disruption in the homophilic interaction between N-cadherin molecules of adjacent granulosa cells.

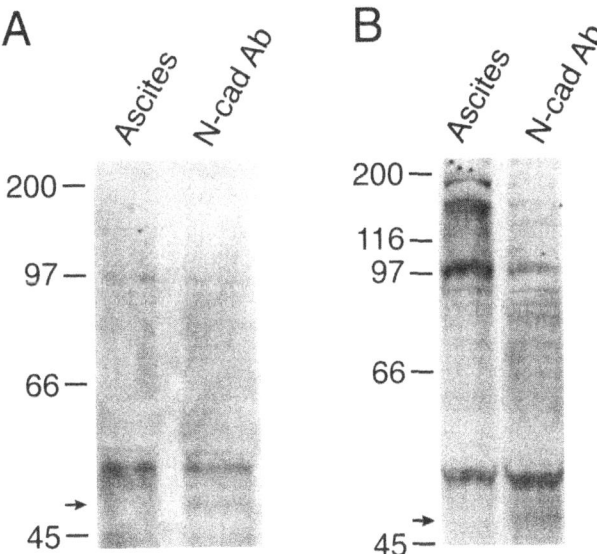

FIGURE 11.6. Electrophoretic profile of tyrosine-phosphorylated proteins. Large granulosa cells were cultured for 24 h with ascites fluid or N-cadherin antibody. Cell lysates were prepared by incubating the cells with boiling lysate buffer as described by Hames (36). Equal amounts of lysate protein were loaded into 8% acrylamide gels, electrophoresed, transferred to nitrocellulose, and tyrosine-phosphorylated proteins were identified by Western blot using an anti-phosphotyrosine antibody (UBI, Lake Placid, NY). These bands were specific as determined by omitting the primary antibody from the Western blot procedure.

Since ovarian surface epithelial cells also die via an apoptotic mechanism (1), studies were conducted using a rat ovarian surface epithelial cell line (i.e. ROSE cells) that were provided by Dr. Burghardt of Texas A & M University. ROSE cells are similar to granulosa cells in that they express N-cadherin, undergo apoptosis in serum-free medium, and bFGF maintains their viability under serum-free conditions (unpublished observations). In addition, aggregated ROSE cells are less likely to be apoptotic after culture in serum-free medium. Importantly, plating ROSE cells with an N-cadherin antibody, even in the presence of serum, increases the percentage of apoptotic aggregated ROSE cells compared to ascites controls (Fig. 11.5B). N-cadherin antibody treatment does not alter the percentage of single apoptotic ROSE cells. Treatment with N-cadherin antibody also alters the profile of tyrosine phosphorylated proteins in a manner similar to granulosa cells (i.e., a decrease in 150 to 180 kDa tyrosine-phosphorylated proteins and the appearance of an ≈50 kDa tyrosine-phosphorylated protein) (Fig. 11.6B). Immunoprecipitation studies using a FGF receptor antibody demonstrate that in ROSE cells these 150 to 180 kDa proteins represent FGF receptors and that exposure to N-cadherin antibody prevents their tyrosine phosphorylation. These studies support the hypothesis that N-cadherin is involved in regulating the FGF receptor signal transduction cascade. Although this has been proposed by others (27), these are the first data to directly show this action.

Overview of Cell Contact-Mediated Cell Survival

The concept that cadherin-mediated contact is part of the mechanism which regulates granulosa cell and ovarian surface epithelial cell viability has important implications to ovarian function. Although very little is known about cadherin expression within the ovary (21, 28), the finding that N-cadherin allows these cells to form functional contacts indicates that cadherin plays a central role in regulating ovarian function. In addition, it is important to appreciate that these findings are one of just a few studies to suggest that cell contact inhibits apoptosis. Recently, it has been demonstrated that the disruption of epithelial cell-matrix interactions by treatment with RGD peptides (30) or antiintegrin antibodies (31) results in apoptosis. This latter study also demonstrates that antiintegrin antibodies prevented intracellular adhesions. Similarly, work by Koopman et al. (32) have shown that intracellular adhesion antibodies inhibit the interaction between follicular dendritic cells and germinal center cells, and subsequently results in apoptosis of the germinal center cells. Frisch and Francis (30) have also shown that once epithelial cells lose contact they undergo apoptosis. They refer to this as "anoikis." It is likely that "anoikis" also occurs in granulosa and surface epithelial cells. A more recent study has specifically shown that N-cadherin is essential in maintaining cell viability (33). This study proposes that N-cadherin mediates its actions through an association with catenins. The catenins link the intracellular domain of N-cadherin to the actin cytoskeleton and are also involved in intracellular signal transduction (34). These studies support the hypothesis that the cadherin-catenin complex plays an important role in regulating granulosa and ROSE cell viability. Since cadherin-catenin complexes are found in all epithelial cells, this concept may have a broader application.

Acknowledgments. The author is grateful to Dr. Bruce A. White for his thoughtful advice throughout the course of these studies and to Ms. Anna Pappalardo and Dr. Mark P. Trolice for conducting the various studies presented in this chapter.

References

1. Ackerman RC, Murdoch WJ. Prostaglandin-induced apoptosis of ovarian surface epithelial cells. Prostaglandins 1993;45:475–85.
2. Tilly JL, Kowalski KI, Johnson AL, Hsueh A. Involvement of apoptosis in ovarian follicular atresia and postovulatory regression. Endocrinology 1991;129:2799–2801.
3. Greenwald G, Terranova P. Follicular selection and its control. In: Knobil E, Neill JD, Ewing LL, Greenwald GS, Markert CL, W PD, eds. The physiology of reproduction. New York: Raven Press, 1988:387–446.
4. Hughes FJ, Gorospe WC. Biochemical identification of apoptosis (programmed cell death) in granulosa cells: evidence for a potential mechanism underlying follicular atresia. Endocrinology 1991;129:2415–22.
5. Schwartzman RA, Cidlowski JA. Apoptosis: the biochemistry and molecular biology of programmed cell death. Endocr Rev 1993;14:133–51.
6. Walker PR, Sikorska M. Endonuclease activities, chromatin structure, and DNA degradation in apoptosis. Biochem Cell Biol 1994;72:615–23.
7. Zeleznik A, Ihrig L, Bassett S. Developmental expression of Ca^{++}/Mg^{++}-dependent endonuclease activity in rat granulosa and luteal cells. Endocrinology 1989;125:2218–20.
8. Billig H, Furuta I, Hsueh A. Estrogens inhibit and androgens enhance ovarian granulosa cell apoptosis. Endocrinology 1993;133:2204–12.
9. Tilly JL, Kowalski KI, Schomberg DW, Hsueh A. Apoptosis in atretic ovarian follicles is associated with selective decreases in messenger ribonucleic acid transcripts for gonadotropin receptors and cytochrome P450 aromatase. Endocrinology 1992;131:1670–76.
10. Peluso JJ, Charlesworth J, Egland-Charlesworth C. Role of estrogen and androgen in maintaining the preovulatory follicle. Cell Tissue Res 1981;216:615–24.
11. Wyllie AH. Cell death. Int Rev Cytol 1987;17:755–85.
12. Tilly JL, Billig H, Kowalski KI, Hsueh A. Epidermal growth factor and basic fibroblast growth factor suppress the spontaneous onset of apoptosis in cultured rat ovarian granulosa cells and follicles by a tyrosine kinase-dependent mechanism. Mol Endocrinol 1992; 6:1942–50.
13. Chun SY, Billig H, Tilly JL, Furuta I, Tsafriri A, Hsueh A. Gonadotropin suppression of apoptosis in cultured preovulatory follicles: mediatory role of endogenous insulin-like growth factor I. Endocrinology 1994;135:1845–53.
14. Morley P, Whitfield J, Vanderhyden B, Tsang B, Schwartz J. A new, nongenomic estrogen action: the rapid release of intracellular calcium. Endocrinology 1992;131:1305–12.
15. Lederer KJ, Luciano AM, Pappalardo A, Peluso JJ. Proliferative and steroidogenic capabilities of rat granulosa cells of different sizes. J Reprod Fertil 1995;103:47–54.

16. Amsterdam A, Plehn DD, Suh BS. Structure-function relationships during differentiation of normal and oncogene-transformed granulosa cells. Biol Reprod 1992;46:513–22.
17. Rao IM, Mills TM, Anderson E, Mahesh VB. Heterogeneity in granulosa cells of developing rat follicles. Anat Rec 1991;229:177–85.
18. Sanbuissho A, Lee GY, Anderson E. Functional and ultrastructural characteristics of two types of rat granulosa cell cultured in the presence of FSH or transforming growth factor alpha (TGF-alpha). J Reprod Fertil 1993;98:367–76.
19. Thompson EB. Apoptosis and steroid hormones. Mol Endocrinol 1994;8:665–73.
20. Luciano AM, Pappalardo A, Ray C, Peluso JJ. Epidermal growth factor inhibits large granulosa cell apoptosis by stimulating progesterone synthesis and regulating the distribution of intracellular free calcium. Biol Reprod 1994;51:646–54.
21. Peluso JJ, Pappalardo A, Trolice MP. N-cadherin-mediated cell contact inhibits granulosa cell apoptosis in a progesterone-independent manner. Endocrinology 1996;137:1196–1203.
22. Peluso JJ, Pappalardo A. Progesterone and cell-cell adhesion interact to regulate rat granulosa cell apoptosis. Biochem Cell Biol 1995;72:547–51.
23. Farookhi R, Desjardins J. Luteinizing hormone receptor induction in dispersed granulosa cells requires estrogen. Mol Cell Endocr 1986;47:13–24.
24. Goldberg GS, Bechberger JF, Naus CG. A pre-loading method of evaluating gap junctional communication by fluorescent dye transfer. BioTechniques 1995;18:490–7.
25. Lawrence TS, Beers WH, Gilula NB. Transmission of hormonal stimulation by cell-to-cell communication. Nature 1978;272:501–6.
26. MacCalman CC, Farookhi R, Blaschuk OW. Estradiol regulates N-cadherin mRNA levels in the mouse ovary. Dev Genet 1995;16:20–4.
27. Williams EJ, Furness J, Walsh FS, Doherty P. Activation of the FGF receptor underlies neurite outgrowth stimulated by L1, N-CAM and N-cadherin. Neuron 1994;13:583–94.
28. Farookhi R, Blaschuk OW. Cadherins and ovarian follicular development. In: Gibori G, ed. Signaling mechanisms and gene expression in the ovary. New York: Springer-Verlag, 1991:254–60.
29. Fantl WJ, Johnson DE, Williams LT. Signaling by receptor tyrosine kinases. Annu Rev Biochem 1993;62:453–81.
30. Frisch SM, Francis H. Disruption of epithelial cell-matrix interactions induces apoptosis. J Cell Biol 1994;124:619–26.
31. Bates RC, Buret A, Van HD, Horton MA, Burns GF. Apoptosis induced by inhibition of intercellular contact. J Cell Biol 1994;125:403–15.
32. Koopman G, Keehnen R, Lindhout E, et al. Adhesion through the LFA-1 (CD11a/CD18)-ICAM-1 (CD54) and the VLA-4 (CD49d)-VCAM-1 (CD106) pathways prevents apoptosis of germinal center B cells. J Immunol 1994;152:3760–7.
33. Hermiston ML, Gordon JI. In vivo analysis of cadherin function in mouse intestinal epithelium: essential roles in adhesion, maintenance of differentiation, and regulation of programmed cell death. J Cell Biol 1995;129:489–506.
34. Gumbiner BM. Proteins associated with cytoplasmic surface of adhesion molecules. Neuron 1993;11:551–64.
35. Barry MA, Reynolds JE, Eastman A. Etoposide-induced apoptosis in human HL-60 cells is associated with intracellular acidification. Cancer Res 1993;53:2349–57.
36. Hames BD, Rickwood D. Gel electrophoresis of proteins: a practical approach, 2nd ed. New York: IRL Press, 1990.

12

The Central Role of Basement Membrane in Functional Differentiation, Apoptosis, and Cancer*

Mina J. Bissell

More than a decade and a half ago, based both on the existing literature and experience at the bench, I concluded that, "if there is one generalization that can be made from all the tissue and cell culture studies with regard to the differentiated state, it is this: since most, if not all, functions are changed in culture, quantitatively and/or qualitatively, there is little or no 'constitutive' regulation in higher organisms; i.e., the differentiated state of normal cells is unstable and the environment regulates gene expression" (1). This concept, more recently referred to as the "plasticity" of the differentiated state (2), has gained some credence as literature has accumulated that differentiation may not be as terminal or fixed as once thought. Nevertheless, much of the literature still, and in my opinion without much justification, uses the term "constitutive gene expression" to define many basic cellular functions. This is not a semantic and literary argument about the use of words: it affects how we think about cells and function, and it has fundamental consequences for how we do experiments and how we interpret our results. There is ample evidence, if we look with an open mind, that all cells retain the ability to modulate most, if not all, of their functions; that even enucleated red blood cells still regulate their behavior depending on where they are and what they encounter. I believe there is enough evidence to support the notion that in order to make a cell live or function properly in a tissue-specific way, it has to be actively *prevented* from growing or dying; i.e., it has to be directed at all times as to how to behave correctly. Thus, if these active signals are withdrawn from a resting, differentiated cell, it will do one of three things: it will die, grow, or function inappropriately. Having come to appreciate this paradigm, I then posed the following question: what are the proper cues in vivo that keep a cell, or more appropriately, a tissue, differentiated?

*For Elizabeth Hay, who saw the importance of extracellular matrix before many others and whose important work provided much inspiration for many of us.

Two of my very early postdoctoral fellows, Richard (Rick) Schwarz and Glenn Hall, taught me many things, including my first lessons on collagens and basement membranes (BM). Learning also from the vast literature of developmental biology and the then recent work of a few groups in cell biology, Glenn Hall and I proposed that the extracellular matrix (ECM) in general, and the BM in particular, have information, and that this solid state information, along with hormones and growth factors (the "soluble" signals), are the final arbitrators of tissue-specificity (3). Furthermore, that the unit of function in higher organisms was not the cell, but the normal cell plus its ECM or in the last analysis, the tissue and the organism itself (4). The model is reproduced in Figure 12.1.

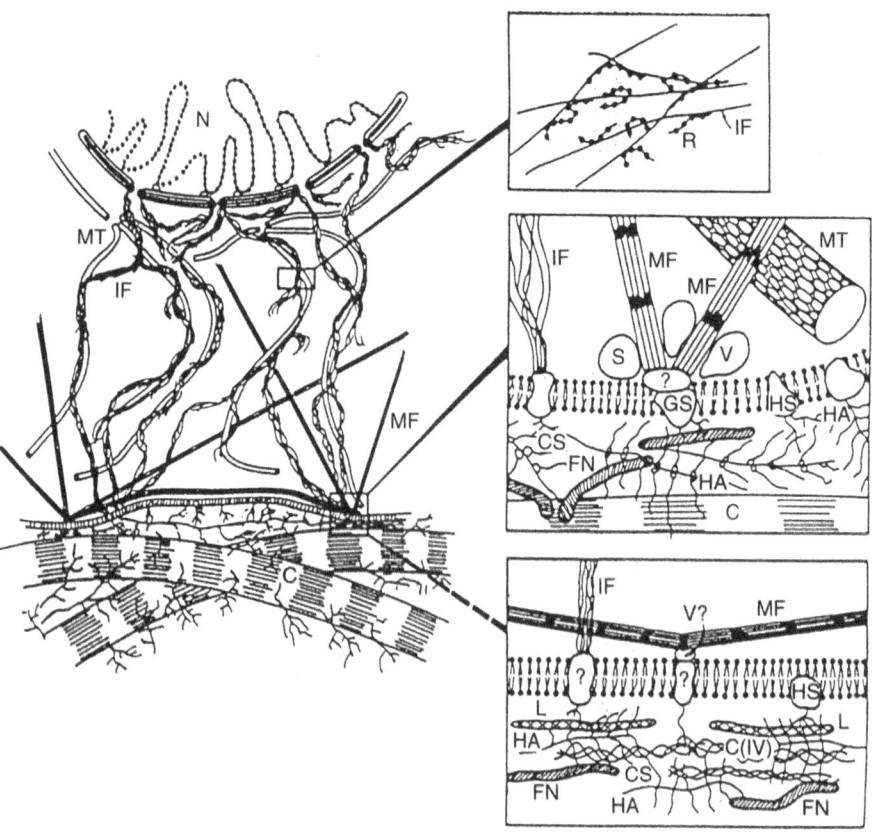

FIGURE 12.1. The minimum required unit for tissue-specific functions. The postulated overall scheme for ECM-cell interactions. N = nucleus; MT = microtubules; IF = intermediate filaments; MF = microfilaments; C = collagen. Top inset: Polyribosome attachment to cyotskeleton. R = ribosomes. Middle inset: V = Vinculin; S = *src* coded protein kinase (?); GS = Ganglioside (attaching fibronectin to membrane?); FN = fibronectin; HA = hyaluronic acid; CS = chondroitin sulfate; HS = heparan sulfate. Bottom inset: Possible attachment site to membranes in epithelial cells. L = laminin; C(IV) = collagen type (IV). (Reproduced with permission from Bissell et al. (3).)

A great body of work in almost every cell type and tissue has shown the validity of this argument (for reviews see refs. 5–8), an unorthodox, if not heretical, point of view only a decade ago. Despite the remarkable advances, including the important discovery of integrins, the largest class of ECM-receptors (9–12), there is much that needs to be understood about how the individual molecules of the ECM (or the ECM as a structural unit) signal via integrins, how the ECM collaborates with soluble factors such as hormones to bring about tissue-specific gene expression, and how it interacts with growth factors to actively (also possibly passively as a reservoir) to override uncontrolled growth, and cell death.

Form and Function in Culture

In a quest to show the importance of BM in the regulation of form and function in vivo, we chose the epithelial cells of the mammary gland as a model since it undergoes remarkable cycles of growth, morphogenesis, differentiation, and involution even after birth, processes that need to be understood before we unravel the mechanisms underlying the high incidence of breast cancer. The pioneering work of Emerman and Pitelka (13) using mouse mammary epithelial cells and floating collagen gels, had paved the way for culturing these cells, and for investigating the role of the ECM in the regulation of gene expression. After Joanne Emerman joined my laboratory as a postdoctoral fellow, we established primary cultures of mammary cells from mid-pregnant mice and began a series of in depth studies, which took off with the help of my talented graduate student, Eva Lee and my colleague, Gordon Parry, first using floating collagen gels (13–14), later using reconstituted BM (15, 16; Figure 12.2), and primary cultures, cell strains, and clonal cell lines. A decade and a half later, a succession of bright and hard-working graduate students and postdoctoral fellows has allowed us to put a pathway together as represented in Figure 12.2. Although many details need to be worked out, one can appreciate the complexity of even this simplified version.

The diagram shown in Figure 12.2 summarizes our culture studies to date on the various levels at which BM regulates the three milk protein genes studied in our laboratory. We have recognized, as have others who study growth and differentiation, that the cells must withdraw from the cell cycle in order to begin the process of tissue-specific gene expression (17–19). Calvin Roskelley, through a series of simple and thoughtful studies (8, 20), demonstrated that the cells had to reorganize their cytoskeleton and change "shape" before they could receive a specific ECM signal. This signal is not obvious in primary mammary epithelial cells in that cells respond to laminin without necessarily appearing to change shape (personal communication with Charles Streuli, a former postdoctoral fellow who is now in Manchester, UK, and who elucidated the role of laminin). However, how a biochemical signal is translated into a "shape" change may indeed be subtle and the P2 cells (a mouse

128   M.J. Bissell

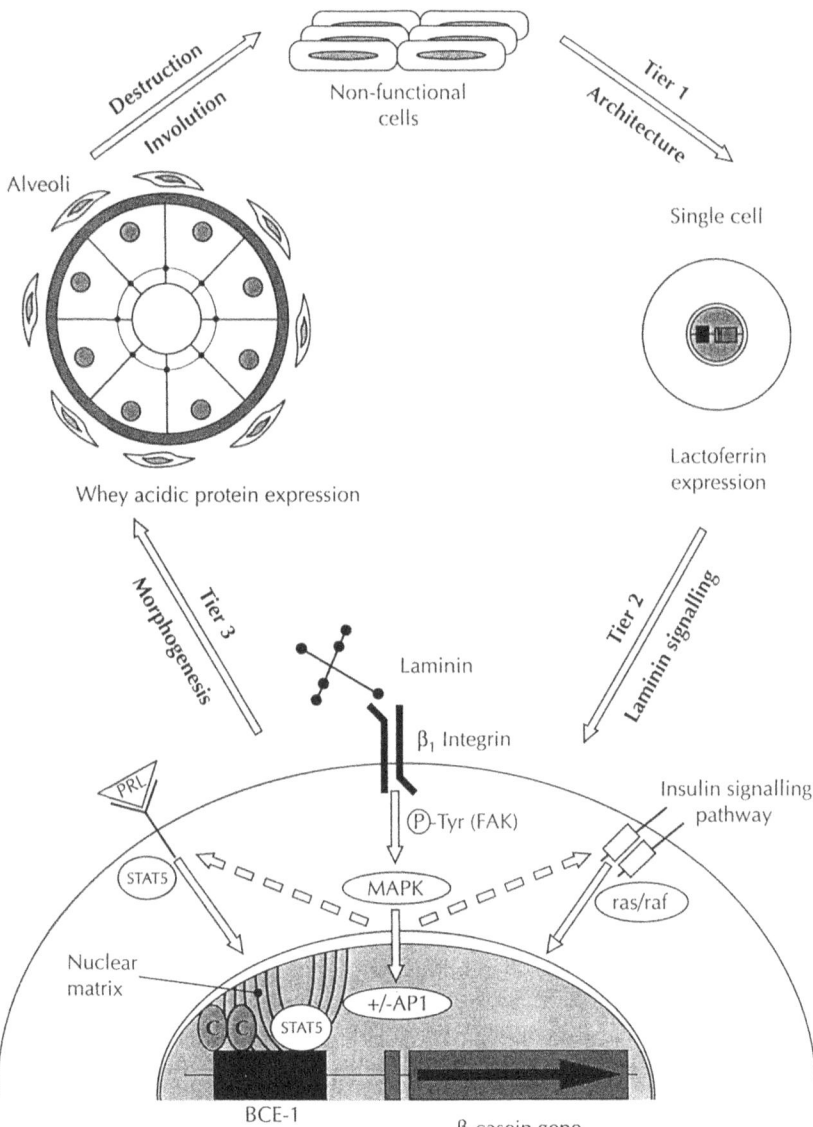

FIGURE 12.2. A hierarchy of ECM-dependent signals regulates mammary gland development. The first tier of the hierarchy is mediated by architectural changes in cell shape that result in lactoferrin expression. The second tier of the hierarchy is mediated by laminin-specific biochemical signals that activate an ECM-responsive element (BCE-1) and induce endogenous β-casein expression. The third tier in the hierarchy is assembled during morphogenesis and serves to regulate whey acidic protein (WAP) expression and the formation of "alveoli" in tissue culture models. Destruction of the hierarchy is mediated by matrix metalloproteases and results in involution. C = CCAATT/enhancer binding protein; FAK = focal adhesion kinase; MAPK = mitogen-activated protein kinase; PRL = prolactin. (Reproduced with permission from Roskelley et al. (8).)

mammary cell line isolated by Pierre Desprez which we use extensively; 21),
being immortal and clonal, undoubtedly contain mutations that would force
an exaggeration of the "shape" change. As such, they provide an opportunity
to measure and "track" this response and eventually to understand it. The
change in cell shape, which in vivo is most probably brought about as a result
of stromal-epithelial interactions (via the cell-BM contact), and in culture by
laminin, can also be brought about using polyhema, a nonadhesive but mal-
leable substratum (Fig. 12.3) (20, 22). This change by itself is sufficient to
result in the expression of lactoferrin, an iron-binding constituent of milk
(see tier 1 in Fig. 12.2). The expression of this gene is independent of the
presence of the lactogenic hormone, prolactin. Once the cells are "rounded,"
they can then receive an additional and specific laminin signal, most probably from the E_3 domain of the α subunit of laminin I to give rise to β-casein
(23) (see tier 2 in Fig. 12.2). Recently, John Muschler has obtained evidence
to indicate that the receptor for cell "shape" change is different from the
laminin receptor for β-casein gene expression and that the latter is most probably an α_6-containing laminin (24).

The discovery of the first "ECM-response element" (BCE-1 in Fig. 12.2)
was made possible through a series of collaborative studies between Chris-
tian Schmidhauser and Connie Meyers of my laboratory and Gerry Casperson
of Monsanto Company, who had cloned the bovine β-casein gene. A 1600-
nucleotide stretch of this gene's promoter and 5' sequences was attached to
a reporter gene (chloramphenicol acetyl transferase, CAT). This was transfected into CID-9 cells, a functional mammary cell line which Schmidhauser
had isolated from COMMA1-D cultures (25), the latter isolated in Dan
Medina's laboratory (26). To overcome a position effect, he combined 300
to 500 stably transfected clones and asked whether or not the reporter gene
will "fire" in response to cultivation on the reconstituted BM. The result
was an astounding "yes!" The reporter gene was 50- to 150-fold more active when cells were placed on a reconstituted BM. Subsequent promoter
deletion analysis identified a 160-nucleotide stretch that defined an ECM-
response element in the regulatory sequences of a tissue-specific gene (27).
Using site-specific mutagenesis, the response element, referred to as BCE-1
(bovine casein element-1) was shown to contain C/EBPβ and STAT-5—binding elements that are essential for the activity of the enhancer (28, 29; C.
Meyers et al., submitted) (the involvement of FAK, MAP kinase, AP1, and
other factors depicted in Scheme 1 is discussed in Roskelly et al. (8) and
Roskelley et al. in preparation). Thus, the expression of β-casein is more
stringent than lactoferrin and is completely dependent on the presence of
both prolactin and laminin.

To achieve whey acidic protein expression (a milk protein that is expressed in high levels in the middle of pregnancy), the cells need to down
regulate transforming growth factor-alpha (TGF-α) and perhaps other unknown factors and form a 3-dimensional organotypic structure (7, 30, 31).
Thus, there is a hierarchy of regulation and requirements to achieve a functional tissue (Fig. 12.4).

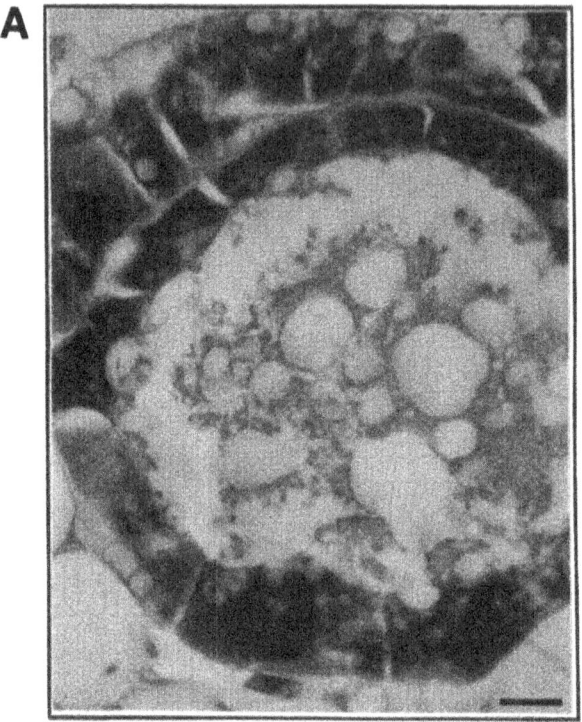

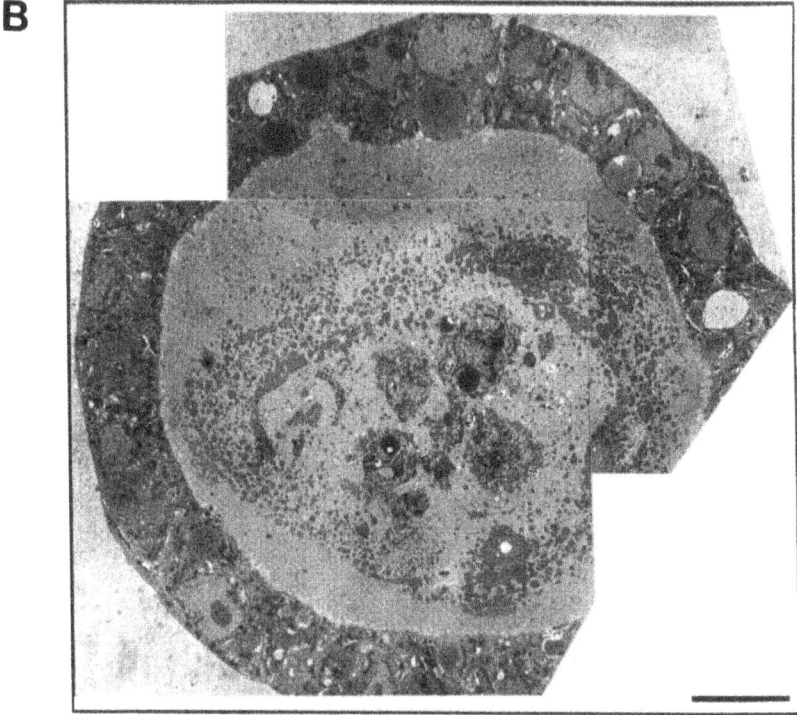

The Importance of Basement Membrane in Maintenance of Form and Function in the Mammary Gland In Vivo

Given the pivotal importance of the BM molecules in culture, how does one prove that BM is indeed a signaling entity in vivo as well? Barring the creation of a "BM knock-out" that either will kill the mouse or create an entirely different beast, I reasoned that if BM is crucial for the mammary gland to make milk during lactation (gain of function), then the loss of milk in involution (loss of function) would have to be preceded by destruction of BM. Thus, we used the involuting gland as a complex "mutant!" A long and fruitful collaboration with Zena Werb of the University of California at San Francisco has laid the foundation for testing this hypothesis and for understanding the role of ECM and ECM-degrading enzymes in vivo. We showed that ECM-degrading enzymes are indeed elaborated before the bulk of milk production is lost (32). We showed that the ratio of ECM-degrading enzymes (e.g., stromelysin-1, SL-1) to tissue inhibitor of metalloproteases (TIMP) (e.g., TIMP-1) was a determining factor for whether the gland would involute or remain functional. Using Elvax pellets containing TIMP-1 or buffer inserted in opposite glands of the mice, Rabih Talhouk, a joint fellow in the laboratory of Zena Werb and my laboratory (33), showed that in the same mouse, the gland that contained the TIMP-1 did not involute much and resembled a late pregnant gland, whereas the other glands that received buffer involuted completely. We also showed that the mammary gland involution has at least two phases: one that is SL-1 independent (34) and occurs immediately (and may indeed be a stress response from lack of suckling and thus gorging by milk), and a second phase that is SL-1 dependent and will result in the complete remodeling of the gland and a loss of the bulk of the lactation phenotype. It is this latter phase that can be inhibited by TIMP-1. Since the mammary gland involutes with massive apoptosis (35), the dominant inhibitory action of TIMP-1 on involution (and thus apoptosis) indicated to us, as early as 1992, that ECM and ECM-degrading enzymes must be involved in regulation of apoptosis. Furthermore, Carolyn Sympson, another joint fellow, showed that mice transgenic for rat SL-1, driven by the whey acidic protein (WAP) promoter (in effect creating a conditional transgenic for the milk producing mammary gland), had impaired BM and an appreciable reduction in gene expression in middle and late pregnancy (36). Wicha et al., using inhibitors of proteolysis, had shown a decade earlier that budding and organization of the rat mammary gland was affected if inhibitors of collagen synthesis were used (37). However, our study proved for the first time in vivo that an intact BM is necessary

___

FIGURE 12.3. Similarity of morphogenesis by mammary epithelial cells in culture and in vivo. Transverse section of an alveolus derived (A) from a lactating mammary gland in vivo and (B) from an alveolar-like sphere formed by primary mouse epithelial cells cultured on a reconstituted basement membrane (see Streuli and Bissell (57)).

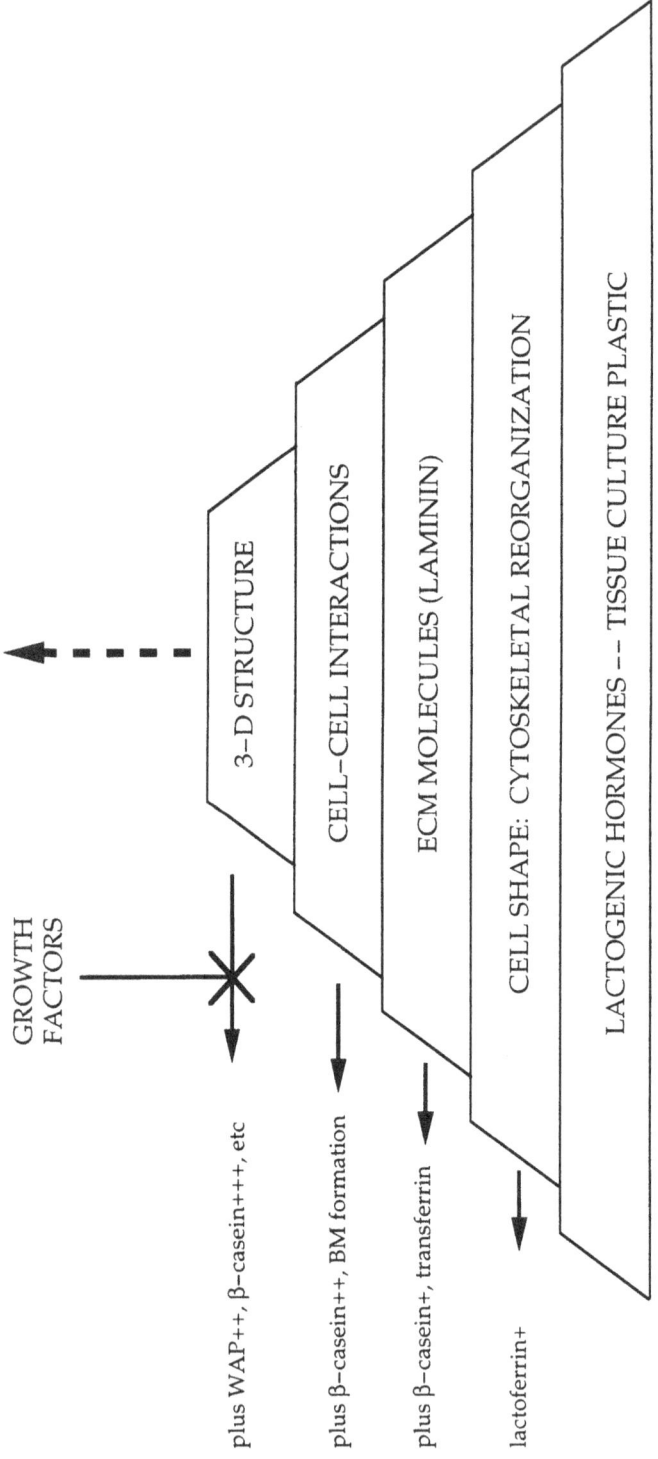

FIGURE 12.4. Hierarchy in regulation of tissue-specific gene expression.

for functional differentiation of the mammary gland. Surprisingly, a very low level of SL-1 expression in the gland of the virgin animal led to remodeling of the gland: the ducts would bud (and contain some β-casein) as if the animals were five to ten days pregnant (36)! It appears that SL-1, and other ECM-degrading enzymes, may be involved in the formation of budding morphogenesis. Thus, a high level of SL-1 expression (which destroys the BM in the pregnant and lactating gland) will lead to loss of function, whereas a small level of SL-1, expressed cyclically with the estrous cycle (WAP comes on every four days during the estrous cycle) will lead to budding and accumulation of some β-casein (36). Finally, and perhaps not so surprisingly in the light of the above discussions, these animals develop both mammary tumors and other forms of tumors such as lymphomas (38, and Z. Werb et al., in preparation). One has to conclude that expression of an activated metalloprotease in mammary epithelial cells, when and where it is not supposed to be expressed, acts as a carcinogen with serious consequences for the mammary gland and the mouse. How is this possible without the introduction of carcinogens and viruses or oncogenes?

We had shown previously that BM is also a suppressor of growth (17, 18) and that its formation inhibits growth factors, such as TGF-β (39) and TGF-α (31). Using in situ hybridization, Nicole Thomasett, a French visiting scientist in my laboratory found that the endogenous SL-1 is expressed in the stroma in both wild-type and transgenic animals, but that the endogenous SL-1 is upregulated in the transgenic animal. A number of other proteases are also induced, blood vessels are increased and the stroma of these mice, as a rule, become reactive with increased collagen deposition and tenasin expression. Indeed, the glands of transgenic animals during later stages of pregnancy resemble those of the involuting mice (40; N. Thomasett et al., submitted).

This phenomenon of a "reactive stroma" (see also ref. 52), coupled with the fact that loss of BM would increase growth factor expression and secretion, must then be postulated to destabilize the epithelial cells and lead to deregulated growth, genomic instability (unpublished data on the mammary tumors with Joe Gray and Dan Pinkel) and cancer. Given that these mice were not constructed with an overexpressed oncogene or a knocked-out suppressor gene, but with a slightly increased metalloprotease expression, the formation of tumors is clearly a surprising phenomenon. The cellular and molecular mechanisms involved await careful examination.

The Importance of Basement Membrane for Integrity and Survival of the Differentiated Tissue

In the last few years, a number of laboratories have shown that interruption of integrin-mediated adhesion could lead to apoptosis of endothelial and epithelial cells (41–43). Nancy Boudreau, at the time a postdoctoral fellow in my laboratory, showed that adhesion per se (e.g., to fibronectin or collagens) did not provide protection against apoptosis beyond 48 hours. Above and beyond adhesion, the long term survival of the mammary cells was shown to be BM-dependent (44). Using functional cell lines in culture, she showed that cells that had formed an endogenous BM and were expressing β-casein when SL-1 was silent, would undergo apoptosis within a few days if the BM was destroyed by activation of an inducible SL-1-transfected gene. She further showed that apoptosis was correlated with the expression of interleukin 1β-converting enzyme (ICE), a homologue of the *C. elegans ced*-3 gene, previously shown to cause apoptosis in other cells (45). Inhibitors of ICE and/or its homologues (these inhibitors do not distinguish between ICE and its homologues) inhibited apoptosis (44). Since cells from the ICE knock-out mouse can apparently undergo apoptosis, there has to be either a compensatory pathway, or the ICE homologues and not ICE per se are the culprits. More recently, Charles Streuli and his colleagues have independently shown a role for BM in apoptosis of primary mammary cells (46). These investigators and others have also produced data to implicate Bax and Bcl-x in mammary gland apoptosis (46, 47). Whether Bax and a homologue of ICE are both involved and, if so, which regulates the other, remain to be determined. Unless the expression of these pro-apoptotic genes are selectively (in a tissue-specific manner) and conditionally knocked out, it would be difficult to sort out the exact molecules involved in different tissues in vivo. Nevertheless, the rules of ECM-cell interactions may be applicable broadly. The recent conditional knock-out studies from Henninghausen's laboratory indicate that these technologies are now within reach for the mammary gland (48). On the basis of the above studies, both in culture and in vivo, the pathway for functional differentiation and apoptosis is depicted in Figure 12.5.

How Does Basement Membrane Protect Against Apoptosis?

The effect of BM on apoptosis may be mediated by down-regulation of genes involved in growth regulation and up-regulation of genes involved in growth arrest, as well as additional signals generated by cell-cell interactions. On the other hand, if the conflicting signals are expressed simultaneously, the cell will undergo apoptosis (49). We showed that upon removal of serum from CID-9 or P2 cells, the cells continue to grow on plastic (myc and cyclin D1 are both present). As they become crowded, they also express genes such as p21 that normally signals arrest and thus the cells undergo apoptosis (50). With the cell lines used in our studies, individual components of the BM, including laminin, do not appear to significantly prevent apoptosis, thus more signals are required to prevent cell death in this model system.

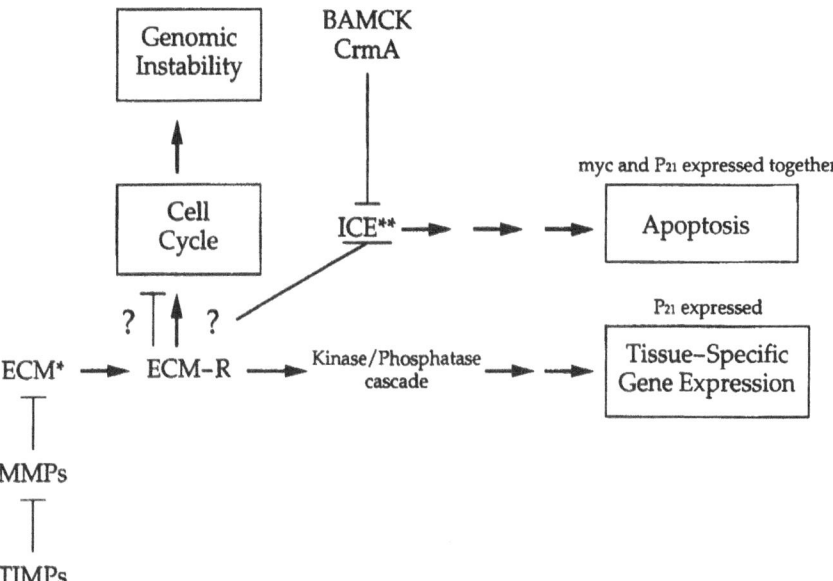

FIGURE 12.5. Integration of apoptosis and tissue specific gene expression in the mammary gland. * Abbreviations: ECM = extracellular matrix; TIMPs = tissue inhibitors of metaloproteases; MMPs = metaloproteases; ICE = interleukin converting enzyme. **ICE or its homologues (BAMCK, CrmA do not distinguish between the members of this family).

Human Breast Epithelial Cells Are Governed by Similar Rules

In the past five years, in a highly productive collaboration with Ole Petersen of the Panum Institute, Denmark, we have extended these studies to human breast epithelial cells and have developed a three-dimensional model for distinguishing normal and malignant breast cells in short term culture (17). This versatile assay system has allowed us to rapidly distinguish between normal and malignant epithelial cells. The results have led us to postulate that the ability to sense the BM appropriately and to form three-dimensional organotypic structures may be the function of a class of "suppressor" genes that are lost as cells become malignant (17). Using this and other three-dimensional assays, we have reported on a possible role for nm23, a putative metastatic suppressor gene (51), a role for $\beta 1$ integrins in growth and survival (51), the origin of α-smooth muscle actin in breast cancer stroma (53), and other observations (54–56).

In conclusion, given the multiple roles of the ECM in the mammary gland, I would propose that the ECM and its receptors are the dominant regulators of

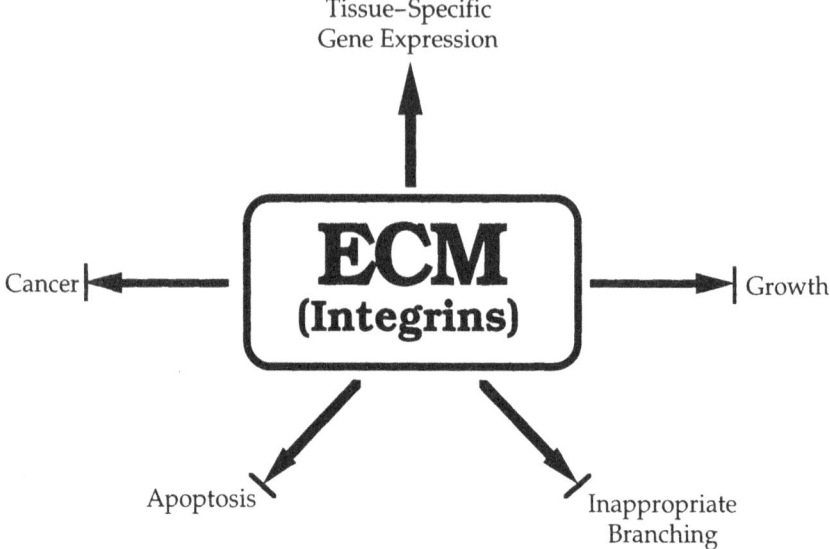

FIGURE 12.6. In the mammary gland, ECM-mediated signals inhibit inappropriate branching, apoptosis, growth, and the development of cancer, but positively influence tissue-specific gene expression.

tissue-specific gene expression and growth. These receptors act via a complex cytoplasmic junction with the cytoskeleton, and a resulting kinase/phospatase-based cascade of signal transduction. A loss of such regulation leads to apoptosis and possibly cancer (Fig. 12.6). ECM and its receptors are attractive candidates because to be a central switch, the candidate must not only be responsible for the stability of the differentiated state, but it also must suppress those events that are contrary to such stability (inappropriate growth, apoptosis, and cancer). At this point, ECM and its receptors appear to fit the bill!

Acknowledgments. I would like to thank the many postdoctoral fellows, graduate students, colleagues, and collaborators whose names I have already mentioned as well as the many others whom I have not mentioned by name but without whose hard work, much of this research would not have been possible. I also thank Bill Johansen of my administrative staff for all his help in the preparation of this personal account. This work was started by a modest NSF grant in 1976, and has been supported primarily by the Office of Health and Environmental Research, U.S. Department of Energy under contract DE-AC03-76SF00098 since then. More recently, the work has also been supported by NIH grants CA-64786 (MJB) and CA-57621 (Zena Werb and MJB).

References

1. Bissell MJ. The differentiated state of normal and malignant cells or how to define a "normal" cell in culture. Int Rev Cytol 1981;70:27–100.
2. Blau HM, Baltimore D. Differentiation requires continuous regulation. J Cell Biol 1991;112:781–3.
3. Bissell MJ, Hall HG, Parry G. How does the extracellular matrix direct gene expression. J Theor Biol 1982;99:31–68.
4. Bissell MJ, Hall HG. Form and function in the mammary gland: the role of extracellular matrix. In: Neville MC, Daniel CW, eds. The mammary gland, Plenum Publishing Corp. 1987:97–146.
5. Adams JM, Watt FM. Regulation of development and differentiation by the extracellular matrix. Development 1993;117:1183–98.
6. Hay ED. Extracellular matrix alters epithelial differentiation. Curr Opin Cell Biol 1993; 5:1029–35.
7. Lin CQ, Bissell MJ. Multi-faceted regulation of cell differentiation by extracellular matrix. FASEB J 1993;7:737–43.
8. Roskelley CD, Srebrow A, Bissell MJ. A hierarchy of ECM-mediated signaling regulates tissue-specific gene expression. Curr Opin Cell Biol 1995;7:736–47.
9. Hynes RO. Integrins: versatility, modulation, and signaling in cell adhesion. Cell 1992; 69:11–25
10. Clarke AS, Lotz MM, Chao C, Mercurio AM. Activation of the p21 pathway of growth arrest and apoptosis by the beta 4 integrin cytoplasmic domain. J Biol Chem 1995; 270:22673–6.
11. Clark EA, Brugge JS. Integrins and signal transduction pathways: the road not taken. Science 1995;268:233–9.
12. Dedhar S. Integrin mediated signal transduction in oncogenesis: an overview. Cancer Metastasis Rev 1995;14:165–72.
13. Emerman JT, Enami J, Pitelka DR, Nandi S. Hormonal effects on intracellular and secreted casein in cultures of mouse mammary epithelial cells on floating collagen membranes. Proc Natl Acad Sci USA 1977;74:4466–70.
14. Lee EY, Lee WH, Kaetzel CS, Parry G, Bissell MJ. Interaction of mouse mammary epithelial cells with collagen substrata: regulation of casein gene expression and secretion. Proc Natl Acad Sci USA 1985;82:1419–23.
15. Li ML, Aggeler J, Farson DA, Hatier C, Hassell J, Bissell MJ. Influence of a reconstituted basement membrane and its components on casein gene expression and secretion in mouse mammary epithelial cells. Proc Natl Acad Sci USA 1987;84:136–40.
16. Barcellos-Hoff MH, Aggeler J, Ram TG, Bissell MJ. Functional differentiation and alveolar morphogenesis of primary mammary cultures on reconstituted basement membrane. Development 1989;105:223–35.
17. Petersen OW, Ronnov-Jessen L, Howlett AR, Bissell MJ. Interaction with basement membrane serves to rapidly distinguish growth and differentiation pattern of normal and malignant human breast epithelial cells. Proc Nat Acad Sci USA 1992;89:9064–8.
18. Desprez PY, Hara E, Bissell MJ, Campisi J. Suppression of mammary epithelial cell differentiation by the helix-loop-helix protein, Id-1. Mol Cell Biol 1995;15:3398–404.
19. Rana B, Mischoulon D, Xie Y, Bucher NL, Farmer SR. Cell-extracellular matrix interactions can regulate the switch between growth and differentiation in rat hepatocytes:

reciprocal expression of C/EBP alpha and immediate-early growth response transcription factors. Mol Cell Biol 1994;14:5858–69.

20. Roskelly CD, Bissell MJ. Dynamic reciprocity revisited: a continuous, bidirectional flow of information between cells and the extracellular matrix regulates mammary epithelial cell function. Biochem Cell Biol 1995;73:391–7.

21. Desprez PY, Roskelley C, Campisi J, Bissell MJ. Isolation of functional clones of mouse mammary epithelial cells: the importance of basement membrane and cell-cell interaction. Mol Cell Diff 1993;1:99–110.

22. Folkman J, Moscona A. Role of cell shape in growth control. Nature 1978; 273:345–9.

23. Streuli CH, Schmidhauser C, Bailey N, Yurchenco P, Skubitz A, Bissell MJ. A domain within laminin that mediates tissue-specific gene expression in mammary epithelia. J Cell Biol 1995;120:253–260.

24. Muschler J, Levy D, Bissell MJ. The integrin $\alpha6\beta1$ mediates signals for survival and tissue-specific gene expression in mammary epithelial cells. Mol Biol Cell 1996;7:240a.

25. Schmidhauser C, Bissell MJ, Myers C, Casperon GF. Extracellular matrix and hormones transcriptionally regulate bovine β-casein 5' sequences in stably transfected mouse mammary cells. Proc Natl Acad Sci USA 1990;87:9118–22

26. Danielson KG, Oborn CJ, Durban EM, Buetel JS, Medina D. Epithelial mouse mammary cell line exhibiting normal morphogenesis in vivo and functional differentiation in vitro. Proc Natl Acad Sci USA 1984;110:1405–15.

27. Schmidhauser C, Casperon GF, Myers CA, Sanzo KT, Bolten S, Bissell MJ. A novel transcriptional enhancer is involved in the prolactin- and extracellular matrix-dependent regulation of β-casein gene expression. Mol Biol Cell 1992;3:699–709

28. Schmidhauser C, Myers CA, Mossi R, Casperson GF, Bissell MJ. Extracellular matrix dependent gene regulation in mammary epithelial cells. Wilde CJ, et al., eds. Intercellular signaling in the mammary gland. New York: Plenum Press 1995:107–19.

29. Myers CA, Casperson GF, Bissell MJ, Schmidhauser C. Identification of functionally important regions in BCE-1: a transcriptional enhancer regulated by the presence of extracellular matrix and prolactin. Mol Biol Cell 1995;5:90a.

30. Chen L-H, Bissell MJ. A novel regulatory mechanism for whey acidic protein gene expression. Cell Regul 1989;1:45–54.

31. Lin CQ, Dempsey P, Coffey C, Bissell MJ. Extracellular matrix regulates whey acidic protein gene expression by suppression of TGF-α in mouse mammary epithelial cells: studies in culture and in transgenic mice. J Cell Biol 1995;129:1115–26.

32. Talhouk RS, Neiswander RL, Schanbacher FL. In vitro culture of cryopreserved bovine mammary cells on collagen gels: synthesis and secretion of casein and lactoferrin. Tissue Cell 1990;22:583–99.

33. Talhouk RS, Bissell MJ, Werb Z. Coordinated expression of extracellular matrix-degrading proteinases and their inhibitors regulates mammary epithelial function during involution. J Cell Biol 1992;118:1271–82.

34. Lund L, Romer J, Thomasset N, Solberg H, Pyke C, Bissell MJ, Dano K, Werb Z. Two distinct phases of apoptosis in mammary gland involution: proteinase-independent and -dependent pathways. Development 1996;122:181–93.

35. Strange R, Li F, Saurer S, Burkhardt A, Friis RR. Apoptotic cell death and tissue remodelling during mouse mammary gland involution. Development 1992; 115:49–58.

36. Sympson CJ, Talhouk RS, Alexander CM, Chin JR, Bissell MJ, Werb Z. Targeted

expression of stromelysin to the mouse mammary gland provides evidence for a role of proteinases in branching morphogenesis and the requirement for an intact basement membrane for tissue-specific gene expression. J Cell Biol 1994;125:681–3.

37. Wicha MS, Liotta LA, Vonderhaar, Kidwell WR. Effects of inhibition of basement membrane collagen deposition on rat mammary gland development. Develop Biol 1980;80:253–66.

38. Sympson CJ, Bissell MJ, Werb Z. Mammary gland tumor formation in transgenic mice overexpressing stromelysin-1. Semin Cancer Biol 1995;6:159–63.

39. Streuli CH, Schmidhauser C, Kobrin M, Bissell MJ, Derynck R. Extracellular matrix regulates expression of the TGF-β gene. J Cell Biol 1993;120:253–60.

40. Thomasset N, Sympson CJ, Lund L, Werb Z, Bissell MJ. Overexpression of stromelysin-1 modifies mammary gland stroma in transgenic mice. Mol Biol Cell 1994;5:181a.

41. Meredith JE Jr, Fazeli B, Schwartz MA. The extracellular matrix as a cell survival factor. Mol Biol Cell 1993;4:953–61.

42. Frisch SM, Francis M. Disruption of epithelial cell-matrix interactions induces apoptosis. J Cell Biol 1994;124:619–26.

43. Bates RC, Lincz LF, Burns GF. Involvement of integrins in cell survival. Cancer Metastasis Rev 1995;14:191–203.

44. Boudreau N, Sympson CJ, Werb Z, Bissell MJ. Suppression of ICE and apoptosis in mammary epithelial cells by extracellular matrix. Science 1995;267:891–3.

45. Kumar S. ICE-like proteases in apoptosis. Trends Biochem Sci 1995;20:198–202.

46. Pullan S, Wilson J, Metcalfe A, Edwards GM, Goberdhan N, Tilly JL, Hickman JA, Dive C, Streuli, CH. Requirement of basement membrane for the suppression of programmed cell death in mammary epithelium. J Cell Sci 1996;109:631–42.

47. Heermeier K, Benedict M, Li ML, Furth P, et al. Bax and Bcl-x(s) are induced at the onset of apoptosis in involuting mammary epithelial cells. Mech Develop 1996; 56:197–207.

48. Ewald D, Li M, Efrat S, Auer G, Wall RJ, Furth PA, Henninghausen L. Time-sensitive reversal of hyperplasia in transgenic mice expressing SV40 T antigen. Science 1996; 273:1384–6.

49. Walker NI, Bennett RE, Kerr JF. Cell death by apoptosis during involution of the lactating breast in mice and rats. Am J Anat 1989;185:19.

50. Boudreau N, Werb Z, Bissell MJ. Suppression of apoptosis by basement membrane requires three-dimensional tissue organization and withdrawal from the cell cycle. Proc Natl Acad Sci USA 1996;93:3509–13.

51. Howlett AR, Petersen OW, Steeg PS, Bissell MJ. A novel function for the nm21-H1 gene: overexpression in human breast carcinoma cells leads to the formation of basement membrane and growth arrest. J Natl Cancer Inst 1994;86:1838–44.

52. Howlett AR, Bailey N, Damsky C, Petersen OW, Bissell MJ. Cellular growth and survival are mediated by β1 integrins in normal human breast epithelium but not in breast carcinoma. J Cell Sci 1995;108:1945–57.

53. Ronnov-Jessen L, Petersen OW, Koteliansky VE, Bissell MJ. The origin of the myofibroblasts in breast cancer: recapitulation of tumor environment in culture unravels diversity and implicates converted fibroblasts and recruited smooth muscle cells. J Clin Invest 1995;95:859–73.

54. Ronnov-Jessen L, Petersen OW, Bissell MJ. Cellular changes involved in conversion of normal to malignant breast: importance of the stromal reaction. Physiol Rev 1996; 76:69–125.

55. Petersen OW, Ronnov-Jessen L, Bissell MJ. The microenvironment of the breast: three-dimensional models to study the roles of the stroma and the extracellular matrix in function and dysfunction. Breast J 1995;1:22–35.
56. Lelievre S, Weaver VM, Larabell CA, Bissell MJ. Extracellular matrix and nuclear matrix interactions may regulate apoptosis and tissue-specific gene expression: a concept whose time has come. In: Getzenberg RH, ed. Advances in molecular and cell biology: cell structure and signaling. Greenwich, CT: JAI Press Inc, in press.
57. Streuli CH, Bissell MJ. Mammary epithelial cells, extracellular matrix, and gene expression. In: Lippman M, Dickson R, eds. Regulatory mechanisms in breast cancer. 1991;pp 365–80.

13

Reactive Oxygen Species and Antioxidants in Luteal Cell Demise

HAROLD R. BEHRMAN, RAYMOND F. ATEN, PINAR KODAMAN,
TONY G. ZREIK, AND PAOLO RINAUDO

The corpus luteum is the end-differentiated state of the primordial follicle and one of the major endocrine substructures of the ovary. The formation and involution of the corpus luteum occurs by processes that are regulated by endocrine, paracrine, and intracrine agents of the pituitary, the uterus (in nonprimates), the ovary, and the immune system. The physiological role of the corpus luteum is the secretion of products necessary for the establishment and maintenance of pregnancy. The more prominent luteal products are progesterone, estrogens, androgens, and peptides such as relaxin and oxytocin. Other products include inhibin and its family members, eicosanoids, cytokines, growth factors, and oxygen radicals. It is now evident that the historical perspective of the corpus luteum as an organ of progesterone and estradiol secretion governed solely by feedback regulation of luteinizing hormone (LH) is inadequate. In this chapter, we offer one view of corpus luteum regression and its regulation based on recent findings of this lab and others using the pseudopregnant rat model.

Luteal regression is an indispensable process because products of the corpus luteum inhibit ovulation. The rapid return of ovulation in the absence of pregnancy serves a vital role in survival of a species and emphasizes the physiological importance of luteal regression. Regression of the corpus luteum occurs in two phases. The first is rapid and is defined by a loss of function, while the second and slower phase is characterized by tissue involution. These processes are referred to as functional and structural regression, respectively, with cellular demise only evident in structural regression. Time-dependent events associated with each stage of luteal regression are summarized in Figure 13.1. Specific and discrete endocrine agents regulate functional and structural regression of the rat corpus luteum. Increasing evidence points to a role of eicosanoids and prolactin as inducers of functional and structural regression, respectively, and to reactive oxygen species as mediators of both processes.

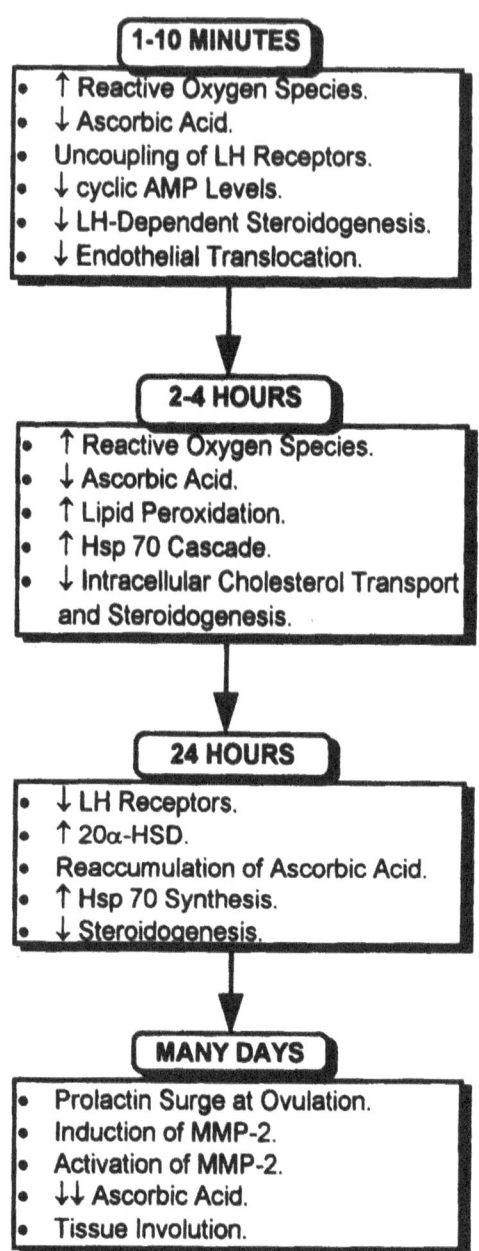

FIGURE 13.1. Time-dependent events in luteal regression of the rat (see text for abbreviations and details).

Nature and Interactions of Reactive Oxygen Species

Superoxide anion (O_2^-), hydrogen peroxide (H_2O_2), and hydroxyl radical (OH) are well known species of reactive oxygen that are produced by the sequential addition of electrons to oxygen. Production of O_2^- generates other reactive oxygen species by dismutation of O_2^- into H_2O_2, and by the reduction of ferric or cupric ions to ferrous and cuprous ions, respectively, thereby initiating a cycle of OH production by Fenton reactions. Hypohalous acids (e.g., HClO) are generated by leukocyte peroxidases in the presence of H_2O_2, and a combination of HClO and H_2O_2 generates singlet oxygen (O_2^1) (1, 2). Highly reactive O_2^1 is also generated by photoexcitation (3). Nitric oxide (NO), a small gaseous radical first described as the endothelium-derived relaxation factor, is produced by endothelium, leukocytes, and other cells (4, 5). The interaction of O_2^- with NO generates the highly reactive peroxynitrite radical, which has properties similar to OH (6). Lipid hydroperoxides, formed by interaction of reactive oxygen species with polyunsaturated fatty acids, propagate peroxidation of other polyunsaturated fatty acids through a chain reaction. Species that induce lipid peroxidation include OH, peroxynitrite, HClO, and O_2^1; peroxidase-coupled reactions also result in lipid peroxidation (7, 8). Lipid peroxides, O_2^1, peroxynitrite, HClO, and OH are directly cytotoxic. These oxidants also breakdown the unstable NO radical and by this process contribute to tissue ischemia (9).

A characteristic feature of reactive oxygen species is their short existence in biological systems. For example, OH, O_2^1, and NO have half-lives of 1 ns, 2 ms, and 4–6 ms, respectively (10, 11). Analysis of these products in biological systems is consequently difficult. At neutral pH, O_2, H_2O_2, and O_2^- are weak oxidizing agents, whereas OH, peroxynitrite, O_2^1, and HClO, are powerful oxidants; the NO radical is a weak reducing agent (12). Although the appearance of reactive oxygen species may be short-lived, their interactions initiate events that are self-propagating and long-lasting.

Biological Origins of Reactive Oxygen Species

The production of reactive oxygen species is a fact of aerobic life. Some of these reactive species escape from enzymes within cellular compartments that utilize oxygen during metabolism. But this leakage of reactive oxygen species is not a major risk to cells under normal conditions and does not represent a regulated process. Prominent and regulated cellular sources of reactive oxygen species include phagocytic leukocytes (13), thyrocytes (14), and endothelium (15). Both phagocytes and the thyroid contain enzymes that utilize NADPH as the electron donor to generate O_2^-, which quickly becomes reduced to H_2O_2 by ubiquitous superoxide dismutases

(SODs). NADPH oxidase is located in the plasma membrane and releases O_2^- into the extracellular space and phagosomes. In phagocytes, the enzyme is under complex regulation by several cytosolic and membrane proteins, which become organized in the plasma membrane after activation of the respiratory burst (13). In the thyroid, an NADPH oxidase-like enzyme is activated by thyroid-stimulating hormone (TSH), and represents the rate-limiting step in thyroid hormonogenesis (16). Increasing evidence indicates that many cells generate reactive oxygen species by mechanisms similar to the thyroid and phagocytes. Examples of such cells identified so far include endothelial and B-cells (17), fibroblasts (18), adipocytes (19), sperm (20, 21), and oocytes (22).

Reactive Oxygen Species in the Corpus Luteum

That the rat corpus luteum produces reactive oxygen species at luteal regression is now well established, as shown by the generation of O_2^- (23, 24), H_2O_2 (25), and lipid peroxides (26–28). The cellular origin of these reactive products is not known. As activated leukocytes are the most powerful cellular generators of reactive oxygen species and are resident in virtually all tissues, these cells probably represent a significant source of reactive oxygen species. In fact, a hallmark of luteal regression in a host of species is the infiltration of leukocytes (29–34). Moreover, activated leukocytes block gonadotropin action in isolated luteal cells (35). These observations, however, do not preclude direct and regulated production of reactive oxygen species by luteal cells. Interestingly, the depletion of ascorbic acid in the rat corpus luteum by LH, which can be viewed as an indirect response to reactive oxygen species production, is blocked by an inhibitor of steroidogenesis (36). Ascorbic acid depletion in response to prostaglandin-$F_{2\alpha}$ ($PGF_{2\alpha}$) is unaffected by inhibitors of steroidogenesis (36), which implies that LH may increase reactive oxygen species by stimulation of steroidogenesis, whereas the actions of $PGF_{2\alpha}$ seem to be mediated by an alternate process. If leukocytes generate reactive oxygen species in response to $PGF_{2\alpha}$ in luteal tissue, this action must be mediated by secondary factors released by luteal cells as there is no evidence that $PGF_{2\alpha}$ directly activates leukocytes. We were unable to detect specific binding of $PGF_{2\alpha}$ to rat peritoneal leukocytes (Koladecik and Behrman, unpublished findings) under conditions where luteal cells and luteal tissue avidly bind this eicosanoid (37, 38). A possible mediator for activation of leukocytes by $PGF_{2\alpha}$ is LTB_4, a leukotriene known to be generated in luteal tissue by $PGF_{2\alpha}$, which is a potent chemotactic factor and activator of leukocytes (34, 39). Research in this area may result in the identification of other ovarian products that regulate leukocyte function.

Protective Mechanisms Against Reactive Oxygen Species

Protection against reactive oxygen species is afforded by specific degradative enzymes, antioxidant vitamins, and other radical scavengers, as well as by cellular repair mechanisms invoked to reverse damage caused by reactive oxygen species. Superoxide dismutases are present both within and outside cells and they rapidly transform O_2^- to H_2O_2. Cytosolic and extracellular SODs are Cu-Zn metalloenzymes, whereas the mitochondrial form is a Mn metalloenzyme. Each SOD is encoded by a separate gene, and mitochondrial SOD appears to be the most readily inducible form in response to oxidative stress (40).

Hydrogen peroxide is degraded by catalase, glutathione (GSH) peroxidase, and other peroxidases under certain circumstances. Catalase is widely distributed in tissues and highly specific for H_2O_2, whereas GSH peroxidase degrades both H_2O_2 and lipid hydroperoxides (41, 42). Glutathione peroxidase has exclusive specificity for GSH as the H-donor substrate, but shows broad specificity for the peroxide substrate (41). A large family of other peroxidases also utilize H_2O_2 as a substrate. These include leukocyte peroxidases (eosinophil peroxidase and myeloperoxidases of neutrophils, monocytes, and macrophages), thyroid peroxidase, soluble salivary and mammary peroxidases (lactoperoxidase), and membrane-bound peroxidases, which are present in most cells (41, 43). Peroxidases specifically utilize H_2O_2, but unlike catalase they must use a secondary H-donor substrate other than H_2O_2, as H_2O_2 alone causes irreversible inactivation (44, 45). Secondary H-donor substrates utilized by these peroxidases include halides, ascorbic acid, polyunsaturated lipids, and others. Clearly, these peroxidases could induce oxidation of secondary substrates like polyunsaturated fatty acids in membranes that would be cytotoxic if protective substrates were not present. We found that lactoperoxidase, myeloperoxidase, and eosinophil peroxidase are very effective scavengers of H_2O_2 but only in the presence of ascorbic acid or thiocyanate;these secondary H-donors alone show no ability to scavenge H_2O_2 (46). Extracellular peroxidases are most likely present in the corpus luteum, as both peroxidase and H_2O_2 are released into the extracellular space by activated leukocytes, and these cells are known constituents of the corpus luteum. Since ascorbic acid is preeminent in the pecking order of H-donor antioxidants, its abundance in the aqueous environment of the corpus luteum places this antioxidant in a key position to serve a vital protective role against reactive oxygen species. Moreover, we recently found that marked and rapid secretion of ascorbic acid occurs by luteal cells in response to LH and $PGF_{2\alpha}$ (36). All of the components for scavenging of H_2O_2 by peroxidase and ascorbic acid are therefore present in both the intracellular and extracellular milieu of the corpus luteum. On the other hand, if ascorbic acid is depleted, as that seen in luteal tissue after $PGF_{2\alpha}$ treatment, protection would be compromised. Perhaps it is for this reason that we found a reciprocal change in $PGF_{2\alpha}$-induced ascorbic acid depletion and lipid peroxidation in the corpus luteum (27). We propose here that $PGF_{2\alpha}$ causes lipid peroxidation like that seen in natural luteal regression (26–28) because unsaturated fatty acids become secondary H-donors in peroxidase-coupled reactions as a result of the excessive and prolonged depletion of ascorbic acid, the preferred substrate in such reactions.

Ascorbic acid is transported from the blood into many peripheral tissues via an energy and sodium-dependent process (47). A similar transporter is present in the rat corpus luteum and luteal cells (36, 48). The time-course for half-maximal depletion of ascorbic acid in vivo in response to $PGF_{2\alpha}$ is extremely rapid (2 to 3 min) (36) compared to that of LH (27). $PGF_{2\alpha}$ is a potent and rapid inhibitor of ascorbic acid uptake, whereas ascorbic acid uptake is unaffected by LH, PGE_2, glucose, cyclic adenosine monophosphate (cAMP), progesterone, phorbol ester, ionomycin, and H_2O_2 (36). Interestingly, secretion of ascorbic acid in luteal cells is also stimulated by $PGF_{2\alpha}$, an effect mimicked by LH, H_2O_2, generators of reactive oxygen species, calcium ionophore, and disruption of the cytoskeleton (36). The ability of $PGF_{2\alpha}$ to inhibit ascorbic acid uptake and to stimulate ascorbic acid secretion implicates these processes as the basis for the rapid depletion of ascorbic acid in the corpus luteum at luteal regression. These findings emphasize the fact that a particular tissue can become deficient in ascorbic acid in the face of normal dietary and blood levels as only hepatic tissue produces ascorbic acid *de novo* in mammals that have no dietary requirement for this vitamin (49, 50). Ascorbic acid is required in the diet of primates and guinea pigs (49). On the other hand, recent evidence shows that ascorbic acid uptake is increased in rat luteal cells by treatment with insulin-like growth factor-I IGF-I (Kodaman and Behrman, unpublished findings). Interestingly, we also found that a similar ascorbic acid transporter is induced at the gene level by follicle-stimulating hormone (FSH) and IGF-I in immature rat granulosa cells, the progenitors of luteal cells (51). These findings show that both accumulation and secretion of ascorbic acid is a hormone-regulated process.

Vitamin E is a hydrophobic molecule and is therefore present in high levels in cell membranes and other lipid-rich environments. It is an excellent scavenger of lipid radicals and serves an important role in the termination of lipid peroxidation reactions. Oxidized vitamin E is recycled to the reduced form by ascorbic acid (52, 53). Consequently, tissue levels of vitamin E remain relatively constant, whereas ascorbic acid levels fluctuate markedly as shown for the corpus luteum (27, 36). Vitamin E is present in blood predominantly within lipoprotein complexes and is transported into cells via the lipoprotein receptor (52, 54). A similar transport process for vitamin E is present in the corpus luteum (55).

Carotenoids are also lipid soluble and serve as antioxidants (56–58). Lutein is a member of this family and is conspicuously concentrated in the corpus luteum of some species. Carotenoids are very effective scavengers of O_2^1 (57).

Luteal Cell Responses to Reactive Oxygen Species

In rat luteal cells, O_2^-, H_2O_2, and lipid hydroperoxide cause an extremely rapid abrogation of gonadotropin-sensitive AMP accumulation and progesterone secretion (59–63), a decrease in membrane fluidity, and an increase in phospholipase A_2 activity (64, 65). While O_2^- evokes antigonadotropic and antisteroidogenic actions in luteal cells, this response is mediated by the production of H_2O_2 (61). In turn, the action of H_2O_2 in luteal cells appears to be mediated in part by the OH radical, which induces DNA and RNA damage, depletes ATP, and inhibits protein synthesis (59, 66, 67). These same responses are produced by lipid hydroperoxide (63). Depletion of ATP is a normal consequence of corpus luteum regression in vivo (68). The acute blockade of steroidogenesis by H_2O_2 and lipid hydroperoxide is due to uncoupling of the LH receptor from adenylate cyclase, as well as to impairment of cholesterol utilization by mitochondrial cholesterol side-chain cleavage ($P450_{scc}$) via inhibition of protein synthesis (63, 67, 69, 70). Gonadotropins and cyclic AMP stimulate steroidogenesis by induction of a protein that mediates the translocation of cholesterol across the inner mitochondrial membrane to the site of $P450_{scc}$. This is the rate-limiting step in stimulation of steroidogenesis by gonadotropin (71). Protein synthesis, not RNA synthesis, is necessary for this response (67, 72, 73). The hormone-sensitive "rapidly turning over protein" of steroidogenic cells was recently identified by Stocco's lab (74, 75). It should be noted that O_2^-, H_2O_2, and lipid hydroperoxide do not cause demise of luteal cells, but irreversibly inhibit hormone-sensitive steroidogenesis.

The Stress Response Mediates Functional Luteal Regression

The half-lives of O_2^-, H_2O_2, and lipid peroxides in tissues are short, and the actions of these agents are rapid, yet luteal regression must persist to permit the resumption of ovulation. It is likely that reactive oxygen species initiate other effector systems that have a longer duration of action. We therefore determined if the regressing corpus luteum evokes a stress protein or heat shock protein (Hsp), as prostaglandins and reactive oxygen species do in other cells (76–82). Moreover, agents like phorbol ester and calcium ionophore induce Hsp (83, 84) and block progesterone secretion in luteal cells (85–88).

The plethora of functions of heat shock proteins have been identified only recently. They serve as molecular chaperones as they bind proteins during translation and prevent folding, aid in the correct folding of proteins, bind to denatured proteins and target them for degradation, and present antigens on the cell surface (89–95). Some Hsps are expressed constitutively, while others are induced by stress. Hsp 90 is constitutively expressed in the cytosol of most cells. Hsp 90 binds steroid receptors, blocks protein synthesis, and it is induced by heat, anoxia, glucose starvation, and ionophores (84, 89). In rat granulosa cells, Hsp 90 is induced by FSH and cyclic AMP, as well as by heat (96). Hsp 70 is a family of proteins whose members are present in the endoplasmic reticulum (Grp 78 or BiP), cytosol (constitutive Hsp 73 and inducible Hsp 72), mitochondria, and plasma membrane (84, 89). Interestingly, Grp 78 contains a C-terminal peptide sequence identical to that of steroidogenesis activating polypeptide (97). Hsp 60 is located in mitochondria and mediates, in association with Hsp 10, the proper folding of imported proteins (84, 89, 91, 98). Hsp 28 is normally found in close association with the Golgi and is markedly induced by heat (89). Interestingly, both Hsp 70 and Hsp 28 are induced during apoptosis (99).

The induction of Hsps is initiated by trans-activating heat shock factors (HSF1 and HSF2). Following activation by stress, HSF1 trimerizes, translocates to the nucleus, binds to heat shock elements (HSE), and initiates transcription of Hsp (100, 101). The initiation of transcriptional activity by HSF is detected within minutes in response to a host of activators (76, 77, 101, 102), and several non-Hsp genes contain HSE (101, 103, 104). The latter finding has significance because activation of HSF could influence the transcription of other genes such as those involved in luteal regression.

We found that induction of Hsp 70 interrupts LH and cyclic AMP-sensitive progesterone secretion in rat luteal cells (105). Inhibition of progesterone secretion by heat shock protein induction is not associated with blockade of the LH receptor, but it is reversed completely by 22-hydroxycholesterol, an analog of cholesterol that is a substrate for steroidogenesis that rapidly penetrates the cell and mitochondria. An early response in natural and $PGF_{2\alpha}$-induced luteal regression is the abrogation of LH-sensitive steroidogenesis (106, 107) similar to that produced by Hsp induction and H_2O_2 in luteal cells (69). In addition, $PGF_{2\alpha}$, phorbol ester, ionomycin, and tumor necrosis factor-alpha (TNFα) induce Hsp in luteal cells (108). McPherson et al. (109) also showed that $PGF_{2\alpha}$ in vivo induced Hsp 70 in large luteal cells of the ewe prior to the induction of luteal regression.

More recent and direct evidence indicates that induction of Hsp 70 is an important mediator of functional luteal regression in the rat (110). Gel-retardation assays show that HSF activation occurs within 7.5 min after induction of luteal regression by $PGF_{2\alpha}$ treatment of the rat. Western analysis revealed Hsp 70 synthesis by 1 h with higher levels seen at 2 h in luteal tissue after regression was induced by $PGF_{2\alpha}$. The stress response cascade was similarly activated during natural regression of the corpus luteum in rats (110). Gel retardation assays demonstrated maximal HSF activation 10 days after ovulation, a period of incipient luteal regression in the pseudopregnant rat (111). Western blotting showed that Hsp 70 levels increased dramatically on this day and were sustained for several days afterwards (110). We also investigated the inhibition of Hsp 70 synthesis in isolated luteal cells using a cholesteryl-conjugated phosphorothioate antisense oligodeoxynucleotide. The control was an oligodeoxynucleotide with the same base composition, but a scrambled sequence. Incubation with antisense oligodeoxynucleotide for 2 h before heat shock prevented synthesis of Hsp 70. Preincubation with antisense oligodeoxynucleotide partially reversed heat stress-induced inhibition of LH-stimulated steroidogenesis and reversed the inhibition of cAMP-stimulated steroidogenesis induced by $PGF_{2\alpha}$. Treatment with control oligodeoxynucleotide did not reverse heat shock or $PGF_{2\alpha}$ inhibition of hormone-dependent steroidogenesis. The findings that the synthesis of Hsp 70 coincides with the loss of luteal function, and that blocking Hsp 70 synthesis reverses inhibition of hormone-dependent steroidogenesis, strongly suggest a role for Hsp as a physiological mediator of luteal regression.

Structural Luteal Regression

There is sound evidence in species such as the rat that structural luteolysis is distinct from functional luteolysis (111, 112). Corpora lutea of the rat persist for several estrous cycles (113), and a similar phenomenon is seen in the human. Although luteal function is completely dependent on pituitary gonadotropins in the nonpregnant rat (114, 115), hypophysectomy does not cause structural luteolysis. On the contrary, hypophysectomy prevents structural involution of corpora lutea in rats for months (114). The histologic features seen during structural luteolysis in the rat show that the diameter of the corpus luteum decreases and connective tissue or stromal elements greatly increase (116). In many cases, large numbers of these fibroblast-like cells are seen near the periphery of the corpus luteum as they invade the lutein tissue. At complete luteolysis, scars remain and occasionally, the connective tissue cells of the scar surround a small number of lutein cells. At the end of the luteal phase in the human corpus luteum, Corner (117) describes "mulberry" cells that have several large vacuoles in the granulosa compartment. A few cells have large empty spaces occupying almost all of the cytoplasm. Small, densely-staining pyknotic nuclei are evident, and an occasional nucleus has degenerated into a dense irregular mass. Degeneration is even more advanced in the theca interna, and nearly every cell is heavily vacuolated. Cells in which the cytoplasm is virtually obscured by vacuoles and cells with pyknotic or degenerated nuclei are common. Hemorrhage into the central cavity is frequent. Adams and Hertig (32) describe the dissolution and phagocytosis of luteal cells during luteal regression in which portions of disrupted cells, present between intact cells, appear to be in the process of engulfment, possibly by thecal cells. In structural degeneration of the human corpus luteum, release of cellular blebs referred to as apoptotic bodies, nuclear condensation, and phagocytosis occurs and is consistent with the process of apoptosis (118, 119). In the ewe, similar autophagocytosis of endothelial cells is associated with the marked decrease in blood flow at regression (120). Biochemical markers of apoptosis, which include oligonucleosome formation, characteristic of endonuclease activation, were recently identified during structural luteolysis of the bovine corpus luteum (121).

Functional luteolysis is a prerequisite condition for structural luteolysis (122). Although $PGF_{2\alpha}$ induces functional luteolysis, it does not directly cause cellular degeneration in bovine or rat luteal cells (87, 123). This indicates that secondary actions of $PGF_{2\alpha}$ in vivo cause luteal involution. In the rat, prolactin induces structural involution of the corpus luteum (116, 124), and levels of prolactin rise precipitously at ovulation in this species. Mechanisms and agents that induce structural luteolysis in the primate corpus luteum are completely unknown, although prolactin is known to impair ovarian and luteal function in the human. Paradoxically, in the rat, prolactin is luteotropic when administered immediately after hypophysectomy and can maintain luteal function for months (125, 126), whereas LH is ineffective under these conditions (126). Structural luteolysis can only be induced by prolactin if administered after functional luteolysis has occurred (122) or if replacement is delayed for about 24 h after hypophysectomy. Functional luteolysis and hypophysectomy cause depletion of luteal LH receptors (127, 128), as described earlier, and this may permit induction of structural luteolysis by prolactin. Indomethacin, an inhibitor of prostaglandin synthesis, was reported to block the luteolytic response to prolactin (129). Immune cells do not appear to mediate prolactin-induced structural involution of the rat corpus luteum, since immunosuppressive agents such as dexamethasone or cyclosporine do not prevent luteal involution (122).

A characteristic feature of luteal cells in culture is their gonadotropin-sensitive loss of differentiated function, which may be in part due to the absence of an appropriate extracellular matrix. Abundant evidence shows that the matrix, which is composed mostly of collagen proteoglycans, fibronectin, and laminin, serves a vital role in modulation of cell function and differentiation (130–133). The rate of collagen synthesis is dependent on collagen-specific mRNA levels (134, 135). There is direct evidence that shows the importance of the extracellular matrix in luteal function and differentiation (136). We recently found that the inclusion of extracellular matrix components, like laminin or fibronectin, greatly increases the steroidogenic potential of granulosa cells as they differentiate into luteal cells, while a cell adhesion receptor antiserum abolishes differentiation of these cells (136). Interestingly, extracellular matrix components block apoptosis in cultured mammary epithelial cells (137).

Ascorbic acid is an essential cofactor for collagen synthesis via hydroxylation of proline and lysine, and also serves a role in this process at the level of gene transcription, mRNA stabilization, translation, and secretion (138). The rat corpus luteum contains enormous levels of ascorbic acid that are transiently depleted at functional regression (27, 122), but severely and persistently depleted during structural involution (122). Such a response would be expected to not only reduce protection against reactive oxygen species, but also impair collagen synthesis. The levels of ascorbic acid or other antioxidant vitamins in the human corpus luteum are not known, nor is it known whether or not depletion of ascorbic acid occurs during involution of the primate corpus luteum, but such studies certainly merit investigation. Possibly, impairment of the luteal ascorbic acid transporter, as seen in other tissues (139–141) as well as luteal cells (36), may cause a localized scorbutic condition in the luteolytic process. In other tissues, IGF-I and TGFβ stimulate collagen synthesis (142), but it is not known if they serve a similar role in the corpus luteum. Interestingly, collagen synthesis is inhibited by TNFα and interferon-alpha (IFNα) in other tissues (142), which seems highly pertinent in view of the actions of such cytokines on ovarian function. Investigation of matrix synthesis not only during luteinization but also in luteal involution would add measurably to our understanding of these processes.

Matrix metalloendopeptidases (MMPs) control matrix dissolution, and at least seven different members of this family have been identified (130). The enzymes are secreted as latent zymogens and are activated by cleavage of an approximately 10 kDa peptide by proteolytic attack or by exposure to oxidants such as HOCl. Regulation of latent MMP secretion is achieved largely by control of gene transcription. Important exceptions are neutrophils and macrophages, in which storage and release of MMP occurs (130). Notable agents that induce MMP activity include those that disrupt the actin cytoskeleton, cytokines/growth factors (IFNs, ILs, PDGF, TNFα), cytotoxic agents, and inflammatory mediators. Repressive agents include glucocorticoids and TGFβ. We demonstrated that structural luteolysis in the rat results in the induction of at least two MMPs, but by far the most predominant forms are a latent collagenase of 72 kDa and an activated 66 kDa form of this enzyme that has the characteristics of MMP-2 (122). The 66 kDa form of activated collagenase increases consistently during prolactin-induced structural luteolysis in the rat. No information is available on the nature of MMPs in the primate corpus luteum or if these collagenases are activated by prolactin, but such studies may shed light on potential mechanisms of prolactin-induced ovarian dysfunction in women.

Major control over extracellular proteolysis is afforded by tissue inhibitors of metalloproteinase (TIMP). Two major forms of TIMP are known, TIMP-1 (a 28 kDa glycosylated peptide) and TIMP-2 (a 20 kDa nonglycosylated peptide), which are coordinately secreted with collagenases. These inhibitors bind stoichiometrically with high affinity to the collagenases and block activity. Other proteinase inhibitors include α2-macroglobulin, α1-macroglobulin, and α1-proteinase inhibitor-3 (130). Interestingly, the highest levels of TIMP mRNA in the mouse are found in the ovaries, particularly in functional corpora lutea. It is suggested that TIMP may serve to prevent luteolysis (143). Recently, TIMP was localized in porcine corpora lutea between luteal cells, around blood vessels in connective tissue, and in the capsule (144). These are highly intriguing findings in view of the histological features of structural luteolysis discussed earlier (145). A decrease in expression of TIMP mRNA occurs at the time of functional luteal regression (144). These recent findings are interesting because they support the postulate that LH is necessary for TIMP synthesis and secretion; receptors for LH are lost at functional luteal regression (127, 128). Interestingly, LH and human chorionic gonadotropin (hCG) stimulate TIMP expression in follicular granulosa cells (146, 147). Thus, structural luteolysis in the rat is associated with ascorbic acid depletion, which probably reflects a decrease in collagen synthesis, an increase in collagenase secretion and activation, and a decrease in mRNA expression of TIMP, the endogenous collagenase inhibitor. Interestingly, another metalloproteinase inhibitor, α2-macroglobulin and its mRNA, is highly expressed in the rat corpus luteum in response to prolactin, but severely depleted at functional luteolysis (148).

Summary

The historical perspective of corpus luteum regression as a passive and unregulated process is inadequate to explain the myriad of events that are now known to occur, which include rapid cellular signaling, gene activation, and induction of proteins. Substantial evidence indicates a role of eicosanoids, reactive oxygen species, and heat shock proteins as well as the degradation and remodeling of the extracellular matrix, in corpus luteum regression. Yet, important questions beg answers for a complete understanding of luteal cell demise. We have no idea what signals initiate luteal regression in primates. Unlike the primate, luteal regression in lower species is initiated by uterine secretion of $PGF_{2\alpha}$ induced by the rising levels of estrogen from the growing follicle (149). Clearly no increase in estrogen secretion precedes luteal regression in the primate, and hysterectomy has no effect on the duration of luteal function in this species (150). Another question that begs research is how luteal regression is completely prevented in early pregnancy. Hormones produced by the implanting conceptus are certainly involved in this process, but their mechanisms and nature remain largely elusive. Finally, there is no information on whether or not antioxidants like ascorbic acid are present in such high and variable quantities in the primate like that seen in lower species where they seem to play such a vital role. Research on these and other questions may eventually lead to a more complete understanding of luteal function and regression.

Acknowledgments. The authors gratefully acknowledge the excellent technical assistance of Sandra L. Preston, Thomas R. Kolodecik, and Tanya A. Fatima, the outstanding editorial assistance of Rebecca E. Morley, and the exceptional administrative assistance of Mrs. Cindy Kolodecik. Supported by NIH-HD-10718.

References

1. Aune TM, Thomas EL. Accumulation of hypothiocyanite ion during peroxidase-catalyzed oxidation of thiocyanate ion. Eur J Biochem 1977;80:209–14.
2. Test ST, Weiss SJ. The generation of utilization of chlorinated oxidants by human neutrophils. Adv Free Rad Biol Med 1986;2:91–116.
3. Uehara K, Hori K, Nakano M, Koga S. Highly sensitive chemiluminescence method for determining myeloperoxidase in human polymorphonuclear leukocytes. Anal Biochem 1991;199:191–6.
4. Palmer RMJ, Ferrige AG, Moncada S. Nitric oxide accounts for the biological activity of endothelium-derived relaxing factor. Nature 1987;327:524–6.
5. Schmidt HHW, Warner TD, Ishii K, Sheng H, Murad F. Insulin secretion from pancreatic B cells caused by L-arginine-derived nitrogen oxides. Science 1992;255:721–3.
6. Floris R, Piersma SR, Yang G, Jones P, Wever R. Interaction of myeloperoxidase with peroxynitrite. A comparison with lactoperoxidase, horseradish peroxidase and catalase. Eur J Biochem 1993;215:767–75.
7. Stelmaszynska T, Kukovetz E, Egger G, Schaur RJ. Possible involvement of myeloperoxidase in lipid peroxidation. Int J Biochem 1992;24:121–8.
8. Wiedau-Pazos M, Goto JJ, Rabizadeh S, et al. Altered reactivity of superoxide dismutase in familial amyotrophic lateral sclerosis. Science 1996;271:515–8.
9. Gryglewski RJ, Palmer RM, Moncada S. Superoxide anion is involved in the breakdown of endothelium-derived vascular relaxing factor. Nature 1986;320:454–6.
10. Marklund S, Marklund G. Involvement of the superoxide anion radical in the autoxidation of pyrogallol and a convenient assay for superoxide dismutase. Eur J Biochem 1974;47:469–74.
11. Schuman EM, Madison DV. A requirement for the intracellular messenger nitric oxide in long-term potentiation. Science 1991;254:1503–6.
12. Halliwell B, Gutteridge JMC. Free radicals in biology and medicine. Oxford: Oxford University Press, 1989.
13. Morel F, Doussiere J, Vignais PV. The superoxide-generating oxidase of phagocytic cells. Eur J Biochem 1991;201:523–46.
14. Nunez J, Pommier J. Formation of thyroid hormones. Vitamins and hormones 1982;39:175-229.
15. Zweier JL, Kuppusamy P, Lutty GA. Measurement of endothelial cell free radical generation: evidence for a central mechanism of free radical injury in postischemic tissues. Proc Natl Acad Sci USA 1988;85:4046–50.
16. Dupuy C, Virion A, Ohayon R, Kaniewski J, Deme D, Pommier J. Mechanism of hydrogen peroxide formation catalyzed by NADPH oxidase in thyroid plasma membrane. J Biol Chem 1991;266:3739–43.
17. Maly F-E. The B lymphocyte: a newly recognized source of reactive oxygen species with immunoregulatory potential. Free Rad Res Comm 1990;8:143–8.
18. Meier B, Cross AR, Hancock JT, Kaup FJ, Jones OTG. Identification of a superoxide-generating NADPH oxidase system in human fibroblasts. Biochem J 1991;275:241–5.
19. Krieger-Brauer HI, Kather H. Human fat cells possess a plasma membrane-bound H_2O_2-generating system that is activated by insulin via a mechanism bypassing the receptor kinase. J Clin Invest 1992;89:1006–13.

20. Alvarez JG, Touchstone JC, Blasco L, Storey BT. Spontaneous lipid peroxidation and production of hydrogen peroxide and superoxide in human spermatozoa: superoxide dismutase as major enzyme protectant against oxygen toxicity. J Androl 1987;8:338–48.

21. Aitken RJ, Clarkson JS, Fishel S. Generation of reactive oxygen species, lipid peroxidation, and human sperm function. Biol Reprod 1989;40:183–97.

22. Shapiro BM. The control of oxidant stress at fertilization. Science 1991;252: 533–6.

23. Sawada M, Carlson JC. Superoxide radical production in plasma membrane samples from regressing corpora lutea. Can J Physiol Pharmacol 1989;67:465–71.

24. Sawada M, Carlson JC. Studies on the mechanism controlling generation of superoxide radical in luteinized rat ovaries during regression. Endocrinology 1994; 135:1645–50.

25. Riley JCM, Behrman HR. In vivo generation of hydrogen peroxide in the rat corpus luteum during luteolysis. Endocrinology 1991;128:1749–53.

26. Sawada M, Carlson JC. Association of lipid peroxidation during luteal regression in the rat and natural aging in the rotifer. Exp Gerontol 1985;20:179–86.

27. Aten RF, Duarte KM, Behrman HR. Regulation of ovarian antioxidant vitamins, reduced glutathione, and lipid peroxidation by luteinizing hormone and prostaglandin $F_{2\alpha}$. Biol Reprod 1992;46:401–7.

28. Sugino M, Nakamura Y, Takeda O, Ishimatsu M, Kato H. Changes in activities of superoxide dismutase and lipid peroxide in corpus luteum during pregnancy in rats. J Reprod Fertil 1993;97:347–51.

29. Bucco RA, Melner MH, Gordon DS, Leers-Sucheta S, Ong DE. Inducible expression of cellular retinoic acid-binding protein II in rat ovary: gonadotropin regulation during luteal development. Endocrinology 1995;136:2730–40.

30. Bulmer D. The histochemistry of ovarian macrophages in the rat. J Anat Lond 1964;98:313–9.

31. Kirsch TM, Friedman AC, Vogel RL, Flickinger GL. Macrophages in corpora lutea of mice: characterization and effects on steroid secretion. Biol Reprod 1981;25:629–38.

32. Adams EC, Hertig AT. Studies on the human corpus luteum. I. Observations on the ultrastructure of development and regression of the luteal cells during the menstrual cycle. J Cell Biol 1969;41:696–715.

33. Bagavandoss P, Wiggins RC, Kunkel SL, Remick DG, Keyes PL. Tumor necrosis factor production and accumulation of inflammatory cells in the corpus luteum of pseudopregnancy and pregnancy in rabbits. Biol Reprod 1990;42:367–76.

34. Murdoch WJ. Treatment of sheep with prostaglandin $F_{2\alpha}$ enhances production of a luteal chemoattractant for eosinophils. Am J Reprod Immunol 1987;15:52–6.

35. Pepperell JR, Wolcott K, Behrman HR. Luteolytic effect of neutrophils in rat luteal cells. Endocrinology 1992;130:1001–8.

36. Musicki B, Kodaman PH, Aten RF, Behrman HR. Endocrine regulation of ascorbic acid transport and secretion in luteal cells. Biol Reprod 1996;54:399–406.

37. Wright K, Luborsky-Moore JL, Behrman HR. Specific binding of prostaglandin $F_{2\alpha}$ to membranes of rat corpora lutea. Mol Cell Endocrinol 1979;13:25–34.

38. Wright K, Pang CY, Behrman HR. Luteal membrane binding to $PGF_{2\alpha}$ and sensitivity of corpora lutea to $PGF_{2\alpha}$-induced luteolysis in pseudopregnant rats. Endocrinology 1980;106:1333–7.

39. Steadman LE, Murdock WJ. Production of leukotriene B4 by luteal tissue of sheep treated with prostaglandin F2 alpha. Prostaglandins 1988;36:741–5.
40. Visner GA, Dougall WC, Wilson JM, Burr IA, Nick HS. Regulation of manganese superoxide dismutase by lipopolysaccharide, interleukin-1, and tumor necrosis factor. J Biol Chem 1990;265:2856–64.
41. Chance B, Sies H, Boveris A. Hydroperoxide metabolism in mammalian organs. Physiol Rev 1979;59:527–605.
42. Meister A, Anderson ME. Glutathione. Annu Rev Biochem 1983;52:711–60.
43. Prabir KD. Tissue distribution of constitutive and induced soluble peroxidase in the rat. Eur J Biochem 1992;206:59–67.
44. Edwards SW, Nurcombe HL, Hart CA. Oxidative inactivation of myeloperoxidase released from human neutrophils. Biochem J 1987;245:925–8.
45. Arnao MB, Acosta M, del Rio JA, Garcia-Canovas F. Inactivation of peroxidase by hydrogen peroxide and its protection by a reductant agent. Biochim Biophys Acta 1990;1038:85–9.
46. Kolodecik TR, Aten RF, Behrman HR. Interaction of ascorbic acid or thiocyanate with peroxidase blocks the action of hydrogen peroxide in rat luteal cells. Biol Reprod 1994;50(Suppl 1):152.
47. Rose RC. Transport of ascorbic acid and other water-soluble vitamins. Biochim Biophys Acta 1988;947:335–66.
48. Stansfield DA, Flint AP. The entry of ascorbic acid into the corpus luteum in vivo and in vitro and the effect of luteinizing hormone. J Endocrinol 1967;39:27–35.
49. Levine M, Morita K. Ascorbic acid in endocrine systems. Vitamins and hormones. Vol. 42. New York: Academic Press, Inc., 1985:1–64.
50. Grollman AP, Lehninger AL. Enzymic synthesis of L-ascorbic acid in different animal species. Arch Biochem Biophys 1957;69:458–67.
51. Behrman HR, Preston SL, Aten RF, Rinaudo P, Zreik T. Hormone induction of ascorbic acid transport in immature granulosa cells. Endocrinology 1996;137:4316–21.
52. Drevon CA. Absorption, transport and metabolism of vitamin E. Free Rad Res Comm 1991;14:229–46.
53. Packer L. Protective role of vitamin E in biological systems. Am J Clin Nutr 1991;53:1050S–5S.
54. Traber MG, Kayden HJ. Vitamin E is delivered to cells via the high affinity receptor for low-density lipoprotein. Am J Clin Nutr 1984;40:747–51.
55. Aten RF, Kolodecik TR, Behrman HR. Ovarian vitamin E accumulation: evidence for a role of lipoproteins. Endocrinology 1994;135:533–9.
56. Yuting C, Rongliang Z, Zhongjian J, Yong J. Flavonoids as superoxide scavengers and antioxidants. Free Rad Biol Med 1990;9:19–21.
57. Bendich A, Olson JA. Biological actions of carotenoids. FASEB J 1989;3:1927–32.
58. Krinsky NI. Antioxidant functions of carotenoids. Free Rad Biol Med 1989;7:617–35.
59. Behrman HR, Preston SL. Luteolytic actions of peroxide in rat ovarian cells. Endocrinology 1989;124:2895–900.
60. Margolin Y, Aten RF, Behrman HR. Antigonadotropic and antisteroidogenic actions of peroxide in rat granulosa cells. Endocrinology 1990;127:245–50.
61. Gatzuli E, Aten RF, Behrman HR. Inhibition of gonadotropin action and progester-

one synthesis by xanthine oxidase in rat luteal cells. Endocrinology 1991;128: 2253–8.

62. Endo T, Aten RF, Leykin L, Behrman HR. Hydrogen peroxide evokes antisteroidogenic and antigonadotropic actions in human granulosa lutein cells. J Clin Endocrinol Metab 1993;76:337–42.

63. Kodaman PH, Aten RF, Behrman HR. Lipid hydroperoxides evoke antigonadotropic and antisteroidogenic activity in rat luteal cells. Endocrinology 1994;135: 2723–30.

64. Wu XM, Sawada M, Carlson JC. Stimulation of phospholipase A_2 by xanthine oxidase in the rat corpus luteum. Biol Reprod 1992;47:1053–8.

65. Wu X, Yao K, Carlson J. Plasma membrane changes in the rat corpus luteum induced by oxygen radical generation. Endocrinology 1993;133:491–5.

66. Musicki B, Behrman HR. Metal chelators reverse the action of hydrogen peroxide in rat luteal cells. Mol Cell Endocr 1993;92:215–20.

67. Musicki B, Aten RF, Behrman HR. Inhibition of protein synthesis and hormone-sensitive steroidogenesis in response to hydrogen peroxide in rat luteal cells. Endocrinology 1994;134:588–95.

68. Soodak LK, Macdonald GJ, Behrman HR. Luteolysis is linked to LH-induced depletion of ATP in vivo. Endocrinology 1988;122:187–93.

69. Behrman HR, Aten RF. Evidence that hydrogen peroxide blocks hormone-sensitive cholestrol transport into mitochondria of rat luteal cells. Endocrinology 1991;128:2958–66.

70. Stocco DM, Wells J, Clark BJ. The effects of hydrogen peroxide on steroidogenesis in mouse Leydig tumor cells. Endocrinology 1993;133:2827–32.

71. Hall PF. Trophic stimulation of steroidogenesis: in search of the elusive trigger. Rec Prog Horm Res 1985;41:1–39.

72. Garren LD, Ney RL, Davis WW. Studies on the role of protein synthesis in the regulation of corticosterone production by adrenocorticotropic hormone in vivo. Proc Natl Acad Sci USA 1965;53:1443–50.

73. Toaff ME, Schleyer H, Strauss JF III. Metabolism of 25-hydroxycholesterol by rat luteal mitochondria and dispersed cells. Endocrinology 1982;111:1785–90.

74. Clark BJ, Wells J, King SR, Stocco DM. The purification, cloning, and expression of a novel luteinizing hormone-induced mitochondrial protein in MA-10 mouse Leydig tumor cells. J Biol Chem 1994;269:28314–22.

75. King SR, Ronen-Fuhrmann T, Timberg R, Clark BJ, Orly J, Stocco DM. Steroid production after in vitro transcription, translation, and mitochondrial processing of protein products of complementary deoxyribonucleic acid for steroidogenic acute regulatory protein. Endocrinology 1995;136:5165–76.

76. Becker J, Mezger V, Courgeon A-M, Best-Belpomme M. Hydrogen peroxide activates immediate binding of a *Drosophila* factor to DNA heat-shock regulatory element in vivo and in vitro. Eur J Biochem 1990;189:553–8.

77. Bruce JL, Price BD, Coleman CN, Calderwood SK. Oxidative injury rapidly activates the heat shock transcription factor but fails to increase levels of heat shock proteins. Cancer Res 1993;53:12–15.

78. Donati YRA, Slosman DO, Polla BS. Oxidative injury and the heat shock response. Biochem Pharm 1990;40:2571–7.

79. Healy AM, Mariethoz E, Pizurki L, Polla BS. Heat shock proteins in cellular defense mechanisms and immunity. Ann NY Acad Sci 1992;663:319–30.

80. Heufelder AE, Wenzel BE, Bahn RS. Methimazole and propylthiouracil inhibit the oxygen free radial-induced expression of a 72 kilodalton heat shock protein in Graves' retroocular fibroblasts. J Clin Endocrinol Metab 1992;74:737–42.

81. Koizumi T, Negishi M, Ichikawa A. Activation of heat shock transcription factors by Δ12-prostaglandin J2 and its inhibition by intracellular glutathione. Biochem Pharmacol 1993;45:2457–64.

82. Santoro MG, Garaci E, Amici C. Prostaglandins with antiproliferative activity induce the synthesis of a heat shock protein in human cells. Proc Natl Acad Sci USA 1989;86:8407–11.

83. Nowak TS, Jr. Synthesis of heat shock/stress proteins during cellular injury. Ann NY Acad Sci 1993;679:142–56.

84. Welch WJ. Mammalian stress response: cell physiology, structure/function of stress proteins, and implications for medicine and disease. Physiol Rev 1992;72: 1063–81.

85. Behrman HR, Riley JCM, Aten RF. Reactive oxygen species and ovarian function. In: Adashi E, Leung P, eds. The ovary. New York: Raven Press, 1993:455–71.

86. Sender Baum M, Rosberg S. A phorbol ester, phorbol 12-myristate 13-acetate, and a calcium ionophore, A23187, can mimic the luteolytic effect of prostaglandin $F_{2\alpha}$ in isolated rat luteal cells. Endocrinology 1987;120:1019–26.

87. Musicki B, Aten RF, Behrman HR. The antigonadotropic actions of $PGF_{2\alpha}$ and phorbol ester are mediated by separate processes in rat luteal cells. Endocrinology 1990;126:1388–95.

88. Dorflinger LJ, Albert PJ, Williams AT, Behrman HR. Calcium is an inhibitor of luteinizing hormone-sensitive adenylate cyclase in the luteal cell. Endocrinology 1984;114:1208–15.

89. Welch WJ, Kang HS, Beckmann RP, Mizzen LA. Response of mammalian cells to metabolic stress: changes in cell physiology and structure/function of stress proteins. Curr Top Microbiol Immunol 1991;167:31–55.

90. Craig EA. Chaperones: helpers along the pathways to protein folding. Science 1993;260:1902–4.

91. Horwich AL, Willison KR. Protein folding in the cell: functions of two families of molecular chaperone, hsp60, and TF55-TCP1. Phil Trans R Soc Lond B 1993;339:57–70.

92. McKay DB. Structure and mechanism of 70-kDa heat-shock-related proteins. Adv Protein Chem 1993;44:67–98.

93. Aquino DA, Brosnan CF. Heat-shock proteins and immunopathology. In: Brosnan CF, ed. Heat shock proteins and gamma-delta T cells. Vol. 53. Basel: Karger, 1992: 1–16.

94. Hartl FU, Martin J. Protein folding in the cell: the role of molecular chaperones Hsp 70 and Hsp 60. Annu Rev Biophys Biomol Struct 1992;21:293–322.

95. Winfield JB, Jarjour WN. Stress proteins, autoimmunity, and autoimmune disease. Curr Top Microbiol Immunol 1991;167:161–89.

96. Ben-Ze'ev A, Amsterdam A. Regulation of heat shock protein synthesis by gonadotropins in cultured granulosa cells. Endocrinology 1989;124:2584–94.

97. Li X, Warren DW, Gregoire J, Pedersen RC, Lee AS. The rat 78000 dalton glucose regulated protein [GRP78] as a precursor for the rat steroidogenesis-activator polypeptide [SAP]: the sap coding sequence is homologous with the terminal end of GRP78. Mol Endocrinol 1989;3:1944–52.

98. Neupert W, Hartl F-U, Craig EA, Pfanner N. How do polypeptides cross the mito-chondrial membranes? Cell 1990;63:447–50.
99. Tenniswood MP, Guenette RS, Lakins J, Mooibroek M, Wong P, Welsh J. Active cell death in hormone-dependent tissues. Cancer Metastasis Rev 1992;11: 197–220.
100. Sarge KD, Murphy SP, Morimoto RI. Activation of heat shock gene transcription by heat shock factor 1 involves oligomerization, acquisition of DNA-binding ac-tivity, and nuclear localization and can occur in the absence of stress. Mol Cell Biol 1993;13:1392–407.
101. Morimoto RI, Sarge KD, Abravaya K. Transcriptional regulation of heat shock genes. J Biol Chem 1992;267:21987–90.
102. Benjamin IJ, Kroger B, Williams RS. Activation of the heat shock transcription factor by hypoxia in mammalian cells. Proc Natl Acad Sci USA 1990;87:6263–7.
103. Shibahara S, Muller RM, Taguchi H. Transcriptional control of rat heme oxyge-nase by heat shock. J Biol Chem 1987;262:12889–92.
104. Colotta F, Polentarutti N, Staffico M, Fincato G, Mantovani A. Heat shock in-duces the transcriptional activation of c-fos protooncogene. Biochem Biophys Res Commun 1990;168:1013–9.
105. Khanna A, Aten RF, Behrman HR. Heat shock protein induction blocks hormone-sensitive steroidogenesis in rat luteal cells. Steroids 1994;59:4–9.
106. Thomas JP, Dorflinger LJ, Behrman HR. Mechanism of the rapid antigonadotropic action of prostaglandins in cultured luteual cells. Proc Natl Acad Sci USA 1978;75:1344–8.
107. Jordan AW. Effects of prostaglandin $F_{2\alpha}$ treatment on LH and dibutyryl cyclic AMP-stimulated progesterone secretion by isolated rat luteal cells. Biol Reprod 1981;25:327–31.
108. Khanna A, Aten RF, Behrman HR. Physiological and pharmacological inhibitors of luteinizing hormone-dependent steroidogenesis induce heat shock protein-70 in rat luteal cells. Endocrinology 1995;136:1775–81.
109. McPherson LA, Van Kirk EA, Murdoch WJ. Localization of stress protein-70 in ovine corpora lutea during prostaglandin-induced luteolysis. Prostaglandins 1993;46:433–40.
110. Khanna A, Aten RF, Behrman HR. Heat shock protein-70 induction mediates luteal regression in the rat. Mol Endocrinol 1995;9:1431–40.
111. Wang F, Riley JCM, Behrman HR. Immunosuppressive glucocorticoid blocks ex-trauterine luteolysins in the rat. Biol Reprod 1993;49:66–75.
112. Malven PV. Hypophysial regulation of luteolysis in the rat. In: McKerns KW, ed. The gonads. New York: Meredith Corporation, 1969:367–82.
113. Long JA, Evans HM. The oestrous cycle in the rat. Memoirs Univ Calif 1922;6: 1–148.
114. Smith PE. Hypophysectomy and a replacement therapy in the rat. Am J Anat 1930;45:205–73.
115. Astwood EB, Greep RO. A corpus luteum-stimulating substance in the rat pla-centa. Proc Soc Exp Biol Med 1958;38:713–6.
116. Malven PV, Sawyer CH. A luteolytic action of prolactin in hypophysectomized rats. Endocrinology 1966;79:268–74.
117. Corner GW. The histological dating of the human corpus luteum of menstruation. Am J Anat 1956;98:377–401.

118. Kerr JFR, Wyllie AH, Currie AR. Apoptosis: a basic biological phenomenon with wide-ranging implications in tissue kinetics. Br J Cancer 1972;26:239–57.

119. Wyllie AH, Kerr JFR, Currie AR. Cell death: the significance of apoptosis. Int Rev Cytol 1980;68:251–306.

120. Azmi TI, O'Shea JD. Mechanism of deletion of endothelial cells during regression of the corpus luteum. Lab Invest 1984;51:206–17.

121. Juengel JL, Garverick HA, Johnson AL, Youngquist RS, Smith MF. Apoptosis during luteal regression in cattle. Endocrinology 1993;132:249–54.

122. Endo T, Aten RF, Wang F, Behrman HR. Coordinate induction and activation of metalloproteinase and ascorbate depletion in structural luteolysis. Endocrinology 1993;133:690–8.

123. Fairchild DL, Pate JL. Modulation of bovine luteal cell synthetic capacity by interferon-gamma. Biol Reprod 1991;44:357–63.

124. Desclin L. Observations sur la structure des ovaires chez des rats soumis a l'influence de la prolactine. Annal D'Endocrinol 1949;10:1–18.

125. Greenwald GS, Rothchild I. Formation and maintenance of corpora lutea in laboratory animals. J Anim Sci (Suppl 1) 1968;27:139–62.

126. Macdonald GJ, Greep RO. Prolactin-induced morphological luteal regression unaffected by LH. Proc Soc Exp Biol Med 1969;131:905–7.

127. Grinwich DL, Hichens M, Behrman HR. Control of the LH receptor by prolactin and prostaglandin $F_{2\alpha}$ in rat corpora lutea. Biol Reprod 1976;14:212–8.

128. Behrman HR, Grinwich DL, Hichens M, Macdonald GJ. Effect of hypophysectomy, prolactin, and prostaglandin $F_{2\alpha}$ on gonadotropin binding in vivo and in vitro in the corpus luteum. Endocrinology 1978;103:349–57.

129. Sanchez-Criado JE, Ochiai K, Rothchild I. Indomethacin treatment prevents prolactin-induced luteolysis in the rat. J Endocrinol 1987;112:317–22.

130. Woessner JF, Jr. Matrix metalloproteinases and their inhibitors in connective tissue remodeling. FASEB J 1991;5:2145–54.

131. Shimizu Y, Shaw S. Lymphocyte interactions with extracellular matrix. FASEB J 1991;5:2292–9.

132. Springer TA. Adhesion receptors of the immune system. Nature 1990;346: 425–34.

133. Tinker D, Rucker RB. Role of selected nutrients in synthesis, accumulation, and chemical modification of connective tissue proteins. Physiol Rev 1985;65: 607–57.

134. Kucharz E. The collagens: biochemistry and pathophysiology. New York: Springer-Verlag, 1992:430.

135. Geesin JC, Darr D, Kaufman R, Murad S, Pinnell SR. Ascorbic acid specifically increases type I and type III procollagen messenger RNA levels in human skin fibroblasts. J Invest Dermatol 1988;90:420–4.

136. Aten RF, Kolodecik TR, Behrman HR. A cell adhesion receptor antiserum abolishes, wheras laminin and fibronectin glycoprotein components of extracellular matrix promote, luteinization of cultured rat granulosa cells. Endocrinology 1995;136:1753–8.

137. Boudreau N, Sympson CJ, Werb Z, Bissell MJ. Suppression of ICE and apoptosis in mammary epithelial cells by extracellular matrix. Science 1995;267:891–3.

138. Franceschi RT. The role of ascorbic acid in mesenchymal differentiation. Nutrition Rev 1992;50:65–70.

139. Padh H. Cellular functions of ascorbic acid. Biochem Cell Biol 1990;68:1166–73.
140. Cullen EI, May V, Eipper BA. Transport and stability of ascorbic acid in pituitary cultures. Mol Cell Endocrinol 1986;48:239–50.
141. Finn FM, Johns PA. Ascorbic acid transport by isolated bovine adrenal cortical cells. Endocrinology 1980;106:811–7.
142. Oikarinen A. Dermal connective tissue modulated by pharmacologic agents. Int J Derm 1992;31:149–56.
143. Nomura S, Hogan BLM, Wills AJ, Heath JK, Edwards DR. Developmental expression of tissue inhibitor of metalloproteinase (TIMP) RNA. Development 1989;105:575–83.
144. Tanaka T, Andoh N, Takeya T, Sato E. Differential screening of ovarian cDNA libraries detected the expression of porcine collagenase inhibitor gene in functional corpora lutea. Mol Cell Endocrinol 1992;83:65–71.
145. Malven PV, Cousar GJ, Row EH. Structural luteolysis in hypophysectomized rats. Emer J Physiol 1969;216:421–4.
146. Mann JS, Kinkey MS, Edwards DR, Curry TE Jr. Hormonal regulation of matrix metalloproteinase inhibitors in rat granulosa cells and ovaries. Endocrinology 1991;128:1825–32.
147. Reich R, Daphna-Iden D, Chun SY, et al. Preovulatory changes in ovarian expression of collagenases and tissue metalloproteinase inhibitor messenger ribonucleic acid: role of eicosanoids. Endocrinology 1991;129:1869–75.
148. Gaddy-Kurten D, Hickey GJ, Fey GH, Gauldie J, Richards JS. Hormonal regulation and tissue-specific localization of α2-macroglobulin in rat ovarian follicles and corpora lutea. Endocrinology 1989;125:2985–95.
149. Caldwell BV, Behrman HR. Prostaglandins in reproductive processes. Med Clin N/A 1981;65:927–35.
150. Behrman HR, Caldwell BV. Role of prostaglandins in reproductive processes. In: Greep RP, ed. Reproductive physiology. Medical and Technical Publishing Co. Ltd, 1973.

14

Potential Regulators of Physiological Cell Death in the Corpus Luteum

Bo R. Rueda, Debora L. Hamernik, Patricia B. Hoyer, and Jonathan L. Tilly

The corpus luteum (CL) is formed within the mammalian ovary from the remaining follicular tissue following ovulation of the dominant follicle(s). The primary function of the CL is to synthesize and secrete progesterone. Circulating levels of progesterone rise during luteinization of the ovulated follicle and are maintained in order to provide uterine quiescence for the establishment and maintenance of pregnancy. If, however, fertilization fails or does not occur, the CL regresses by a process referred to as luteolysis. The CL, like most female reproductive endocrine tissues, goes through a phase of rapid cellular growth, proliferation and neovascularization that, by many criteria, is analogous to tumor development. Unlike a tumor, however, the lifespan of the luteal cells is defined as removal of the CL is important for resumption of the next estrous or menstrual cycle. As would be expected, luteal regression is likely mediated via apoptosis. Although much is known about CL function, relatively little is known about mechanisms that regulate luteal cell death. Recently, there has been a plethora of information published regarding the role of reactive oxygen species (ROS), protooncogenes, tumor suppressor genes, transcription factors and proteases in regulating the fate of many cell types. The focus of this review is to discuss the potential role of oxidative stress response and cell death genes in regulating the lifespan of the CL. A brief review of the morphological and physiological aspects of luteal function and regression is also presented.

Morphological and Functional Characteristics of the Mammalian CL

The CL is composed of multiple cell types, including steroidogenic and nonsteroidogenic cells (1–3). Morphologically and functionally distinct populations of steroidogenic cells have been identified in many mammalian species (4–13). The two steroidogenic cell types comprise approximately 40% of the total cell population. These cells are designated small and large, based on their size, and are believed to develop from the steroid-producing granulosa and theca-interstial cells of follicular origin (3, 14–19). Both the large and small luteal cells are capable of producing progesterone (10, 13, 20); however, progesterone production is differentially regulated in the two cell types. On a per cell basis, the basal secretion of progesterone by large cells is 5 to 10 times greater than that by small cells (12, 13). The nonsteroidogenic cells are primarily vascular-derived endothelial cells which make up approximately 50% of the CL (1, 21, 22). Fibroblasts (1, 3) and immune cells (23–26) comprise the remainder (<10%) of the luteal cell population.

Our laboratories, as well as others, have used the ovine and bovine CL as models for studying how each cell type contributes to the regulation of luteal function. Ovine small luteal cells contain the majority of receptors for luteinizing hormone (LH) (13) and LH stimulates progesterone synthesis through a cyclic adenosine monophospate (cAMP)—dependent second messenger pathway (13, 27, 28). Although some receptors for LH are detectable on large steroidogenic cells, the LH receptors are not functionally coupled to steroidogenesis (13, 27, 29, 30). Large luteal cells are limited in numbers (approximately 29% of the volume of the ovine CL) and are the primary contributors to circulating progesterone levels during the mid-luteal phase (1). The large luteal cells synthesize progesterone through an LH- and cAMP-independent mechanism (29). Despite the fact that the synthesis of progesterone is regulated differently between the two steroidogenic cell types, it seems likely that proper function of the CL requires communication between the large and small steroidogenic cells.

Luteolysis: Dissecting Functional Versus Structural Aspects of Regression

Luteolysis consists of two distinct components: first, a reduction in the synthesis and secretion of progesterone (functional regression), and second, cellular demise (structural regression). The temporal association of these two events has been analyzed in a number of in vitro and in vivo experiments. The decrease in synthesis and secretion of progesterone precedes luteolysis, and is reversible in vitro (31) and in vivo (32–35). Once the structural integrity of the CL has been challenged, however, cell death is inevitable. Therefore, since functional and structural regression appear to be distinct events, it seems likely that they are regulated by different intercellular and intracellular mechanisms.

The Role of Prostaglandin $F_{2\alpha}$ in Luteolysis

In the absence of fertilization, luteolysis begins on approximately day 14 of the ovine estrous cycle. Once initiated, luteal regression then rapidly proceeds to days 16 to 17 (36), marking the end of that estrous cycle and the beginning of the next. Prostaglandin $F_{2\alpha}$ ($PGF_{2\alpha}$) has been identified as the physiological signal that initiates luteal regression in many species, including the ewe (37, 38), cow (39), and rat (40). In humans, however, the involvement of $PGF_{2\alpha}$ in luteolysis is unclear (41, 42). Although the daily mean concentrations of $PGF_{2\alpha}$ are similar, or even slightly higher, in pregnant than nonpregnant ewes (43–45), $PGF_{2\alpha}$ secretory patterns are different (46). The pulsatile secretory pattern of $PGF_{2\alpha}$ increases in amplitude and duration at the end of the normal ovine estrous cycle compared to pregnancy (46). Uterine derived $PGF_{2\alpha}$ is transported to the ovary via a countercurrent vascular exchange between the uterine vein and the ovarian artery (38, 47–49). Although there are several hypotheses concerning how $PGF_{2\alpha}$ initiates luteolysis, two specific mechanisms are generally accepted. The first involves direct binding of $PGF_{2\alpha}$ to its receptor (the FP receptor) on large luteal cells which leads to an inhibition of steroidogenesis (reviewed in 50, 51). The second hypothesis involves a decrease in blood flow to the CL which leads to ischemia and possibly tissue hypoxia (22, 52–55). A prolonged hypoxic environment may also inhibit steroid synthesis and ultimately initiate luteal cell death. Thus, although the mechanisms by which $PGF_{2\alpha}$ specifically initiates luteolysis have not been fully elucidated, it is likely that $PGF_{2\alpha}$ directly initiates functional regression and indirectly affects structural regression of the CL.

The Effects of $PGF_{2\alpha}$ on Luteal Cells and Luteal Blood Flow

Large luteal cells directly mediate the $PGF_{2\alpha}$-induced signal for luteal regression. In the ewe, large luteal cells, but not small luteal cells, contain a significant number of high-affinity functional receptors for $PGF_{2\alpha}$ (13, 56–58). Additional studies suggest that $PGF_{2\alpha}$ mediates luteolysis by inhibiting secretion of progesterone (functional regression) in large cells via a calcium-mediated intracellular pathway (59–61). Increased calcium levels have been associated with both pathologic and apoptotic cell death in other systems (62–64). It still remains to be determined if $PGF_{2\alpha}$-initiated increases in intracellular calcium in large luteal cells play a role in the synchronous degeneration of all luteal cell types. Therefore, it is not clear if the effects of $PGF_{2\alpha}$ are limited to functional inhibition or if $PGF_{2\alpha}$ also has a direct role in structural regression of the CL.

   Blood flows through the ovary at a surprisingly high rate are compared to other tissues (65–67), and is largely directed to the functional corpus luteum (53, 65, 66). Prostaglandin $F_{2\alpha}$, a vasoconstrictor, initiates a decrease in blood flow to the corpus luteum (37, 43, 65, 67, 69, 81). Whether the decrease in luteal blood flow precedes (47, 69) or follows (67, 70) the decrease in progesterone production is not clear and may be species specific.

Involvement of Immune Cells in Luteal Regression

Cells and cytokines of the immune system normally associated with inflammation have recently been implicated in luteal regression (reviewed in 51, 71). Furthermore, there is a resident immune cell population, and some cytokines have been detected, within luteal tissue (26, 71). Infiltrating leukocytes may be involved in the removal of cellular debris produced during luteolysis. However, it may be that cytokines released by immune cells directly facilitate luteolysis and, therefore, may also regulate the onset of cell death. The ability of cytokines to regulate human luteal function has recently been reviewed (25). Although there is accumulating evidence that supports a role for cytokine interactions in mediating the structural component of luteal regression, the exact cellular mechanisms have yet to be elucidated.

Physiological Cell Death Associated With Luteolysis

The structural component of luteal regression is irreversible and is evident following the decrease in serum concentrations of progesterone associated with functional luteal regression (22, 72–74). Although information regarding events occurring after functional regression and before cellular destruction is limited, recent studies have implicated apoptosis as an important component of structural regression. Apoptosis has been described in many female reproductive tissues including the uterus (Chapters 5, 20;75–77), the mammary gland (78), ovarian germ cells (Chapter 3;79, 80), ovarian follicles (Chapters 8–11;81–84) and the corpus luteum (72, 73, 85, 86). Of relevance to the present discussion, apoptosis has been identified in the CL by both morphological (85, 87) and biochemical (35, 72–74, 86, 88) parameters at the time of structural regression in many species. In the ovine and bovine CL, the appearance of internucleosomal cleavage of DNA ("laddering") characteristic of apoptosis is associated with natural, as well as $PGF_{2\alpha}$-induced, luteolysis (72, 73, 74).

In addition to apoptosis, induced or naturally occurring luteolysis is accompanied by a decrease in the overall extent of microvascularization within the CL (85, 89, 90), resulting from an involution of the vessels due to orderly loss of endothelial cells rather than from random cellular degeneration (85, 90). Furthermore, uterine-derived $PGF_{2\alpha}$ and luteal oxytocin are believed to act synergistically to amplify vasoconstriction of arterioles within the CL resulting in a severe tissue ischemia (22). Although this evidence supports a universal ischemic effect within the CL at the time of regression, it is not clear if decreased luteal blood flow directly initiates luteal regression or is a resultant effect. Consequently, it is not known if the vascular system plays a passive or active role in luteolysis.

Morphological evaluations of the regressing CL ascertained that the rate at which cell death occurred was dependent on the specific cell type. The small steroidogenic and endothelial cells were the first to decline in numbers followed by a reduction in the number of large luteal cells (22, 91). Therefore, it is possible that vascular regression is the consequence of specific cellular events initiated by changes in external stimuli (hormones) as opposed to a nonspecific response to decreased blood flow. Despite the collapse in the luteal microvasculature system, oxygen availability to luteal cells in the regressing CL may still be adequate since blood flow is reduced but not occluded completely (55). In any case, the conditions generated within the CL at the time of regression are likely associated with alterations in the oxidative state of the luteal cells.

Oxidative Stress and Cell Death

Reactive oxygen species (ROS; superoxide anion radical, hydroxyl radical) and their intermediates (hydrogen peroxide) are generated in all cells as a consequence of normal cellular metabolism and hormone-mediated signaling events involving lipid membrane turnover (92–94). A superoxide radical is formed by the addition of a single electron to molecular oxygen. Addition of a second electron to the newly formed superoxide radical results in the formation of hydrogen peroxide which can be further reduced to form a hydroxyl radical. Reactive oxygen species are known to disrupt cellular function and homeostasis, and thus cells possess enzymatic defense mechanisms that serve to detoxify or metabolize ROS. One such enzyme, superoxide dismutase (SOD), exists in three forms: mitochondrial or manganese-containing (MnSOD), cytosolic or copper/zinc-containing (Cu/ZnSOD), and secreted (SecSOD). The SOD enzymes are the first lines of defense to protect cells from oxidative stress as these enzymes catalyze the conversion of superoxide anions to hydrogen peroxide (95). Once formed, hydrogen peroxide can be neutralized by its conversion to water via the actions of two other intracellular antioxidant enzymes, catalase and glutathione peroxidase (GSHPx) (96). If the function of one or more of the defense enzymes is impaired, the subsequent rise in ROS can destabilize proteins, disrupt the plasma membrane and initiate DNA strand breaks (94, 97–100). The characteristics of cells exposed to prolonged oxidative stress resemble many of the morphological and biochemical features characteristic of apoptosis (101, 102), suggesting that high levels of ROS can initiate this form of cell death (97, 103–106).

Reactive Oxygen Species: A Link Between Functional and Structural Luteolysis

By analogy to other systems, we and others predicted that the events induced by increased ROS may trigger regression of the CL (74, 107–111). Laloraya et al. (112) proposed a role for ROS in luteal cell function after observing changes in the levels of superoxide radicals in rat ovarian tissue during the estrous cycle. Additionally, elevated levels of ROS in rat luteal cells (111, 113, 114) were measured during natural luteolysis or in response to $PGF_{2\alpha}$. The increase in ROS triggered by prostaglandin was biphasic with the first peak occurring within 10 min of $PGF_{2\alpha}$ injection, prior to the fall in plasma progesterone (114). Elevated levels of hydrogen peroxide have also been linked to a loss of steroidogenic capacity (109, 110) and protein synthesis (110) in rat luteal cells. A recent study documented that progesterone inhibited superoxide radical formation in mononuclear phagocytes within the CL (115). Although these results provide evidence of a role for ROS in triggering functional regression of the CL, their involvement in apoptosis induction during structural regression has not been defined. It may be that the initial increase in ROS, and the ensuing decline in steroidogenesis, may prime cells of the CL for further damage by removing the inhibitory effect of progesterone on superoxide radical formation. In view of the tendency for oxidatively-stressed cells to undergo apoptosis (94, 97, 100), similar events initiated by uterine-derived $PGF_{2\alpha}$ in luteal cells could provide a point of passage from functional to structural regression.

Our laboratories have taken another tact to investigate the role of oxidative stress in luteolysis by determining the changes in the relative amounts of mRNA encoding enzymes that provide protection against ROS. Using Northern blot analysis we demonstrated that functional CL of the bovine ovary contained relatively high levels of mRNA encoding SecSOD and MnSOD when compared to their respective levels in regressing CL (74). Similarly, levels of mRNA encoding catalase were also higher in functional CL as compared with regressing CL of the bovine ovary (74). These results, in conjunction with a study that reported a decrease in SOD activity in the human CL during the late luteal phase (116), suggest there is a reduced capacity of the CL to respond to ROS during the time of luteal regression.

A subsequent study by our laboratories was then conducted to determine if levels of mRNAs which encode the oxidative stress response enzymes varied throughout the luteal phase of the estrous cycle in a closely related species (sheep). Similar to the bovine CL, the ovine CL contained less mRNA encoding Cu/ZnSOD during luteolysis when compared to those mRNA levels measured in functional CL collected during the midluteal phase of the estrous cycle. However, no differences were observed in the amount of mRNA encoding either catalase or GSHPx in the ovine CL throughout the estrous cycle and pregnancy (Rueda et al., manuscript submitted). Nevertheless, the decline in mRNAs that encode enzymes responsible for the conversion of superoxide anions to hydrogen peroxide may reduce the ability of the luteal cells to maintain an adequate or sustained response to the damaging effects of ROS. Decreased expression of oxidative stress response genes were similarly reported in granulosa cells undergoing apoptosis during follicular atresia in the rat ovary (106). Furthermore, the addition of exogenous anti-oxidant enzymes to ovarian follicles cultured in vitro suppresses tropic hormone deprivation-induced apoptosis in granulosa cells (106), supporting the hypothesis that ROS can in fact lead to ovarian cell demise.

It should be pointed out that the regulation of luteal regression is not limited to oxidative stress response enzymes. Heat shock protein-70 (HSP-70), another stress-associated protein, may also play a role in mediating luteolysis (Chapter 13;117–119). For example, HSP-70 was increased in the ovine CL within 2 h following administration of $PGF_{2\alpha}$. This increase in HSP-70 was observed prior to the decline in concentrations of progesterone, and was histochemically localized to large luteal cells (117, 118). Furthermore, the induction of heat shock protein has been shown to block hormone-sensitive steroidogenesis in rat luteal cells (119). At this point, however, it has not been determined whether or not heat shock proteins play a role in apoptosis during structural regression of the CL.

Conserved Intracellular Regulators of Cell Death

In recent years, several new genes have been identified that appear to function as key regulators of cell death or cell survival. The gene encoding BCL-2, an inhibitor of apoptosis in a wide variety of cell types, was originally identified at the breakpoint of the translocation between chromosomes 14 and 18 in human B-cell lymphoma (120–122). Over expression of BCL-2 in B cells and certain cytokine-dependent lymphoid cell lines inhibits apoptosis (123–125). The inhibitory effects of BCL-2 on the induction of apoptosis, however, are not observed in all cells in response to all stimuli (126). The BCL-2 protein is found in the outer membranes of the mitochondria, nucleus and endoplasmic reticulum (127–130). Interestingly, these sites are recognized as areas of free radical generation (131), and BCL-2 can inhibit cell death via antioxidant actions (131, 132). However, whether the antioxidative function of BCL-2 is a direct or indirect effect is unknown (133).

Since the identification of bcl-2, several bcl-2-related genes have been identified. The BAX protein, which is a BCL-2 antagonist, can form heterodimers with BCL-2 or homodimers with itself. It is now thought that BAX homodimers, which form in the absence of BCL-2, directly accelerate apoptosis (134). The ratio of BAX to BCL-2 is critical in determining cell fate with BAX believed to be the active determinant in enhancing cell death susceptibility (135). Along with the identification of BAX, a third BCL-2-like protein, termed BCL-X, was also identified to act as a regulator of apoptotic cell death (136). Furthermore, two forms of BCL-X, derived from alternative splicing, are produced in cells. Like BCL-2, expression of BCL-X$_{LONG}$ inhibits apoptosis in some systems. In contrast, BCL-X$_{SHORT}$, a truncated form of the protein, inhibits the ability of BCL-2 to enhance cell survival and thus has been described as a BAX-like protein (136).

There is additional evidence that interactions between the products of tumor suppressor genes and proto-oncogenes determine cell fate (134, 137, 138). Therefore, other regulators of cell death must be considered when discussing physiological cell death. For instance, the p53 protein has a dual role as a cell cycle regulator and cell death inducer. Following cellular DNA damage, increased levels of p53 prevent the progression of the cell cycle during the G1 to S phase transition (139), allowing the cell time to repair the genomic damage (140). If the damage is too extensive, the continued elevation of p53 can initiate apoptosis, presumably to maintain cellular homeostasis and prevent potential tumor development (141–144). Although it is not known how p53 initiates apoptosis, recent data suggest that the p53 tumor suppressor protein binds to specific enhancer response elements in the *bax* gene promoter leading to increased transcription of the *bax* gene (145). In addition to increasing *bax* gene expression, p53 also inhibits the expression of *bcl-2* through transcriptional repression (146).

Many of the genes associated with cell death regulation are expressed in specific cell types of the ovary. Increased expression of the *bax* death gene is associated with the initiation of apoptosis in granulosa cells during follicular atresia (147). In support of these observations, *bax* gene knock-out mice display a defect in the normal induction of apoptosis in granulosa cells of follicles destined for atresia (148). Thus, BAX may be one of the first factors identified that is necessary for granulosa cell death. In addition to BCL-2 family members, recent data have shown that p53 protein accumulates in the nuclei of rat granulosa cells during apoptosis and follicular atresia (138). Consequently, it may be that the translocation of p53 to the nuclei of ovarian cells destined for apoptosis leads to increased expression of the *bax* death-susceptibility gene (149). Therefore, we proposed that p53 may similarly regulate *bax* expression during luteolysis. Likewise, other genes associated with the control of cell death may play a role in determining luteal cell fate.

BCL-2-Related Genes in the CL

Since the discovery of specific genes directly associated with apoptosis induction or repression, several investigators have hypothesized that cell death genes may also encode regulators of cell death in the CL. A *bcl-2* cDNA was recently cloned from an ovine luteal cell cDNA library, and expression of the gene in luteal tissue was verified by ribonuclease protection assay (150). In addition, by using immunohistochemistry, positive staining for BCL-2 protein was found primarily in steroid producing cells of the human CL during the luteal phase of the menstrual cycle (151). However, no differences were observed in the intensity or localization of BCL-2 immunostaining in the CL throughout the luteal phase. Although expression of *bcl-2* in luteal cells suggests a potential role in luteal function, the lack of change in expression of BCL-2 throughout the luteal phase would imply that factors in addition to this one protein are needed for the regulation of luteal cell demise.

Recent studies have discovered that other genes known to accelerate or inhibit apoptosis are expressed in the ovine and bovine CL. Furthermore, some of these genes are differentially expressed in functional versus regressing CL. For instance, during the bovine estrous cycle elevated levels of *bax* mRNA are evident in the CL during luteolysis (152). Although expressed, levels of *bcl-X* were not different based on Northern blot analysis. Since the original probe used to determine *bcl-x* mRNA levels did not distinguish between the long and truncated (short) form, the RT-PCR technique was used and revealed that the long form was the predominant message expressed in the bovine CL; however, there were no changes in the relative levels of $bcl\text{-}x_{LONG}$ versus $bcl\text{-}x_{SHORT}$ during luteolysis (152). These findings are reminiscent of data obtained from analysis of apoptosis in granulosa cells during atresia that have implicated changes in *bax* as a primary determinant of cell fate (81, 147, 148).

Although there were no pronounced differences in the amount of p53 mRNA in the bovine CL when comparing the functional CL of pregnancy and the regressing CL of the estrous cycle, there was a slight (21%) elevation in p53 mRNA levels during luteolysis (15). However, in the rat minimal changes in p53 mRNA levels were associated with dramatic changes in the amount of p53 protein (138). Since elevated p53 is associated with DNA damage and p53 accumulates in the nucleus of the compromised cell in response to DNA damage (142, 153), it is possible that uncontrolled oxidative stress in the CL is associated with p53 translocation to the nucleus and an ensuing increase in *bax* expression as a trigger for luteal cell apoptosis. Although the exact role of BAX and p53 in structural luteal regression remains to be determined, current data support the concept that p53 accumulates in the nuclei of ovarian cells destined for apoptosis (138), and that this increase in p53 is directly associated with an induction of cell death (154).

To complement studies conducted with bovine luteal tissues, we also initiated experiments to characterize changes in p53, *bax,* and *bcl-x* expression in the ovine CL throughout the luteal phase of the estrous cycle and during pregnancy. Based on data obtained from the analysis of the bovine CL, we hypothesized that the levels of p53 and *bax* would be highest, and levels of $bcl\text{-}x_{LONG}$ lowest, during luteolysis. Unexpectedly, however, the highest levels of p53 and *bax* were observed during the mid-luteal phase, a time at which the lowest levels of $bcl\text{-}x_{LONG}$ mRNA were detected (Rueda et al., manuscript submitted). Interestingly, the elevated levels of *bax* and p53, and the decreased levels of $bcl\text{-}x_{LONG}$, mRNA occurred during a period that corresponds to the time of maternal recognition of pregnancy in the ewe. It is, therefore, tempting to speculate that the decision for the initiation of apoptosis in the CL, an event driven by prior alterations in cell death gene expression, is initiated as early as days 12 to 13 of the ovine estrous cycle, a critical period for interactions between the conceptus and the CL.

Role of Ice and Related Proteases in Cell Death

Recent evidence has suggested that specific cytoplasmic proteases are directly involved in the execution of apoptosis. Furthermore, these proteases have been described as fundamental players in the final steps of the "death pathway" (155). Many of the advances in linking proteases to apoptosis were derived from studies using the nematode, *Caenorhabditis elegans*. In *C. elegans*, proteins encoded by the *ced-3* and *ced-4* genes promote apoptosis whereas the *ced-9* gene inhibits their function and protects cells from death (156, 157). Cloning of the *ced-3* gene enabled the identification of its mammalian counterpart, interleukin-1β-converting enzyme (ICE), which shares both sequence and functional homology to CED-3 (156, 158). The cysteine protease, ICE, was previously known for its role in converting prointerleukin -1β to mature interleukin -1β (159). However, consistent with its deduced function from homology comparisons to CED-3, forced expression of either ICE or CED-3 in fibroblasts resulted in apoptosis (160). The activity of ICE is inhibited by the poxvirus, CrmA (161), and overexpression of CrmA prevents apoptosis in certain tumor cell lines following incubation with tumor necrosis factor-alpha (TNFα) or ligation of the FAS antigen, two stimuli previously shown to initiate cell death (162, 163). Although these studies suggest a conserved role for ICE in cell death induction, a recent study demonstrated that *Ice* gene knock-out mice exhibit defects in the induction of apoptosis only in cells of immune origin (164). This finding initially suggested that there may be other CED-3-related enzymes in vertebrates that are involved in apoptosis.

The isolation of additional CED-3/ICE homologs has provided support that ICE-related proteases play a primary role in the initiation of apoptosis. The CED-3 homolog, Nedd2, was first identified in the mouse central nervous system (165). The human homolog of Nedd2 was later isolated and named ICE/CED-3-homolog-1 or ICH-(166). It was also determined that alternative splicing resulted in two forms of Ich-1mRNA (long and short). Overexpression of Ich-1$_{LONG}$ initiates apoptosis in a number of cell lineages, although the degree of apoptosis induction is dependent on cell type (166). In contrast, overexpression of a truncated form of ICH-1, termed Ich-1$_{SHORT}$, antagonizes the function of the long isoform by inhibiting apoptosis initiated following serum deprivation (166). Interestingly, BCL-2, and to a lesser extent CrmA, can inhibit apoptosis induced by forced expression of Ich-1$_{LONG}$ (166). Another ICE homolog, CPP32, was cloned from a human T cell line (167). The CPP32 enzyme is also structurally and functionally similar to ICE, and recent data suggest that CPP32 may be one of the most important members of the ICE family in mediating the induction of apoptosis (168). In addition to gene transfer experiments that demonstrated an induction of cell death following CPP32 transfection, other investigations have shown that poly(ADP)-ribose polymerase, an enzyme involved in DNA repair, is just one of many proteins cleaved by CPP32 in association with apoptosis (169, 170).

Role of Ice-Related Proteases in Luteal Regression

In light of the discussions above, the physiological roles of ICE and related enzymes were evaluated relative to the onset of granulosa cell demise in the rat (171) and mouse (171) ovary. Results of these studies indicated that ICH-1 and CPP32, but not ICE per se, were involved in mediating granulosa cell death during ovarian follicular atresia. Subsequently, we hypothesized that ICE or related enzymes of the CED-3 family were involved in regulating apoptosis during structural luteal regression. Published and unpublished results from our laboratories demonstrated that genes encoding ICE, CPP32, and ICH-1 were expressed in the bovine and ovine CL. Furthermore, higher levels of mRNA for ICE were found in regressed CL when compared to functional CL of cows on the same day of the estrous cycle or pregnancy (152). Similarly, in the ovine ovary, there were significantly higher levels of *Ice* mRNA in CL undergoing luteolysis when compared to functional CL collected from pregnant ewes. In contrast, there were no differences in the relative amounts of mRNA for CPP32 or ICH-1 in the ovine or bovine CL when comparing functional to luteolytic tissue (172). Following the administration of $PGF_{2\alpha}$ on day 10, however, there were elevated levels of CPP32 mRNA in the ovine CL when compared to those levels observed in CL collected from day 10 nontreated ewes. Therefore, ICE and related proteases may play a role in regulating the lifespan of the CL in cattle and sheep. Furthermore, since ICE-like proteases are proposed to be located downstream in the overall sequence of events leading to cell death (173), we believe that these death proteases play an active role in luteal regression at a point distal to p53 and BAX.

Conclusion

The inhibition of steroid synthesis by $PGF_{2\alpha}$ (functional regression) and the eventual disruption of cellular homeostasis (structural regression), collectively described as luteolysis, occur as a result of the initiation of a number of synchronous events. During the initial phase of luteal regression there is a rapid increase in ROS leading to oxidative stress, decreased blood flow and loss of progesterone synthesis, followed by vascular involution and a disruption of cellular integrity. These events are correlated with a rapid influx of immune cells that may play an important role in luteal regression, particularly the "clean-up" of apoptotic bodies via phagocytosis. More recently, expression of genes that encode proteins known to regulate the cell cycle or cell survival have been identified within the CL. We hypothesize that several regulators of cell cycle or cell survival play an active role in luteal maintenance (anti-oxidant enzymes, $BCL-X_{LONG}$), whereas others are involved in apoptosis during luteal regression (p53, BAX, ICE-like proteases). Additional information regarding the cellular events that occur after the inhibition of steroid synthesis but before the disruption of the cellular integrity are critical to our understanding of the mechanisms that are involved in luteolysis. More specifically, it is imperative that further studies are implemented to precisely determine which cell types express the death regulatory genes and to elucidate how their cognate proteins mediate the fate of the luteal cells.

References

1. Farin CE, Moeller CL, Sawyer HR, Gamboni F, Niswender GD. Morphometric analysis of cell types in the ovine corpus luteum throughout the estrous cycle. Biol Reprod 1986;35:1299–1308.
2. O'Shea JD, Rodgers RJ, Wright PJ. Cellular composition of the sheep corpus luteum in the mid- and late-luteal phases of the oestrous cycle. J Reprod Fertil 1986;76: 685–91.
3. O'Shea JD, Rodgers RJ, D'Occhio MJ. Cellular composition of the cyclic corpus luteum of the cow. J Reprod Fertil 1989;85:483–7.
4. Ohara A, Mori T, Taii S, Ban C, Narimoto K. Functional differentiation in steroidogenesis of two types of luteal cells isolated from mature human corpora lutea of menstrual cycle. J Clin Endocr Met 1987;65:1192–1200.
5. Lei ZM, Chegini N, Rao CV. Quantitative cell composition of human and bovine corpora lutea from various reproductive states. Biol Reprod 1991;44:1148–56.
6. Retamales I, Carrasco I, Troncoso JL, Las Heras J, Devato L, Vega M. Morphofunctional study of human luteal cell subpopulations. Human Reprod 1994; 9:591–6.
7. Hild-Petito S, Ottobre AC, Hoyer PB. Comparison of subpopulations of luteal cells obtained from cyclic and superovulated ewes. J Reprod Fertil 1987;80:537–44.
8. Smith CJ, Greer TB, Banks TW, Sridaran R. The response of large and small luteal cells from the pregnant rat to substrates and secretagogues. Biol Reprod 1989;41:1123–32.
9. Lemon M, Mauleon P. Interaction between two luteal cell types from the corpus luteum of the sow in progesterone synthesis in vitro. J Reprod Fertil 1982;64:315–23.
10. Koos RD, Hansel W. The large and small cells of the bovine corpus luteum: ultrastructural and functional differences. In: Schwartz NB, Hunzicker-Dunn M eds. Dynamics of ovarian function. New York: Raven Press, 1982:197–203.
11. O'Shea JD, Cran DG, Hay MF. The small luteal cell of the sheep. J Anat 1979;128:239–51.
12. Rodgers RJ, O'Shea JD. Purification, morphology and progesterone production and content of three cell types isolated from the corpus luteum of the sheep. J Biol Sci 1982;35:441–55.
13. Fitz TA, Mayan HR, Sawyer HR, Niswender GD. Characterization of two steroidogenic cell types in the ovine corpus luteum. Biol Reprod 1982;27:703–11.
14. Corner GW. On the origin of the corpus luteum of the sow from both granulosa and theca internal. Am J Anat 1919;26:117–83.
15. McClellan MC, Diekman MA, Abel JH, Niswender GD. Luteinizing hormone, progesterone and the morphological development of normal and superovulated corpora lutea in sheep. Cell Tissue Res 1975;164:291–307.

16. Wilkinson RF, Anderson G, Aalberg J. Cytological observations of dissociated rat corpus luteum. J Ultrastruct Res 1976;57:168–84.

17. O'Shea JD, Cran DG, Hay MF. Fate of the theca interna following ovulation in the ewe. Cell Tissue Res 1980;210:305–19.

18. Chegini N, Ramani N, Rao CV. Morphological and biochemical characterization of small and large bovine luteal cells during pregnancy. Mol Cell Endocr 1984;37:89–102.

19. Fields MJ, Dubois W, Fields PA. Dynamic features of luteal secretory granules: ultrastructural changes during the course of pregnancy in the cow. Endocrinology 1985;117:1675–81.

20. Rodgers RJ, O'Shea JD, Findlay JK. Do small and large luteal cells of the sheep interact in the production of progesterone? J Reprod Fertil 1985;75:85–94.

21. Meyer GT, Bruce NW. The cellular pattern of corpus luteal growth during pregnancy in the rat. Anat Rec 1979;193:823–30.

22. Sawyer HR, Niswender KD, Braden TD, Niswender GD. Nuclear changes in ovine luteal cells in response to $PGF_2\alpha$. Dom Anim Endocrinol 1990;7:229–38.

23. Lobel BL, Levy E. Enzymatic correlates of development, secretory function and regression of follicles and corpora lutea in the bovine ovary. Acta Endocrinol Suppl 1968;132:5–63.

24. Bagavandoss P, Kunkel SL, Wiggins RC, Keyes PL. Tumor necrosis factor-α (TNFα) production and localization of macrophages and T-lymphocytes in the rabbit corpus luteum. Endocrinology 1988;122:1185–7.

25. Brannstrom M, Norman RJ. Involvement of leukocytes and cytokines in the ovulatory process and corpus luteum function. Human Reprod 1993;8:1762–75.

26. Brannstrom M, Pascoe V, Norman RJ, McClure N. Localization of leukocyte subsets in the follicle wall and in the corpus luteum throughout the human menstrual cycle. Fertil Steril 1994;61:488–95.

27. Hoyer PB, Fitz TA, Niswender GD. Hormone-independent activation of adenylate cyclase in large steroidogenic ovine luteal cells does not result in increased progesterone secretion. Endocrinology 1984;114:604–8.

28. Harrison LM, Kenny N, Niswender GD. Progesterone production: LH receptors and oxytocin secretion by ovine luteal cell types on days 6, 10 and 15 of the estrous cycle and day 25 of pregnancy. J Reprod Fertil 1987;79:539–48.

29. Hoyer PB, Niswender GD. The regulation of steroidogenesis is different in the two types of ovine luteal cells. Can J Physiol Pharmacol 1985;63:240–8.

30. Hoyer PB, Niswender GD. Adenosine 3', 5'-monophosphate-binding capacity in small and large ovine luteal cells. Endocrinology 1986;119:1822–9.

31. Pate JL, Nephew KP. Effects of in vivo and in vitro administration of prostaglandin $F_{2\alpha}$ on lipoprotein utilization in cultured bovine luteal cells. Biol Reprod 1988;38:568–76.

32. Hutchinson JS, Zeleznik AJ. The corpus luteum of the primate menstrual cycle is capable of recovering from a transient withdrawal of pituitary gonadotropin support. Endocrinology 1985;117:1043–9.

33. Keyes PL, Possley RM, Yuh K-CM. Contrasting effects of oestradiol-17β and human chorionic gonadotropin on steroidogenesis in the rabbit corpus luteum. J Reprod Fertil 1983;69:579–86.

34. Fraser HM, Nestor JJ, Vickery BH. Suppression of luteal function by a luteinizing hormone-releasing hormone antagonist during the early luteal phase in the stumptailed macaque monkey, and the effects of subsequent administration of human chorionic gonadotropin. Endocrinology 1987;121:612–8.

35. McGuire WJ, Juengal JL, Niswender GD. Protein kinase C second messenger system mediates the antisteroidogenic effects of prostaglandin $F_{2\alpha}$ in the ovine corpus luteum in vivo. Biol Reprod 1994;51:800–6.

36. Hauger RL, Karsch FJ, Foster DL. A new concept for control of the estrous cycle of the ewe based on the temporal relationships between luteinizing hormone, estradiol and progesterone in peripheral serum and evidence that progesterone inhibits tonic LH secretion. Endocrinology 1977;101:807–17.

37. McCracken JA, Glew ME, Scaramuzzi RJ. Corpus luteum regression induced by prostaglandins F^2_α. J Clin Endocrinol Met 1970;30:544–6.

38. McCracken JA, Carlson JC, Glew ME, Goding JR, Baird DT, Green K, Samuelsson B. Prostaglandin $F_{2\alpha}$ identified as a luteolytic hormone in sheep. Nature New Biol 1972;83:527–36.

39. Kimball FA, Lauderdale JW. Prostaglandin E_1 and F_{2alpha} specific binding in bovine corpora lutea: comparison with luteolytic effects. Prostaglandins 1975;10:313–31.

40. Phariss BB, Wyngarden LJ. The effect of prostaglandin $F_{2\alpha}$ on the progesterone content of ovaries from pseudopregnant rats. Proc Soc Exp Biol Med 1969;130:92–4.

41. Bennegard B, Hahlin M, Wennberg E, Noren H. Local luteolytic effect of prostaglandin F_{2alpha} in the human corpus luteum. Fertil Stertil 1991;56:1070–6.

42. Gelety TJ, Chaudhuri G. Prostaglandins in the ovary and fallopian tube. Baillieres Clin Obstet Gynecol 1992;6:707–29.

43. Pexton JE, Weems CW, Inskeep EK. Prostaglandin F in uterine venous plasma, ovarian arterial and venous plasma and in ovarian and luteal tissue of pregnant and non–pregnant ewes. J Anim Sci 1975;41:154–9.

44. Nett TM, Staigmiller RB, Akbar AM, Diekman NA, Ellinwood WE, Niswender GD. Secretion of prostaglandin F_{2alpha} in cycling and pregnant ewes. J Anim Sci 1976;42:876–80.

45. Lewis GS, Wilson L Jr, Wilks JW, Pexton JE, Fogwell Rl, Ford SP, Butcher RL, Thayne WV, Inskeep EK. $PGF_{2\alpha}$ and its metabolites in uterine and jugular venous plasma and endometrium of ewes during early pregnancy. J Anim Sci 1977; 45:320–27.

46. Zarco L, Stabenfeldt GH, Basu S, Bradford GE, Kindahl H. Modification of prostaglandin $F_{2\alpha}$ synthesis and release in the ewe during the initial establishment of pregnancy. J Reprod Fertil 1988;83:527–36.

47. Nett TM, McClellan MC, Niswender GD. Effects of prostaglandins on the ovine corpus luteum: blood flow, secretion of progesterone and morphology. Biol Reprod 1976;15:66–78.

48. Del Campo CH, Ginther OJ. Vascular anatomy of the uterus and ovaries and the unilateral luteolytic effect of the uterus: horses, sheep and swine. Am J Vet Res 1973;34:300–16.

49. Heap RB, Fleet IR, Hamon M. Prostaglandin $F2\alpha$ is transferred from the uterus to the ovary in the sheep lymphatic and blood vascular pathways. J Reprod Fertil 1985;74:645–56.

50. Michael AE, Abayasekara DRE, Webley GE. Cellular mechanisms of luteolysis. Mol Cell Endocr 1994;99:R1–R9.

51. Pate JL. Cellular components involved in luteolysis. J Anim Sci 1994;72:1884–90.

52. Keyes PL, Wiltbanlk MC. Endocrine regulation of the corpus luteum. Annu Rev Physiol 1988;50:465–82.

53. Knickerbocker JJ, Wiltbank MC, Niswender GD. Mechanisms of luteolysis in domestic livestock. Dom Anim Endocrinol 1988;5:91–107.

54. Wiltbank MC, Gallagher KP, Christensen AK, Brabec RK, Keyes PL. Physiological and immunological cytochemical evidence for a new concept of blood flow regulation in the corpus luteum. Biol Reprod 1990;42:139–49.
55. Wiltbank MC. Cell types and hormonal mechanisms associated with mid-cycle corpus luteum function. J Anim Sci 1994;72:1873–83.
56. Balapure AK, Caicedo IC, Kawada K, Watt DS, Rexroad CE Jr, Fitz TA. Multiple classes of prostaglandin $F_{2\alpha}$ binding sites in subpopulations of ovine luteal cells. Biol Reprod 1989;41:385–92.
57. Wiepz GJ, Wiltbank MC, Nett TM, Niswender GD, Sawyer HR. Receptors for prostaglandins $F_{2\alpha}$ and E_2 in ovine corpora lutea during maternal recognition of pregnancy. Biol Reprod 1992;47:984–91.
58. Powell WS, Hammarstrom S, Samuelsson B. Prostaglandin $F_{2\alpha}$ receptor in ovine corpora lutea. Eur J Biochem 1974;41:103–7.
59. Wiltbank MC, Guthrie PB, Mattson MP, Kater SB, Niswender GD. Hormonal regulation of free intracellular calcium concentrations in small and large ovine luteal cells. Biol Reprod 1989;41:771–78.
60. Wegner JA, Martinez-Zaguilan R, Wise ME, Gillies RJ, Hoyer PB. Prostaglandin $F_{2\alpha}$-induced calcium transient in ovine large luteal cells: I. Alterations in cytosolic free calcium levels and calcium flux. Endocrinology 1990;127:3029–37.
61. Wegner JA, Martinez-Zaguilan R, Gillies RJ, Hoyer PB. Prostaglandin $F_{2\alpha}$-induced calcium transient in ovine large cells: II. Modulation of the transient and resting cytosolic free calcium alters progesterone secretion. Endocrinology 1991;128:929–36.
62. Geeraerts MD, Ronveaux-Dupal MF, Lemasters JJ, Herman B. Cytosolic free Ca^{2+} and proteolysis in lethal oxidative injury in endothelial cells. Am J Physiol 1991;261:C889–96.
63. Richter C, Kass GEN. Oxidative stress in mitochondria: its relationship to cellular Ca^{2+} homeostasis, cell death, proliferation, and differentiation. Chem Biol Interactions 1991;77:1–23.
64. Trump BF, Berezesky IK. The role of cytosolic Ca^{2+} in cell injury, necrosis and apoptosis. Cur Opinion Cell Biol 1992;4:227–32.
65. Abdul-Karim RW, Bruce N. Blood flow to the ovary and corpus luteum at different stages of gestation in the rabbit. Fertil Steril 1973;24:44–47.
66. Wiltbank MC, Gallagher KP, Dysko RC, Keyes PL. Regulation of blood flow to the rabbit corpus luteum: effects of estradiol and human chorionic gonadotropin. Endocrinology 1989;124:605–11.
67. Bruce NW, Meyer GT, Dharmarajan AM. Rate of blood flow and growth of the corpora lutea of pregnancy and of previous cycles throughout pregnancy in the rat. J Reprod Fertil 1984;71:445–52.
68. Novy MJ, Cook MJ. Redistribution of blood flow by prostaglandin $F_{2\alpha}$ in the rabbit ovary. Am J Obstet Gynecol 1973;117:381–5.
69. Nett TM, Niswender GD. Luteal blood flow and receptors for LH during $PGF_{2\alpha}$-induced luteolysis: production of PGE_2 and $PGF_{2\alpha}$ during early pregnancy. Acta Vet Scand, Suppl 1981;77:117–30.
70. Pang CY, Behrman HR. Relationship of luteal blood flow and corpus luteum function in pseudopregnant rats. Am J Physiol 1979;237:E30–E4.
71. Pate JL. Involvement of immune cells in regulation of ovarian function. J Reprod Fertil Suppl 1995;49:365–77.
72. Juengel JL, Garverick HA, Johnson AL, Youngquist RS, Smith MF. Apoptosis during luteal regression in cattle. Endocrinology 1993;132:249–54.

73. Rueda BR, Wegner JA, Marion SL, Wahlen DD, Hoyer PB. Internucleosomal DNA fragmentation in ovine luteal tissue associated with luteolysis: in vivo and in vitro analyses. Biol Reprod 1995;52:305–12.
74. Rueda BR, Tilly KI, Hansen TR, Hoyer PB, Tilly JL. Expression of superoxide dismutase, catalase and glutathione peroxidase in the bovine corpus luteum: evidence supporting a role for oxidative stress in luteolysis. Endocrine 1995;3:227–32.
75. Rotello RJ, Lieberman RC, Lepoff RB, Gerschenson LE. Characterization of uterine epithelium apoptotic cell death kinetics and regulation by progesterone and RU486. Am J Path 1992;140:449–56.
76. Hopwood D, Levison DA. Atrophy and apoptosis in the cyclical human endometrium. J Pathol 1976;119:159–66.
77. Welsh AO. Uterine cell death during implantation and early placentation. Microsc Res Tech 1993;25:223–45.
78. Walker NI, Bennet RE, Kerr JFR. Cell death by apoptosis during involution of the lactating breast in mice and rats. Am J Anat 1989;185:19–32.
79. Pesce M, De Filci M. Apoptosis in mouse primordial germ cells: a study by transmission and scanning electron microscope. Anat Embryo 1994;189:435–40.
80. Ratts VS, Flaws JA, Kolp R, Sorenson CM, Tilly JL. Ablation of *bcl-2* gene expression decreases the numbers of oocytes and primordial follicles established in the post-natal female mouse gonad. Endocrinology 1995;136:3665–8.
81. Tilly JL, Kowalski KI, Johnson AL, Hsueh AJW. Involvement of apoptosis in ovarian follicular atresia and postovulatory regression. Endocrinology 1991;129:2799–801.
82. Hughes FM Jr, Gorospe WC. Biochemical identification of apoptosis (programmed cell death) in granulosa cells: evidence for a potential mechanism underlying follicular atresia. Endocrinology 1991;129:2415–22.
83. Tilly JL, Ratts VS. Biological and clinical importance of ovarian cell death. Contemp Obstet Gynecol 1996;41:59–86.
84. Tilly JL. Apoptosis and ovarian function. Rev Reprod 1996;1:162–72.
85. O'Shea JD, Nightingale MG, Chamley WA. Changes in small blood vessels during cyclical luteal regression in sheep. Biol Reprod 1977;17:162–77.
86. Dharmarajan AM, Goodman SB, Tilly KI, Tilly JL. Apoptosis during functional corpus luteum regression: evidence of a role for chorionic gonadotropin in promoting luteal cell survival. Endocrine 1994;2:295–303.
87. O'Shea JD, McCoy K. Weight, composition, mitosis, cell death and content of progesterone and DNA in the corpus luteum of pregnancy in the ewe. J Reprod Fertil 1988;83:107–17.
88. Zheng J, Fricke PM, Reynolds LP, Redmer DA. Evaluation of growth, cell, proliferation, and cell death in bovine corpora lutea throughout the estrous cycle. Biol Reprod 1994;51:623–32.
89. Niswender GD, Reimers TJ, Diekman MA, Nett TM. Blood flow: a mediator of ovarian function. Biol Reprod 1976;14:64–81.
90. Azmi TI, O'Shea JD. Mechanism of deletion of endothelial cells during regression of the corpus luteum. Lab Invest 1984;51:206–17.
91. Braden TD, Gamboni F, Niswender GD. Effects of prostaglandin F_{2a}-induced luteolysis on the populations of cells in the ovine corpus luteum. Biol Reprod 1988;39:245–53.
92. Flohe L. Glutathione peroxidase brought into focus. In: Pryor WA, ed. Free Radicals in Biology. New York: Academic Press, 1982;5:223–54.

93. Fridovich I. Biological effects of the superoxide radical. Arch Biochem Biophys 1986;247:1–11.
94. Yu BP. Cellular defenses against damage from reactive oxygen species. Physiol Rev 1994;74:139–62.
95. Fridovich I. Superoxide dismutases: defence against endogenous superoxide radical. Ciba Foundation Symposium 1978;65:77–93.
96. Blum J, Fridovich I. Inactivation of glutathione peroxidase by superoxide radical. Arch Biochem Biophys 1985;240:500–8.
97. McConkey DJ, Hartzell P, Nicotera P, Wyllie AH, Orrenius S. Stimulation of endogenous endonuclease in cells exposed to oxidative stress. Toxicol Lett 1989;42:123–30.
98. Wong GHW, Elwell JH, Oberly LW, Goeddel DV. Manganous superoxide dismutase is essential for cellular resistance to cytotoxicity of tumor necrosis factor. Cell 1989;58:923–31.
99. Hockenbery DM, Oltvai ZN, Yin XM, Milliman CL, Korsmeyer SJ. Bcl-2 functions in an antioxidant pathway to prevent apoptosis. Cell 1993;75:241–51.
100. Buttke TM, Sandstrom PA. Oxidative stress as a mediator of apoptosis. Immunol Today 1994;15:7–10.
101. Kerr JFR, Wyllie AH, Currie AR. Apoptosis: a basic biological phenomenon with wide ranging implications in tissue kinetics. Br J Cancer 1972;26:239–57.
102. Wyllie AH, Kerr JFR, Currie AR. Cell death: the significance of apoptosis. Int Rev Cytol 1980;68:251–306.
103. Schraufstatter IU, Hyslop PA, Hinshaw DB, Spragg RG, Sklar LA, Cochrane CG. Hydrogen peroxide-induced injury of cells and its prevention by inhibitors of poly (ADP-ribose) polymerase. Proc Natl Acad Sci USA 1986;83:4908–12.
104. Carson DA, Seto S, Wasson DB, Carrera CJ. DNA strand breaks, NAD metabolism and programmed cell death. Exp Cell Res 1986;164:273–81.
105. Janssen YMW, Van Houten B, Borm PJA, Mossman BT. Biology of disease: cell and tissue responses to oxidative damage. Lab Invest 1993;69:261–74.
106. Tilly JL, Tilly KI. Inhibitors of oxidative stress mimic the ability of follicle-stimulating hormone to suppress apoptosis in cultured rat ovarian follicles. Endocrinology 1995;136:242–52.
107. Gatzuli E, Aten RF, Behrman HR. Inhibition of gonadotropin action and progesterone synthesis by xanthine oxidase in rat luteal cells. Endocrinology 1991;128:2253–58.
108. Carlson JC, Wu XM, Sawada M. Oxygen radicals and the control of the corpus luteum function. Free Rad Biol Med 1993;14:79–84.
109. Behrman HR, Aten RF. Evidence that hydrogen peroxide blocks hormone-sensitive cholesterol transport into mitochondria of rat luteal cells. Endocrinology 1991;128:2958–66.
110. Musicki B, Aten RF, Behrman HR. Inhibition of protein synthesis and hormone-sensitive steroidogenesis in response to hydrogen peroxide in rat luteal cells. Endocrinology 1994;134:588–95.
111. Sawada M, Carlson JC. Studies on the mechanism controlling generation of superoxide radical in luteinized rat ovaries during regression. Endocrinology 1994;135:1645–50.
112. Laloraya M, Pradeep KG, Laloraya MM. Changes in the levels of superoxide anion radical and superoxide dismutase during the estrous cycle of *Rattus norvegicus* and

induction of superoxide dismutase in the rat ovary by lutropin. Biochem Biophys Res Commun 1988;157:146–53.

113. Sawada M, Carlson JC. Superoxide radical production in plasma membrane samples from regressing rat corpora lutea. Can J Physiol Pharmacol 1989;67:465–71.

114. Sawada M, Carlson JC. Rapid plasma membrane changes in superoxide radical formation, fluidity, and phospholipase A_2 activity in the corpus luteum of the rat during induction of luteolysis. Endocrinology 1991;128:2992–98.

115. Sugino N, Shimamura K, Tamura H, Ono M, Nakamura Y, Ogino K, Kato H. Progesterone inhibits superoxide radical production by mononuclear phagocytes in pseudopregnant rats. Endocrinology 1996;137:749.

116. Vega M, Castillo M, Retamales I, Las Heras J, Devoto L, Videla LA. Steroidogenic capacity and oxidative stress-related parameters in human luteal cell regression. Free Rad Biol Med 1994;17:493.

117. Murdoch WJ. Treatment of sheep with prostaglandin $F_{2\ alpha}$ enhances production of a luteal chemoattractant for eosinophils. Am J Reprod Immunol Micro 1987;15:52–56.

118. McPherson LA, VanKirk EA, Murdoch WJ. Localization of stress protein-70 in ovine corpora lutea during prostaglandin-induced luteolysis. Prostaglandins 1993;46: 433–40.

119. Khanna A, Aten RF, Behrman HR. Heat shock protein induction blocks hormone-sensitive steroidogenesis in rat luteal cells. Steroids 1994;59:4–7.

120. Tsujimoto Y, Cossman J, Jaffe E, Croce CM. Involvement of the *bcl-2* gene in human follicular lymphoma. Science 1985;228:1440–43.

121. Bakhshi A, Jensen JP, Goldman P, Wright JJ, McBride OW, Epstein AL, Korsmeyer SJ. Cloning the chromosomal breakpoint of t(14;18) human lymphomas: clustering around J_H on chromosome 14 and near a transcriptional unit on chromosome 18. Cell 1985;41:899–906.

122. Cleary ML, Sklar J. Nucleotide sequence of a t(14;18) chromosomal breakpoint in follicular lymphoma and demonstration of a breakpoint-cluster region near a transcriptionally active locus on chromosome 18. Proc Natl Acad Sci USA 1985; 82;74309.

123. Vaux DL, Weissman IL, Kim SL. Prevention of programmed cell death in *Caenorhabditis elegans* by human *bcl-2*. Science 1992;258:1955–7.

124. Vaux DL, Cory S, Adams JM. *bcl-2* gene promotes haemopoietic cell survival and cooperates with c-*myc* to immortalize pre-B cells. Nature 1988;335: 440–2.

125. Nunez G, London L, Hockenberry D, Alexander M, McKern JP, Korsmeyer SJ. Deregulated *bcl-2* gene expression selectively prolongs survival of growth factor-deprived hematopoietic cell lines. J Immunol 1990;144:3602–10.

126. Reed JC. Bcl-2 and the regulation of programmed cell death. J Cell Biol 1994; 124:1–6.

127. Hockenbery DM, Nunez G, Millman CL, Schreiber RD, Korsmeyer SJ. Bcl-2 is an inner mitochondrial membrane protein that blocks programmed cell death. Nature 1990;348:334–36.

128. Chen-Levy Z, Nourse J, Cleary ML. The Bcl-2 candidate proto-oncogene product is a 24-kilodalton integral-membrane protein highly expressed in lymphoid cell lines and lymphomas carrying the t(14;18). Mol Cell Biol 1989;9:701–10.

129. Jacobson MD, Burne JF, King MP, Miyashita T, Reed JC, Raff MC. 1993 Bcl-2 blocks apoptosis in cells lacking mitochondrial DNA. Nature 1993;361:365–9.

130. Monaghan P, Robertson D, Amos TAS, Dyer MJS, Mason DY, Greaves MF. Ultrastructural localization of Bcl-2 protein. J Histochem Cytochem 1992;40: 1819–25.

131. Hockenbery DM, Oltavi ZN, Yin X-M, Milliman CL, Korsmeyer SJ. Bcl-2 functions in an antioxidant pathway to prevent apoptosis. Cell 1993;75:241–51.

132. Kane DJ, Sarafian TA, Anton R, Hajn H, Gralla EB, Valentine JS, Ord T, Bredesen DE. Bcl-inhibition of neural death: decreased generation of reactive oxygen species. Science 1993;262:1274–78

133. Korsmeyer SJ. Regulators of cell death. Trends Gene 1995;11:101–5.

134. Oltvai ZN, Milliman CL, Korsmeyer SJ. Bcl-2 heterodimerizes in vivo with a conserved homolog, Bax, that accelerates programmed cell death. Cell 1993;74: 609–19.

135. Korsmeyer SJ, Yin X-M, Oltvai ZN, Veis-Novack DJ, Linette GP. Reactive oxygen species and the regulation of cell death by the bcl-2 gene family. Bichim Biophys Acta 1995;1271:63–66.

136. Boise LH, Gonzalez-Garcia M, Postema CE, Ding L, Lindsten T, Turka LA, Mao X, Nunez G, Thompson CB. bcl-x, a bcl-2-related gene that functions as a dominant regulator of apoptotic cell death. Cell 1993;74:597–608.

137. Yin X-M, Oltvai ZN, Korsmeyer SJ. BH1 and BH2 domains of Bcl-2 are required for inhibition of apoptosis and heterodimerization with Bax. Nature 1994;369:321–3.

138. Tilly KI, Banerjee S, Banerjee PP, Tilly JL. Expression of the p53 and Wilm's tumor suppressor genes in the rat ovary: gonadotropin repression in vivo and immunohistochemical localization of nuclear p53 protein to apoptotic granulosa cells of atretic follicles. Endocrinology 1995;136:1394–402.

139. Mercer WE, Sheieds MT, Amin M, Sauve GJ, Appella E, Romano JW, Ullrich SJ. Negative growth regulation in a glioblastoma tumor cell line that conditionally expresses human wild-type p53. Proc Natl Acad Sci USA 1990;87:6166–70.

140. Martin SJ, Green DR, Cotter TG. Dicing with death: dissecting the components of the apoptosis machinery. TIBS 1994;19:26–30.

141. Yonish-Rouach E, Resnitzky D, Lotem J, Sachs L, Kimchi A, Oren M. Wild-type p53 induces apoptosis of myeloid leukaemic cells that is inhibited by interleukin-6. Nature 1991;352:345–7.

142. Fritsche M, Haessler C, Brander G. Induction of nuclear accumulation of the tumor-suppressor protein p53 by DNA-damaging agents. Oncogene 1993;8: 307–18.

143. Hollestein M, Sidransky D, Vogelstein B, Harris C. p53 mutations in human cancers. Science 1991;253:49–53.

144. Vogelstein B, Kinzler KW. p53 function and dysfunction. Cell 1992;70:523–6.

145. Miyashita T, Reed JC. Tumor suppressor p53 is a direct transcriptional activator of the human bax gene. Cell 1995;80:293–9.

146. Miyashita T, Harigai M, Hanada M, Reed JC. Identification of a p53-dependent negative response element in the bcl-2 gene. Cancer Res 1994;54:3131–5.

147. Tilly JL, Tilly KI, Kenton ML, Johnson AL. Expression of members of the bcl-gene family in the immature rat ovary: equine chorionic gonadotropin-mediated inhibition of apoptosis is associated with decreased bax and constitutive bcl-and bcl-x_{LONG} messenger ribonucleic acid levels. Endocrinology 1995;136:232–41.

148. Knudson CM, Tung KSK, Tourtellotte WG, Brown GAJ, Korsmeyer SJ. Bax-deficient mice with lymphoid hyperplasia and male germ cell death. Science 1995;270:96–9.

149. Tilly JL. The molecular basis of ovarian cell death during germ cell attrition, follicular atresia, and luteolysis. Frontiers in Bioscience 1996;1:1–10.

150. Leighr DR, Gentry PC, Smith GW, Smith MF. Cloning and initial characterization of ovine Bcl-2 expression by corpora lutea. Biol Reprod 1995;52(Suppl 1):44.

151. Rodger FE, Fraser HM, Duncan WC, Illingworth PJ. Immunolocalization of Bcl-2 in the human corpus luteum. Mol Human Repr 1995;10:1566–70.

152. Rueda BR, Tilly KI, Botros I, Jolly PD, Hansen TR, Hoyer PB, Tilly JL. Increased *bax* and interleukin-1β-converting enzyme (*Ice*) messenger RNA levels coincide with apoptosis in the bovine corpus luteum during structural regression. Biol Reprod 1997;56:186–93.

153. Kastan MB, Onyekwere O, Sidransky D, Volgelstein B, Craig RW. Participation of p53 protein in the cellular response to DNA damage. Cancer Res 1991;51:6304–11.

154. Keren-Tal I, Suh B-S, Dantes A, Lindner S, Oren M, Amsterdam A. Involvement of p53 expression in cAMP-mediated apoptosis in immortalized granulosa cells. Exp Cell Res 1995;218:283–95.

155. Martin SJ, Green DR. Protease activation during apoptosis: death by a thousand cuts? Cell 1995;82:349–52.

156. Yaun J, Shaham S, Ledoux S, Ellis HM, Horvitz HR. The *C. elegans* death gene ced-3 encodes a protein similar to mammalian interleukin-1β-converting enzyme. Cell 1993;75:653–60.

157. Hengartner MO, Horvitz HR. *C. elegans* survival gene *ced-9* encodes a functional homolog of the mammalian proto-oncogene *bcl-2*. Cell 1994;76:655–76.

158. Yaun J, Horvitz HR. Genetic and mosaic analysis of *ced-3* and *ced-4*, two genes that control programmed cell death in the nematode, *C. elegans*. Dev Biol 1990;138:33–41.

159. Black RA, Kronheim SR, Sleath PR. Activation of interleukin-1β by a coinduced protease. FEBS Lett 1989;247:386–90.

160. Miura M, Zhu H, Rotello R, Hartweig EA, Yaun J. Induction of apoptosis in fibroblasts by IL-1β-converting enzyme, a mammalian homolog of the *C. elegans* cell death gene *ced-3*. Cell 1993;75:653–60.

161. Ray CA, Black RA, Kronheim SR, Greenstreet TA, Sleath PR, Salvesen GS, Pickup DJ. Viral inhibition of inflammation: cowpox virus encodes an inhibitor of the interleukin-1β-converting enzyme. Cell 1992;69:597–604.

162. Enari M, Hug H, Nagata S. Involvement of an ICE-like protease in FAS-mediated apoptosis. Nature. 1995;375:78–81.

163. Tewari M, Dixit VM. *Fas*-and tumor necrosis factor-induced apoptosis is inhibited by the poxvirus crmA gene product. J Biol Chem 1995;270:3255–60.

164. Li P, Allen H, Banerjeee S, Franklin S, Herzog L, Johnston C, McDowell J, Paskind M, Rodman L, Salfield J, Towne E, Tracy D, Wardwell S, Wei F-Y, Wong W, Kamen R, Seshadri T. Mice deficient in IL-1β-converting enzyme are defective in production of mature IL–1β and resistant to endotoxic shock. Cell 1995;80:401–11.

165. Kumar S, Tomooka Y, Noda M. Identification of a set of genes with developmentally down-regulated expression in the mouse brain. Biochem Biophys Res Commun 1992;185:1155–61.

166. Wang L, Miura M, Bergeron L, Zhu H, Yaun J. *Ich-1*, an ICE/*ced-3*-related gene, encodes both positive and negative regulators of programmed cell death. Cell 1994;78:739–50.

167. Fernandes-Alnemri T, Litwack G, Alnemri ES. CPP32, a novel human protein with homology to *Caenorhabditis elegans* cell death protein CED-3 and the mammalian interleukin-1β-converting enzyme. J Biol Chem 1994;269:30761–4.

168. Nicholson DW, Ali A, Thornberry NA, Vaillancourt JP, Ding CK, Gallant M, et al. Identification and inhibition of the ICE/CED-3 protease necessary for mammalian apoptosis. Nature 1995;376:37–43.

169. Tewari M, Quan LT, O'Rourke K, Desnoyers S, Zeng Z, Beidler DR, Poirier GG, Salverson GS, Dixit VM. Yama/CPP32β, a mammalian homolog of CED-3, is a *crmA*-inhibitable protease that cleaves the death substrate poly (ADP-ribose) polymerase. Cell 1995;81:801–9.

170. Lazebnik YA, Kaufman SH, Desnoyers S, Poirer GG, Earnshaw WC. Cleavage of poly(ADP-ribose) polymerase by a proteinase with properties like ICE. Nature 1994;371:346–7.

171. Flaws JA, Kugu K, Trbovich AM, DeSanti A, Tilly KI, Hirshfield AN, Tilly JL. Interleukin-1β-converting enzyme-related proteases (IRPs) and mammalian cell death: dissociation of IRP-induced oligonucleosomal endonuclease activity from morphological apoptosis in granulosa cells of the ovarian follicle. Endocrinology 1995;136:5042–53.

172. Maravei DV, Trbovich AM, Perez GI, Tilly KI, Talanian RV, Banach D, Wong WW, Tilly JL. Cleavage of cytoskeletal proteins by caspases during ovarian cell death: evidence that cell-free systems do no always mimic apoptotic events in intact cells. Cell Death Differen 1997;4(in press).

173. Chinnaiyan AM, Orth K, O'Rourke K, Duan H, Poirier GG, Dixit VM. Molecular ordering of the cell death pathway. Bcl-2 and Bcl-$_{XL}$ function upstream of the CED-3-like apoptotic proteases. J Biol Chem 1996;271:4573–84.

15

ICER and the Nuclear Response to cAMP

CARLOS A. MOLINA

The mechanisms governing the balance between cellular differentiation and proliferation rely upon a complex and versatile array of signal transduction pathways. Among the several existing pathways, the one mediated by the second messenger, cyclic adenosine monophosphate (cAMP), has been clearly linked to cellular growth and differentiation (1–6). Cyclic AMP has a crucial role in the regulation of cell proliferation in the mammalian endocrine system. Several hormones that activate the cAMP-dependent signaling pathway in target endocrine cells also promote growth. The notion that activation of the cAMP transduction pathway results in modulation of gene expression (7) suggests that nuclear effectors of this pathway may be involved in the regulation of the cell cycle and proliferation (8).

Ligand-receptor interactions are known to stimulate the enzyme adenylyl cyclase via interaction with G proteins (9). The subsequent rise in intracellular cAMP concentrations results in the activation of protein kinase A (PKA) and the translocation of active catalytic subunits of PKA into the nucleus (7, 10, 11). PKA phosphorylates and, thereby, stimulates transcriptional activators that bind as dimers to cAMP-responsive elements (CREs). These nuclear factors, encoded by genes such as CREB, CREM, and ATF-1 (12–15), then induce transcription of cAMP-responsive genes. At least 10 genes have been cloned that encode CRE binding nuclear factors (CRE-binding proteins, CREBs, and Activator Transcription Factors or ATFs). By the formation of heterodimers as well as homodimers, these transcriptional regulators exhibit great functional diversity. Studies on CREB and CREM have established that induced transcription results from phosphorylation of these activators by PKA (12). The expression of most CREBs and ATFs is uninducible and ubiquitous (9, 11), confirming the notion that their function is regulated predominantly via differential phosphorylation.

The CREM Gene

The discovery of the CREM (cAMP-response element modulator) gene opened a new dimension in the study of transcriptional response to cAMP (15). This is due to the remarkable dynamic and modular genomic structure of the gene, which offers clues to the understanding of the generation of functional diversity in transcription factors. CREM is a multi-exonic gene that encodes a family of both activators and antagonists of cAMP inducible transcription by differential splicing (15–23). (Fig. 15.1). The most striking feature of the CREM cDNA is the presence of two DNA-binding domains (Fig. 15.1). The first is complete and contains a leucine zipper and basic region very similar to CREB; the second is located in the 3' untranslated region of the gene, out

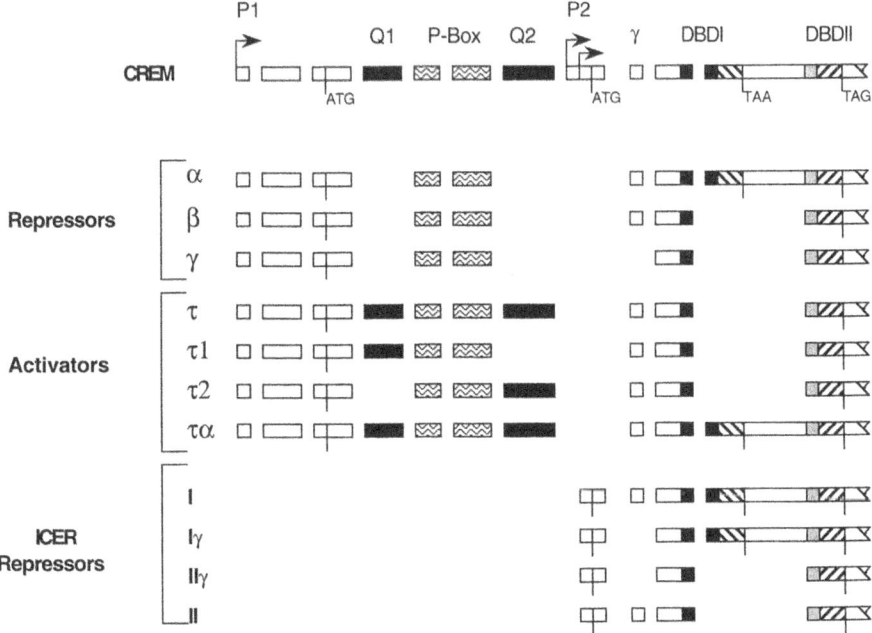

FIGURE 15.1. Activators and repressors from the same gene. The top line indicates a schematic representation of the CREM gene. Exons encoding the glutamine-rich domains (Q1 and Q2), the P-box, the γ-domain (γ), and the two alternative DNA binding domains (DBDI and DBDII) are shown. Below are represented the various activator and repressor isoforms which have been described to date. The P1 promoter is GC-rich and directs a noninducible pattern of expression. Also shown is the ICER family. All of the ICER transcripts are derived from an internal start-site of transcription (P2) located between the Q2 and the γ-exon. A family of four types of ICER transcripts is generated by alternative splicing of the DBDs and γ-domain exons: ICERI, ICERIγ, ICERII, and ICERIIγ.

of phase with the main coding region, and contains a half basic region and a lucine zipper more divergent from CREB. Various mRNA isoforms have been demonstrated in several tissues and cell types, where different patterns of expression have been found (15–18, 20, 24).

The CREM antagonists α, β, and γ lack two glutamine-rich domains and block cAMP-induced transcription (15). These isoforms reveal alternative usage of two DNA binding domains (α and β isoforms, see Fig. 15.1), as well as a small deletion of 12 amino acids (γ isoform). Another isoform, CREMτ includes these glutamine-rich domains that convert CREM into a transcriptional activator (17). The production of CREMτ during spermatogenesis is modulated by the pituitary-hypothalamic axis via a posttranscriptional mechanism. The specific role of CREM in spermiogenesis was addressed using CREM-mutant mice generated by homologous recombination. Analysis of the seminiferous epithelium in mutant male mice reveals postmeiotic arrest at the first step of spermiogenesis. Late spermatids are completely absent, and there is a significant increase in apoptotic germ cells (25). It thus appears that cell-specific splicing is a crucial mechanism of CREM regulation which modulates the DNA-binding specificity and the activity of the final CREM products.

The CREM proteins specifically recognize CREs and show the same binding properties as CREB. This is not surprising, considering the high homology in the DNA-binding domains between these proteins. CREM proteins containing either DNA-binding domain I or II, heterodimerize with CREB (16, 19), although it appears that CREMα-CREB heterodimer formation is more favored than CREMβ-CREB. These notions suggest that CREM proteins might occupy CRE sites as CREM dimers or as CREM-CREB heterodimers, thus generating complexes with altered transcriptional functions. In fact, CREM repressors act by imparing CRE-mediated transcription, and as such are considered antagonists of cAMP-induced expression. In transfection experiments, using CRE reporter plasmids, it has been demonstrated that CREMα, β, and γ antagonist proteins block the transcriptional activation obtained by the joint action of CREB and the catalytic subunit of the cAMP-dependent PKA (26). These observations strongly support the notion that CREM antagonistic proteins negatively modulate CRE promoter elements in vivo. An important question is how CREM proteins work. The two most likely hypotheses are as follows. According to the first scenario, CREM proteins dimerize and bind to CRE sites. Down-regulation is achieved by occupation of these sites, which are unavailable for CREB. Similarly, if CREB is already bound, CREM proteins might squelch them because of their possible higher affinity for a specific site. According to the second model, CREM proteins are able to dimerize with CREB to generate nonfunctional heterodimers. Negative regulation is achieved by titrating active CREB molecules, and CREM dimers and CREB-CREM heterodimers bind to CRE sites. Both hypotheses are justified, and both mechanisms may operate. However, results by Laoide et al. (1993) indicate that the production of nonfunctional heterodimers is the most likely mechanism operating to obtain CREM-mediated antagonism of cAMP-induced transcription.

ICER: A Remarkable Repressor of cAMP-Mediated Gene Expression

During studies of CREM expression within the endocrine system, an unexpected new facet emerged: transcription of the CREM gene is inducible by cAMP (21–23). Furthermore, the kinetics of this induction resemble that of an early response gene. This important finding further reinforces the notion that CREM products play a pivotal role in the nuclear response to cAMP, since to date the expression of no other CRE-binding factor has been shown to be inducible. For example, the recently characterized CREB promoter is GC-rich and reminiscent of the promoter of constitutively-expressed, housekeeping genes (27, 28). Upon detailed analysis of the induced CREM products, there have been more surprises. The promoter that directs expression of the previously characterized CREM isoforms (P1) is not cAMP inducible. Instead, an alternative promoter lying within an intron near the 3' end of the gene directs cAMP induced transcription of a novel truncated CREM product, termed ICER (Induced cAMP Early Repressor) (22, 23) (Fig. 15.1). ICER is the smallest bZip factor yet described, functions as a powerful repressor of cAMP-induced transcription and, furthermore, negatively autoregulates the ICER promoter (22). The expression of ICER was first described in the pineal gland where it exhibits a dramatic circadian pattern of expression (23). Dynamic ICER expression is a general feature of endocrine systems (23).

Inducibility of the CREM Gene: Use of an Alternative Intronic Promoter

Clues that the CREM gene is cAMP inducible first came from the demonstration that adrenergic signals direct CREM transcription in the pineal gland (23). The inducibility phenomenon has been characterized in detail in the pituitary corticotroph cell line, AtT20 (22). In unstimulated cells the level of CREM transcript is below the threshold of detectability. However, upon treatment with forskolin (or cAMP analogs), within 30 min there is a rapid increase in CREM transcript levels that peak after 2 h and then progressively decline to basal levels by 5 h. This characteristic kinetic classifies CREM as an early response gene and thus, for the first time, directly implicates the cAMP pathway in the cell's early response. CREM inducibility is specific for the cAMP pathway since it is not inducible by PKC activators (e.g., phorbol esters) or dexamethasone treatment (22).

In order to further characterize the cAMP-induced CREM transcript, a battery of exon-specific probes have been used in a systematic Northern, RNase protection and RT-PCR analysis. In this way it has been demonstrated that all previously characterized exons located 5' to the γ-exon are absent from the induced CREM transcript. The 5' boundary of the γ-exon is defined by a consensus splicing acceptor site. Thus, in order to identify the remaining exons constituting the 5' end, a RACE-PCR cloning strategy was used. By this approach a series of short, overlapping cDNA clones define a novel cDNA which is termed ICER (22, 23). An 82 basepair (bp) sequence lying 5' of the γ-exon boundary extends the CREM open reading frame upstream by eight amino acids to a consensus Kozak ATG codon and includes a short 5' untranslated region. Full-length cDNA clones, isolated from a pineal cDNA library, together with RNase protection assays, confirmed the results of RACE-PCR while primer extension analysis demonstrated that the 5' end of the ICER clones correspond to a transcription start site (22, 23). The ICER cDNA clones also reveal different splicing of the two alternative DNA binding domains and of the γ-exon (Fig. 15.1), as previously described for the CREMα and β isoforms. In addition, by Northern blot and RNase protection analyses it is apparent that the ICER transcripts employ the various polyadenylation sites in a cell-specific fashion, thus cAMP induction of the CREM gene generates a family of transcripts (22).

In order to locate the start point of transcription of the ICER transcripts relative to the promoter of the previously described isoforms, an overlapping series of phage clones encompassing the entire CREM gene, have been screened with a probe for the 5' ICER-specific exon. Hybridization locates the start of transcription (P2) within the 10 kilobasepair (kb) intron that lies between the Q2 glutamine-rich domain exon and the γ-exon (Fig. 15.1) (22, 23).

In contrast to the promoter (P1) which generates all of the previously characterized CREM isoforms, is GC-rich and not inducible (N. Foulkes, personal communication), the P2 promoter has a normal A-T and G-C content and is strongly inducible by cAMP. It contains two pairs of closely-spaced CRE elements (cAMP-autoregulatory responsive elements, CARE) (Fig. 15.2) organized in tandem. Furthermore, the distance between the CAREs in each pair is only three nucleotides. These features make P2 unique among cAMP-regulated promoters and are suggestive of cooperative interactions among the factors binding to these sites (22). Previously, tandemly repeated pairs of CRE elements have been described in the promoters of the Pit-1 and α-CG genes (29, 30), but the individual elements are more widely spaced. The promoter directs transcription from alternative start points (S1 and S2), and 23 bp upstream from each start site lie A-T rich elements that presumably function as TATA elements.

ICER, the Smallest bZip Factor

The ICER open reading frame is constituted by the C-terminal segment of CREM. The predicted open reading frame encodes a small protein of 120 amino acids with an expected molecular weight of 13.4 kD. This protein,

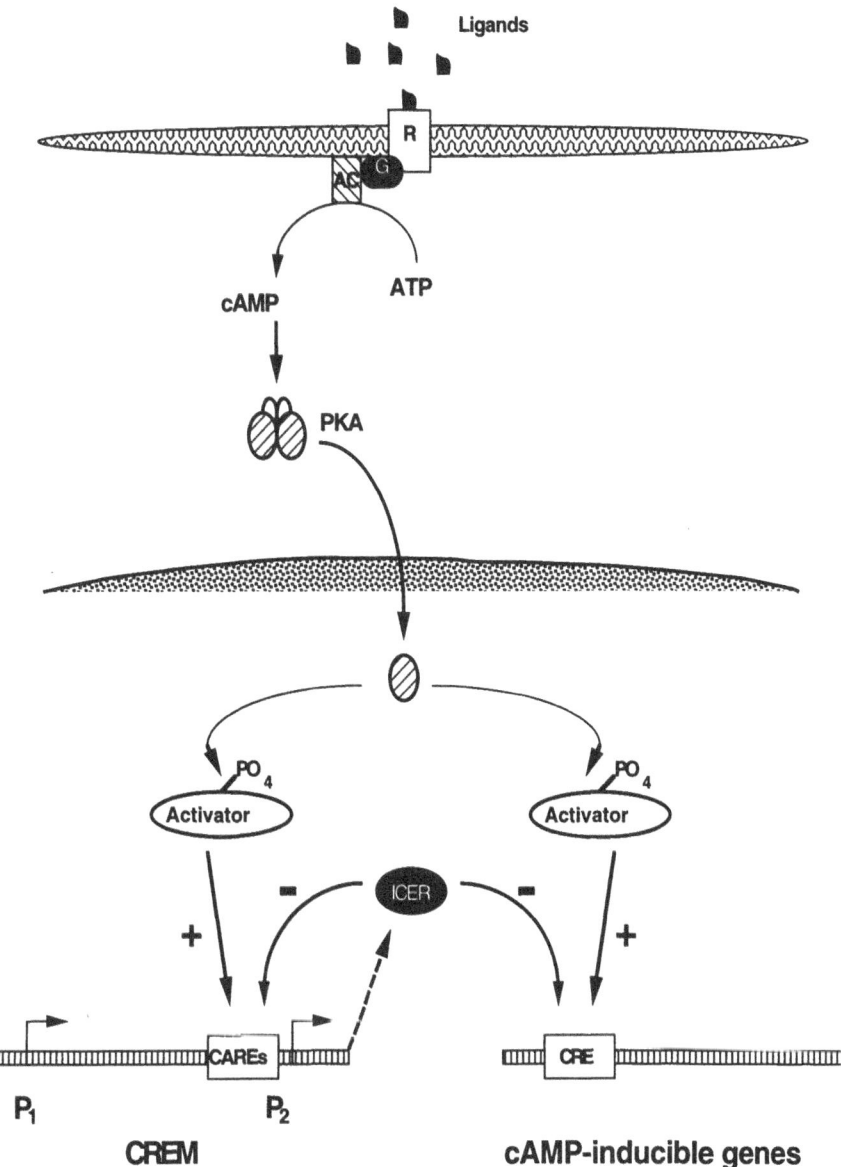

FIGURE 15.2. The role of ICER in the regulation of gene expression by cAMP. Schematic representation of the cAMP signal transduction pathway operating from the cell membrane, through the cytoplasm and into the nucleus. Ligands interacting with transmembrane receptors (R) stimulate the enzyme, adenylyl cyclase (AC), via interaction with G-proteins (G). The subsequent rise in intracellular cAMP concentration results in the dissociation of the regulatory and catalytic subunits of PKA and the translocation of active catalytic subunits into the nucleus. PKA phosphorylates and thereby stimulates transcriptional activators binding to CREs (activators, e.g. CREB and CREMτ) that induce transcription from the promoters of cAMP-responsive genes. These factors activate transcription from the CREM P2 promoter via the CRE elements and ultimately lead to a rapid increase in ICER protein levels. ICER represses cAMP-induced transcription, including that from its own promoter. The consequent fall in ICER protein levels eventually leads to a release of repression and permits a new cycle of transcriptional activation.

compared with the previously described CREM isoforms, essentially consists of only the DNA binding domain, which is constituted by the leucine zipper and basic domain. The provocative stucture of ICER is suggestive of its function and makes it one of the smallest transcription factors described (22).

The intact DNA binding domain directs specific ICER binding to a consensus CRE element. Consistent with previous analysis of the other CREM isoforms, ICER fails to bind to Sp1 sites. Importantly, ICER is able to heterodimerize with CREMτ, as well as with the other CREM proteins and with CREB. ICER functions as a powerful repressor of cAMP-induced transcription in transfection experiments using an extensive range of reporter plasmids carrying induvidual CREs of cAMP-inducible promoter fragments (23). Interestingly, ICER-mediated repression is obtained at substoichiometric concentrations, similarly to the previously described CREM antagonists (15). However, it should be noted that ICER is significantly more potent when compared to CREMα or CREMβ (23).

The small sizes of the ICER products are striking and have possibly masked their presence in previously performed analyses. The smallest CRE-binding factor described previously is the liver regenerating factor (LRF) (Hsu et al. 1991). Based on homology, it is thought that LRF represents a splicing variant of ATF3 (31), although it is also conceivable that it could be generated by the use of an alternative promoter as in the case of ICER. This 21 kD bZip protein functions as an activator of transcription in a heterodimeric complex with c-Jun, and its expression is also stimulated by mitogens. The major feature of ICER proteins is that they lack the N-terminal domain shared by the other CREM isoforms (Fig. 15.1). This N-terminal domain, also referred to as the P-box domain, is retained in all of the P1 promoter-generated CREM isoforms, even those acting as repressors (CREMα, β, and γ). In the latter case, phosphorylation of the P-box domain by PKA has been shown to modulate the degree of repression activity (19). In contrast, ICER proteins escape from PKA-dependent phosphorylation and thus constitute a new category of CRE binding factors, for which the principle determinant of their activity is their intracellular concentration and not their degree of phosphorylation.

Negative Autoregulation

Upon cotreatment with cycloheximide, the kinetics of CREM gene induction by forskolin are altered showing a significant delay in the post-induction decrease in the transcript (e.g., elevated levels persist for as long as 12 h). This implicates a *de novo* synthesized factor that might down-regulate CREM transcription (22). This observation, combined with the presence of CRE elements in the P2 promoter (see above), suggests, that the transient nature of the inducibility could be due to ICER. Consistently, the CARE elements in the P2 promoter have been shown to bind to the ICER proteins. Detailed studies have demonstrated that the ICER promoter is indeed a target for ICER negative regulation (22). Thus, there exists a negative autoregulatory mechanism controlling ICER expression (Fig. 15.2).

ICER Physiological Functions

Day-Night Switch in ICER Expression in the Pineal Gland

A crucial step in understanding of the physiological significance of ICER came with the observation that it is expressed in a rhythmic fashion in the pineal gland (23). Crucial elements for the synchronization of biological rhythms in mammals are the pineal gland (32, 33) and the suprachiasmatic nucleus (SCN) (34, 35). Environmental lighting conditions are transduced by the pineal gland from a neuronal to an endocrine message, that being the rhythmic secretion of melatonin (32, 33, 35). This hormone synthesis is controlled by the SCN, and is elevated at night and decreased during the day (32, 35). The cAMP-dependent signal transduction pathway serves as a relay to stimulate melatonin (35–37). Thus, from the neuronal pathway, including the retina and the SCN, the pineal gland acts as a temporal regulator for the function of the hypothalamic-pituitary-gonadal axis (32). During the night, ICER constitutes an abundant transcript while during the day it is present at low levels (23). In a series of physiological experiments, the mechanism controlling this pattern of ICER expression has been determined (23). Rhythmic adrenergic signals originating in the SCN direct ICER expression via stimulation of the cAMP signal transduction pathway. The question of possible targets for down-regulation by ICER in the pineal gland can, at the moment, only be a matter for speculation. However, it has been proposed that a reasonable target could be the enzyme that catalyzes the rate-limiting step of melatonin synthesis (N-acetyl transferase; 38) or factors that regulates activity of this enzyme (39).

ICER and Long-Term Desensitization of Receptors in the Thyroid Gland

Thyriod gland function is regulated by the hypothalamic-pituitary axis via the secretion of thyrotropin (thyroid-stimulating hormone; TSH), according to environmental, developmental, and circadian stimuli (40, 41). TSH modulates both the secretion of thyroid hormone and gland growth through interaction with a specific receptor on the thyroid cell membrane (TSH receptor, TSH-R; 42). The cAMP signal-transduction pathway is essential in mediating signalling through the TSH-R (43). In addition to the stimulation of several genes (44–47), the expression of other genes in the thyroid gland is repressed by TSH. The best known example is represented by the TSH-R itself (48). Down-regulation of the TSH-R gene is crucial in the attenuation process of the thyroid cell response to TSH. Using this negative feedback mechanism (long-term desensitization; LTD), it is possible to avoid excessive hormone production and gland hypertrophy after a TSH stimulus. It has been shown that transcriptional down-regulation represents the molecular basis for TSH-R LTD in thyroid cells (48). TSH treatment induces the expression of ICER in the rat thyroid gland and in the differentiated thyroid cell line FRTL-5 (49). ICER displays a characteristic early response inducibility, which correlates with TSH-R transcriptional down-regulation. Furthermore, ICER represses the expression of TSH-R. Thus, ICER inducibility may represent a general paradigm for the molecular mechanism used by pituitary hormones to obtain homologous LTD of their own receptors.

Sertoli Cells

The complex structure and function of the mammalian testis are maintained by cell-cell interactions as well as by endocrine signals originating from the hypothalamic-pituitary axis (50). The gonadotropins, follicle-stimulating hormone (FSH) and luteinizing hormone (LH) released by the gonadotrophs of the pituitary gland, regulate spermatogenesis by directing the function of the somatic cells of the testis, the Sertoli, and Leydig cells, respectively. The cAMP signal-transduction pathway is responsible for converting these hormonal signals into long-term changes in gene expression. In Sertoli cells, activation of the cAMP pathway by FSH leads to induction of ICER expression (51). ICER induction accompanies early down-regulation of the FSH receptor transcript, which leads to LTD. ICER represses FSH receptor expression by binding to CRE-like sequences in the regulatory region of the gene.

Conclusions and Perspectives

Much of the reseach in transcription factor biology has been devoted to understanding the functional and structural relationships of these factors. However, a much greater challenge lies ahead, that being to relate transcriptional control mechanisms to the physiology and biology of the organism. The data presented here clearly link the function of two CREM gene isoforms (CREMτ and ICER) to spermatogenesis and long-term desensitization of receptors. Even though much is known about CREM function in the testis, the role of CREM in the ovary has not been extensively explored. Recently, we have begun to study CREM expression in rat ovaries at different stages of development. We have observed that the expression of ICER correlates with the time of luteolysis (unpublished data). We have also observed a correlation between the down-regulation of 3β-hydroxysteroid dehydrogenase and the up-regulation of ICER. The kinetics of ICER induction suggest that ICER plays a pivotal role in the temporal regulation of the cAMP-programmed gene expression occuring in the ovary, specifically at the time of luteal demise. These observations concerning the function of ICER in the ovary are currently being extended.

References

1. Asa SL, Kovacs K, Hammer GD, Liu B, Roos BA, Low MJ. Pituitary corticotroph hyperplasia in rats implanted with a medullary thyroid carcinoma cell line transfected with a corticotropin-releasing hormone complementary deoxyribonucleic acid expression vector. Endocrinology 1992;131:715–20.
2. Burton FH, Hasel KW, Bloom FE, Sutcliffe JG. Pituitary hyperplasia and gigantism in mice caused by cholera toxin transgene. Nature 1991;350:74–7.
3. Di Blasio AM, Fujii DK, Yamamoto M, Martin MC, Jaffe RB. Maintenance of cell proliferation and steroidogenesis in culture human adrenal cells chronically exposed to adrenocorticotropic hormone: rationalization of in vitro and in vivo findings. Biol Reprod 1990;42:683–91.
4. Dumont JE, Jauniaux J-C, Roger PP. The cyclic AMP-mediated stimulation of cell proliferation. TIBS 1989;14:67–71.
5. Lin C, Lin S-C, Chang C-P, Rosenfeld MG. Pit-1-dependent expression of the receptor for growth hormone releasing factor mediates pituitary cell growth. Nature 1992;360:765–8.
6. Struthers RS, Vale WW, Arias C, Sawchenko PE, Montminy MR. Somatotroph hypoplasia and dwarfism in transgenic mice expressing a non-phosphorylatable CREB mutant. Nature 1991;350:622–4.
7. Lalli E, Sassone-Corsi P. Signal transduction and gene regulation: the nuclear response to cAMP. J Biol Chem 1994;269:17359–62.
8. Desdouets C, Matesic G, Molina CA, et al. Cell cycle regulation of cyclin A gene expression by the cyclic AMP-responsive transcription factors CREB and CREM. Mol Cell Biol 1995;15:3301–9.
9. Borrelli E, Montmayeur JP, Foulkes NS, Sassone-Corsi P. Signal transduction and gene control: the cAMP pathway. CRC Rev Oncogenesis 1992;3:321–38.
10. Nigg EA, Hilz H, Eppenberger HM, Dutly F. Rapid and reversible translocation of the catalytic subunit of cAMP-dependent protein kinase type II from the Golgi complex to the nucleus. EMBO J 1985;4:2801–9.
11. Habener JF. Cyclic AMP-response element binding proteins: a cornucopia of transcription factors. Mol Endocrinol 1990;4:1087–94.
12. De Groot RP, Sassone-Corsi P. Hormonal control of gene expression: multiplicity and versatility of cyclic adenosine 3', 5'-monophosphate-responsive nuclear regulators. Mol Endocrinol 1993;10:145–53.
13. Hoeffler JP, Lustbader JW, Chen CY. Identification of multiple nuclear factors that interact with cyclic adenosine 3',5'-monophosphate response element-binding protein and activating transcription factor-2 by protein-protein interactions. Mol Endocrinol 1991;5:256–66.
14. Gonzalez GA, Menzel P, Leonard J, Fischer WH, Montminy MR. Characterization of motifs which are critical for activity of the cyclic AMP-responsive transcription factor CREB. Mol Cell Biol 1991;11:1306–12.

15. Foulkes NS, Borrelli E, Sassone-Corsi P. CREM gene: use of alternative DNA-binding domains generates multiple antagonists of cAMP-induced transcription. Cell 1991;64:739–49.
16. Foulkes NS, Laoide BM, Schlotter F, Sassone-Corsi P. Transcriptional antagonist cAMP-responsive element modulator (CREM) down-regulates c-fos cAMP-induced expression. Proc Natl Acad Sci USA 1991;88:5448–52.
17. Foulkes NS, Mellstrom B, Benusiglio E, Sassone-Corsi P. Developmental switch of CREM function during spermatogenesis: from antagonist to activator. Nature 1992;355:80–4.
18. Foulkes NS, Schlotter F, Pevet P, Sassone-Corsi P. Pituitary hormone FSH directs the CREM functional switch during spermatogenesis. Nature 1993;362:264–7.
19. Laoide BM, Foulkes NS, Schlotter F, Sassone-Corsi P. The functional versatility of CREM is determined by its modular structure. EMBO J 1993;12:1179–91.
20. Delmas V, Laoide BM, Masquilier D, de Groot RP, Foulkes NS, Sassone-Corsi P. Alternative usage of initiation codons in mRNA encoding the cAMP-responsive-element modulator generates regulators with opposite functions. Proc Natl Acad Sci USA 1992;89:4226–30.
21. Masquilier D, Foulkes NS, Mattei MG, Sassone-Corsi P. Human CREM gene: evolutionary conservation, chromosomal localization, and inducibility of the transcript. Cell Growth Differen 1993;4:931–7.
22. Molina CA, Foulkes NS, Lalli E, Sassone-Corsi P. Inducibility and negative autoregulation of CREM: an alternative promoter directs the expression of ICER, an early response repressor. Cell 1993;75:875–86.
23. Stehle JH, Foulkes NS, Molina CA, Simonneaux V, Pevet P, Sassone-Corsi P. Adrenergic signals direct rhythmic expression of transcriptional repressor CREM in the pineal gland. Nature 1993;365:314–20.
24. Delmas V, van der Hoorn F, Mellstrom B, Jegou B, Sassone-Corsi P. Induction of CREM activator proteins in spermatids: downstream targets and implications for haploid germ cell differentiation. Mol Endocrinol 1993;7:1502–14.
25. Nantel F, Monaco L, Foulkes N, Masquilier D, Lemeur M, Henriksen K, Dierich A, Parvinen M, Sassone-Corsi P. Spermiogenesis deficiency and germ-cell apoptosis in CREM-mutant mice. Nature 1996;380:159–62.
26. Mellon PL, Clegg CH, Correll LA, McKnight SG. Regulation of transcription by cyclic AMP-dependent protein kinase. Proc Natl Acad Sci USA 1989;86:4887–91.
27. Meyer TE, Waeber G, Lin J, Beckmann W, Habener JF. The promoter of the gene encoding 3',5'-cyclic adenosine monophosphate (cAMP) response element binding protein contains cAMP response elements: evidence for positive autoregulation of gene transcription. Endocrinology 1993;132:770–80.
28. Cole TJ, Copeland NG, Gilbert DJ, Jenkins NA, Schutz G, Ruppert S. The mouse CREB (cAMP responsive element binding protein) gene: structure, promoter analysis, and chromosomal localization. Genomics 1992;13:974–82.
29. Delegeane A, Ferland L, Mellon PL. Tissue specific enhancer of the human glycoprotein hormone α-subunit gene: dependence on cyclic AMP-inducible elements. Mol Cell Biol 1987;7:3994–4002.
30. McCormick A, Brady H, Theill L, Karin M. Regulation of the pituitary-specific homeobox gene GHF1 by cell-autonomous and enviromental cues. Nature 1990; 345:829–32.
31. Hai T-Y, Liu F, Coukos WJ, Green MR. Transcription factor ATF cDNA clones: an extensive family of leucine zipper proteins able to selectively form DNA binding heterodimers. Genes Dev 1989;3:2083–90.

32. Reiter RJ. Pineal gland: Interface between the photoperiodic environment and the endocrine system. Trends Endocrinol Metab 1991:1:13.

33. Tamarkin L, Baird CJ, Almeida O. Melatonin: a coordinating signal for mammalian reproduction? Science 1985;227:774.

34. Moore RY. Neuroendocrine regulation of reproduction. In: Yen SSC, Jaffe RB, eds. Reproductive endocrinology. Philadelphia: Saunders 1978:3–33.

35. Klein DC. Photoneural regulation of the mammalian pineal gland. In: Photoperiodism, melatonin, and the pineal gland. London: Pitman, 1985:38–56.

36. Sudgen D, Vanecek J, Klein DC, Thomas TD. Activation of protein kinase C potentiates isoprenaline-induced cyclic AMP accumulation in rat pinealocytes. Nature 1985;314:359.

37. Vanecek J, Sudgen D, Weller J, Klein DC. Atypical synergistic $\alpha 1$ and β-adrenergic regulation of adenosine 3',5' Endocrinology. 1985;116:2167.

38. Takahashi JS. Circadian clock a la CREM. Nature 1993;365:299.

39. Foulkes NS, Duval G, Sassone-Corsi P. Adaptive inducibility of CREM as transcriptional memory of circadian rhythms. Nature 1996;381:83–5.

40. Wong C-C, Döhler K-D, Atkinson MJ, Geerling H, Hesch R-F, von zur Mühlen A. Influence of age, strain and season on diurnal periodicity of thyroid stimulating hormone, thyroxine, triiodothyronine and parathyroid hormone in the serum of male laboratory rats. Acta Endocrinol 1983;102:377–85.

41. Utiger RD. The pathogenesis of autoimmune thyroid disease. In: Felig P, Baxter JD, Broadus AE, Frohman LA, eds. Endocrinology and Metabolism, New York: MacGraw-Hill, 1987;389–472.

42. Akamizu T, Ikuyama S, Saji M, et al. Cloning, chromosomal assignment, and regulation of the rat thyrotropin receptor: expression of the gene is regulated by thyrotropin, agents that increase cAMP levels, and thyroid autoantibodies. Proc Natl Acad Sci USA 1990;87:5677–81.

43. Nagayama Y, Rapoport B. The thyrotropin receptor 25 years after its discovery: new insight after its molecular cloning. Mol Endocrinol 1992;6:145–56.

44. Avvedimento EV, Musti AM, Ueffing M, et al. Reversible inhibition of a thyroid-specific trans-acting factor by Ras. Genes Develop 1991;5:22–8.

45. Civitareale D, Lonigro R, Sinclair AJ, Di Lauro R. A thyroid-specific nuclear protein essential for tissue-specific expression of the thyroglobulin promoter. EMBO J 1989;8:2537–42.

46. Francis-Lang H, Price M, Polycarpou-Schwarz M, Di Lauro R. Multiple mechanisms of interference between transformation and differentiation in thyroid cells. Mol Cell Biol 1992;12:576–88.

47. Hansen C, Javaux F, Juvenal G, Vassart G, Cristophe D. cAMP-dependent binding of a trans-acting factor to the thyroglobulin promoter. Biochem Biophys Res Commun 1989;160:722–31.

48. Saji M, Akamizu T, Sanchez M, et al. Regulation of thyrotropin receptor gene expression in rat FRTL-5 thyroid cells. Endocrinology 1992;130:520–33.

49. Lalli E, Sassone-Corsi P. Thyroid-stimulating hormone (TSH)-directed induction of the CREM gene in the thyroid gland participates in the long-term desensitization of the TSH receptor. Proc Natl Acad Sci USA 1995;92:9633–7.

50. Skinner MK. Cell-cell interactions in the testis. Endocr Rev 1991;12:45–77.

51. Monaco L, Foulkes NS, Sassone-Corsi P. Pituitary follicle-stimulating hormone (FSH) induces CREM gene expression in Sertoli cells: involvement in long-term desensitization of the FSH receptor. Proc Natl Acad Sci USA 1995;92:10673–7.

16

Clinical Ramifications of Apoptosis in the Human Testis

Leo Dunkel, Krista Erkkilä, Seppo Taskinen, Sakari Wikström, Håkan Billig, and Jonathan L. Tilly

Germ Cell Degeneration and Apoptosis

In mammals, multiplication of germ cells in the testis is always accompanied by degeneration of some of the proliferating cells. During spermatogenesis, germ cell death occurs spontaneously at various phases of germ cell development, and in consequence the seminiferous epithelium yields fewer spermatozoa than would be anticipated from spermatogonial proliferations (1–3). Germ cell deletion during normal spermatogenesis has been estimated to result in the loss of up to 75% of the potential numbers of mature sperm cells in the adult testis (2, 4, 5). Detailed analyses of germ cell degeneration have been published (1–3, 5–7); morphometric analyses of semithin sections of perfusion-fixed testes have indicated that the loss of germ cells is greatest during the mitoses of type A2, A3, and A4 spermatogonia and during the first meiotic division. No degeneration occurs in type A1 spermatogonia, or in intermediate of type B spermatogonia (7, 8). The mitotic and meiotic division take place at distinct stages of the cycle of the seminiferous epithelium (often referred to as the spermatogenic cycle). Therefore, spontaneously degenerating cells can normally be found only at certain stages: at stages II to VII few if any cells degenerate, whereas stages XIII to I show considerable germ cell degeneration (3, 7, 8).

Although the degeneration of testicular germ cells was originally characterized more than a century ago (9), it was only recently discovered how the cells die. On the basis of morphologic evidence, it was first suggested that cell death of spermatogonia during normal spermatogenesis takes place through the apoptotic mechanism, with the spermatocytes and spermatids undergoing necrosis (6). Recently, however, quantification of small molecular weight DNA fragments and in situ DNA 3'-end labeling analyses of apoptosis in the testis have shown that spermatocytes and spermatids can also undergo apoptosis (10–12). During normal spermatogenesis, apoptosis has been shown to take place in the very same cells and stages in which germ cell degeneration was previously observed (10, 12), indicating that germ cell degeneration is in fact apoptotic cell death.

The physiological significance of germ cell apoptosis in the testis is as yet unclear. It may be a means for regulating the number of differentiating spermatogonia to meet the capacity of the Sertoli cells to support the developing germ cells. The outcome of this regulation would be a constant number of meiotic cells. Since Sertoli cells do not divide in adulthood and are able to support only a certain number of germ cells at a time, the number of spermatogonia entering meiosis must be limited. Apparently, when too many spermatogonia are formed, the surplus cells degenerate (7), most likely by apoptosis.

Regulation of Male Germ Cell Apoptosis

In the immature rat testis, hypophysectomy or treatment with a gonadotropin-releasing hormone (GnRH) antagonist resulted in a marked increase in apoptosis (13). Supplementation with a follicle stimulating hormone (FSH) agonist, human chorionic gonadotropin (hCG), or testosterone inhibited apoptotic DNA cleavage by 84%, 51%, and 75%, respectively. Hypophysectomy-induced DNA fragmentation was found both in interstitial cells and in seminiferous tubules, but the exact cell type undergoing apoptosis was not identified. Russel and Clermont (14) had shown earlier that the first cells to degenerate after hypophysectomy in adult rats are primary spermatocytes and spermatids at stage VII of the spermatogenic cycle. FSH did not inhibit this degeneration, whereas luteinizing hormone (LH) or, more efficiently, a combination of FSH and LH, was able to prevent the degeneration. These data suggested that androgens (stimulated by LH) are indispensable for the maintenance of spermatogenesis. Extensive evidence supporting this concept has accumulated since the 1970s (for review, see ref. 15).

The role of testosterone in inhibition and induction of apoptosis in the immature rat testis was recently further clarified by in situ quantification of squash preparations (12). Destruction of Leydig cells with ethane dimethane sulfonate (EDS) in vivo resulted in rapid apoptosis of Leydig cells followed by a decrease in both serum and intratesticular testosterone concentrations. The reduction in androgen induced a significant increase in the number of apoptotic cells in the seminiferous tubules at stages I to XI of the spermatogenic cycle, primarily affecting the pachytene spermatocytes in stages II to VIII, and step 16 to 18 spermatids in stages II to VI. A surprising finding was a decrease in apoptosis at stage XII of the cycle, following testosterone reduction, induced by EDS-treatment. In EDS-treated rats supplemented with testosterone, apoptotic germ cell death was prevented at most stages of the spermatogenic cycle. Interestingly, testosterone seemed to be a positive regulator of apoptotic germ cell death at stage XII.

On the basis of quantifications of DNA fragmentation, a gradual increase in apoptosis has been observed in the testis cells of juvenile (16- to 28-day old) rats compared with neonatal (8-day-old) animals, followed by a decrease in adult animals (10).

Taken together, the studies referred to above have shown that at least three factors determine the onset of apoptosis in male germ cells: (1) lack of hormones, especially gonadotropins and androgens; (2) the specific stage in the spermatogenic cycle; (3) the developmental stage of the animal.

Germ Cell Number in the Human Testis During Development

Serum concentrations of FSH, LH (16), and testosterone (17) are transiently elevated during the first 6 months of postnatal life. At the same time, the total number of testicular germ cells increase, closely paralleling the changes in serum gonadotropin and androgen levels (18). Subsequently, the number of testicular germ cells decreases, reaching the lowest count per tubule by the age of 2 years, and this is followed by a gradual increase during the prepubertal period. At the onset of puberty, the number of spermatogonia per tubule rapidly increases concomitantly with an increase in serum gonadotropin and androgen concentrations, an increase in tubular diameter, and the onset of active spermatogenesis (19).

In prepuberty, two maturational steps have been described in the germ cells. The first step is the transformation of gonocytes into spermatogonia, which is usually complete by 6 months of age. In normal testes, the histomorphometric features of this transformation are a steady decrease in the number of gonocytes from birth until they disappear entirely at about 3 months of age. Simultaneously, a steady increase is observed in the number of spermatogonia, so that by 3 months of age most biopsies contain spermatogonia (20, 21). It has been suggested that spermatogonia are the stable pool of stem cells that continuously replenish the supply of germ cells throughout the rest of life (22). The second of the two prepubertal steps in the maturation of germ cells in human testis is the transformation of spermatogonia into primary spermatocytes, which normally begins at approximately 3 years of age. The morphometric feature of this step in normal testes is the appearance of primary spermatocytes while the number of spermatogonia remains stable (23). The changes in germ cell number are closely associated with changes in serum gonadotropin and androgen concentrations. The role of apoptosis in regulating the number of germ cells in the human testis during development has not been examined.

Accelerated Decrease in Germ Cell Number in the Cryptorchid Human Testis

Cryptorchidism is the most common congenital disorder in newborn boys. The incidence of cryptorchidism has increased during the last few decades (24) from about 1% to 1.5%. It is a serious risk factor for testicular cancer (25) and an important cause of infertility (26).

The pathogenesis of the reduced fertility seen in cryptorchidism has not been fully clarified. Several studies have detailed the morphometric features of cryptorchid testes during the early years of life (23, 27, 28). The germ cell counts of cryptorchid testes are within normal limits during the first year of postnatal life. They fall below the normal range between 1 and 2 years of age, reaching the lowest level of germ cells per tubule at approximately 2 years of age (23, 27, 28). The reduced fertility has been linked to the reduced number of germ cells, because the cryptorchid patients with the lowest total germ cell counts have the poorest spermiograms in adulthood (29). The cause of the reduced total germ cell counts is a matter of debate. Some data suggest that the reduced total germ cell counts in cryptorchidism are caused by defects in the two prepubertal steps leading to the maturation of germ cells (23).

The unfavorable temperature affecting the undescended testis may also be an important factor in the occurrence of germ cell loss infertility. The temperature in the scrotum, which is a few degrees lower than body temperature, is believed to maintain an optimal environment for testicular function. Surgical induction of cryptorchidism in experimental animals causes disruption of spermatogenesis, which leads to infertility (30). Furthermore, surgically induced cryptorchidism in the immature rat testis is followed by a time-dependent, 2- to 4-fold increase in DNA cleavage into the low molecular weight fragments characteristic of apoptosis. The cell type affected by induced cryptorchidism appears to be the primary spermatocyte (31).

Various hormonal therapies have been advocated for the treatment of cryptorchidism, usually involving administration of hCG (32, 33) or GnRH agonists (34, 35) for several weeks. Human CG has been known to induce some adverse effects in the cryptorchid testis, including inflammation-like changes, and to stimulate spermatogenesis and increase the diameter of the seminiferous tubules (36). Treatment with hCG also transiently increases serum androgen concentrations by about 150-fold, e.g., to the levels seen in adult men (37), followed by a gradual decrease over several weeks (Fig. 16.1)

Increased Apoptotic Germ Cell Death in the Human Testis After Treatment of Cryptorchidism With hCG

Since it has been clearly shown that the germ cell death in the rat testis takes place through the process of apoptosis (10, 13), the major objective of our first study was to examine the occurrence of apoptotic germ cell death in the

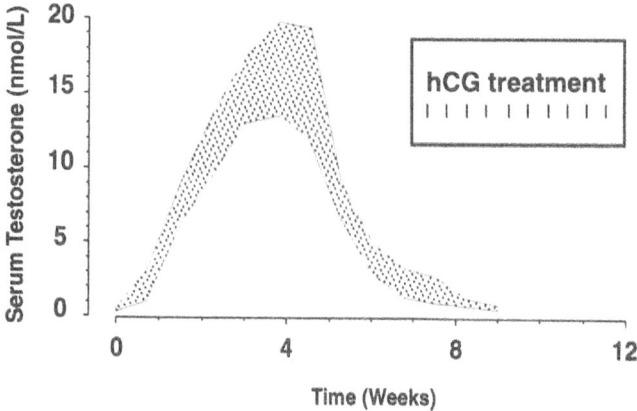

FIGURE 16.1. Serum testosterone levels (± SEM) in cryptorchid boys during and after hCG treatment consisting of 10 intramuscular injections over a period of 5 weeks.

normal and the undescended (cryptorchid) human testis. Furthermore, the findings of gonadotropin- and androgen-mediated control of apoptosis (10–13) prompted us to study the effects of the dramatic changes in serum androgen levels following treatment of cryptorchidism with hCG.

We evaluated the occurrence of apoptosis in the scrotal and cryptorchid testes of 73 prepubertal boys, 43 of whom had received hCH treatment. A total of 102 testis biopsies were examined. In this study, testicular cell apoptosis was investigated mainly by the in situ 3' end-labeling technique in order to discern the specific cells undergoing apoptosis. To quantify our findings, the amount of apoptosis in each section analyzed was recorded as the number of positively-stained cells per square millimeter of tubular area, and the percentage of positively-stained cells was also noted.

With the in situ 3' end-labeling technique, the apoptotic cells in the seminiferous tubules were identified as being exclusively spermatogonia (Fig. 16.2). These apoptotic spermatogonia appeared as single cells in both the scrotal and the cryptorchid testes.

To validate our in situ labeling method, we selected three biopsy specimens containing a great number of apoptotic cells detected in situ and three others containing only a few positive cells per sample. We then isolated the DNA and analyzed the degrees of DNA fragmentation (Fig. 16.3). In those samples in which apoptosis, in situ, was abundant (4 through 6), the ladder pattern indicating the internucleosomal DNA cleavage typical of apoptosis was pronounced, whereas in the samples with little apoptosis, in situ (1 through 3), the ladder pattern was barely or not detectable. This result corroborates the finding that the cells appearing to be stained in situ are indeed apoptotic.

The effect of hCG on spermatogonial apoptosis was investigated by assigning the patients to 4 subgroups according to their hCG treatment. In the

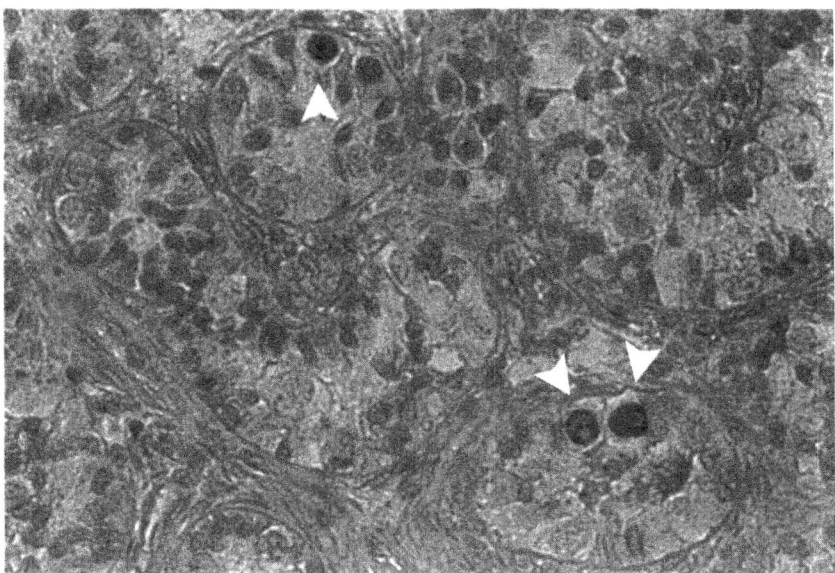

FIGURE 16.2. Apoptotic cell death in situ in the germ cells of the human testis. A representative testicular biopsy from a scrotal testis is shown. The biopsy was taken from a 5-year-old boy 2 weeks after hCG treatment. The histologic sections were stained with an in situ 3' end-labeling technique in which the digoxigenin-ddUTP labels is detected by digoxigenin antibodies conjugated with alkaline phosphatase. Arrows indicate apoptotic cells.

first group, no hormonal treatment had been given prior to the biopsy. In the other three groups, patients had received hCG injections and had a biopsy taken either 0 to 4, 4 to 12, or 12 to 52 weeks after the last hCG and had had an injection (Fig. 16.4). Overall, when the number of apoptotic spermatogonia was expressed per square millimeter of tubular area, the scrotal testes contained more apoptotic spermatogonia than the cryptorchid testes ($p < 0.05$) (Fig. 16.4A). This proved to be true in all four treatment groups. The higher total number of apoptotic spermatogonia in the scrotal, as compared to the cryptorchid, testes was due to the lower total number of spermatogonia in the cryptorchid testes. However, a higher percentage of spermatogonia, out of the total number of spermatogonia, were apoptotic in the cryptorchid testes as compared to the scrotal testes ($p < 0.01$) (Fig. 16.4B).

Analysis of the treatment groups demonstrated that immediately after hCG treatment the number of apoptotic cells had increased significantly as compared with the nontreated group ($p < 0.001$). The effect of hCG was not permanent, since 4 to 12 weeks after the last hCG injection the number of apoptotic spermatogonia was comparable to that in the nontreated group. The difference between the groups less than 4 weeks after treatment and 4 to 12 weeks

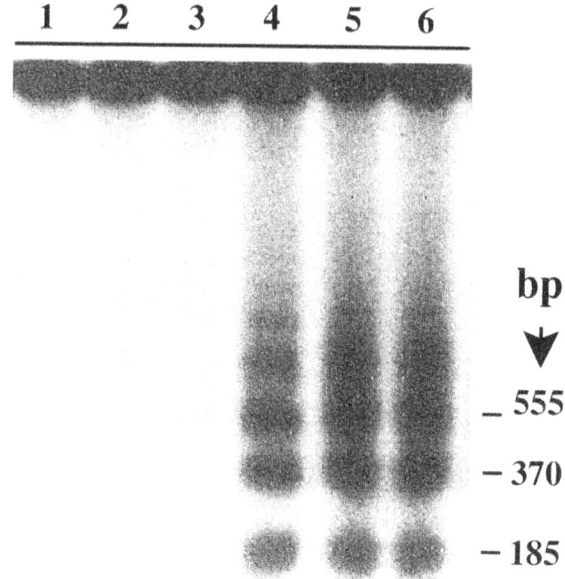

FIGURE 16.3. DNA isolation and analysis for fragmentation. The specificity of the apoptotic staining in situ was validated in selected biopsies, of which three displayed considerably more apoptosis in situ than the other three. The typical ladder pattern indicates the extent of DNA fragmentation, consistent with low (lanes 1 through 3) and high (lanes 4 through 6) numbers of stained cells. For this detection technique, one microgram of DNA from each biopsy sample was isolated before labeling its 3' ends with (α^{32}P)-dideoxy-ATP. The DNA samples were then loaded on agarose gels and separated by electrophoresis. Evidence of apoptosis, indicated by the occurrence of internucleosomal DNA breakdown into 185-basepair (bp) multiples, is observed. (Reproduced with permission from Heiskanen et al. (40).)

after treatment was significant ($p < 0.001$). In the group in which more than 12 weeks had elapsed after the last injection of hCG, the level of apoptosis was similar to the level in the nontreated group (Fig. 16.4). These data suggest that hCG (and/or androgen) withdrawal increases germ cell apoptosis in the human testis.

To compare the levels of testicular apoptosis at different ages, the patients were divided into three age groups (1 to 5 years, 5 to 9 years, and 9 to 16 years). The levels of apoptosis in the spermatogonia of these three groups did not differ significantly.

Number of apoptotic spermatogonia / mm^2

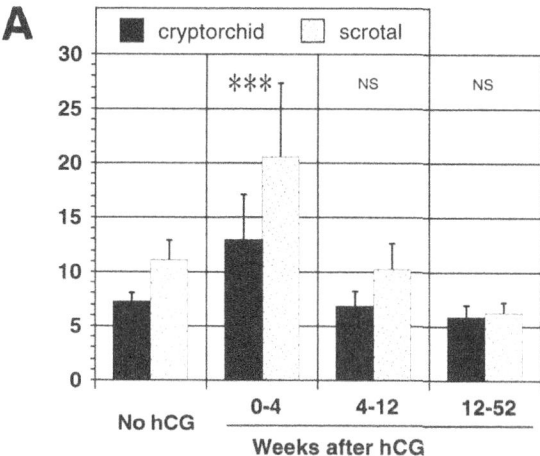

Apoptotic vs. non apoptotic spermatogonia (%)

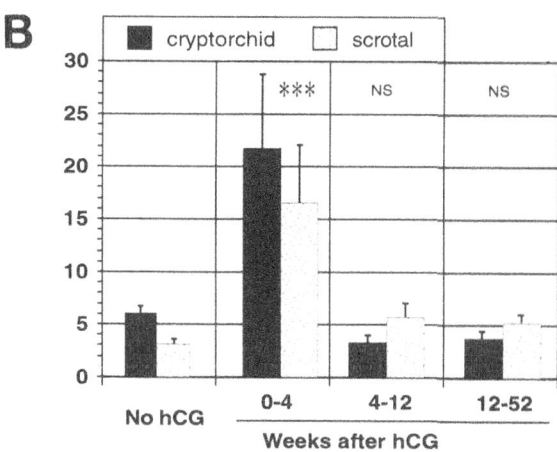

FIGURE 16.4. Effects of hCG treatment on apoptosis in seminiferous tubules of scrotal and cryptorchid testes. Patients were divided into four groups according to hCG treatment. In the first group, no hormone treatment was given. In the other three groups, the biopsy was obtained 0–4, 4–12, or 12–52 weeks after the last injection of hCG. The histological samples were stained by the in situ DNA 3′ end-labeling technique. The seminiferous tubules of the sections were analyzed under the light microscope and the results were recorded as stained cells per square millimeter of the tubular area (A) or as a percentage of the stained cells (B). Asterisks denote significant differences between the nontreated groups and the various treatment groups (*** = $p < 0.001$, NS=nonsignificant difference).

Long-Term Consequences of the hCG
Treatment-Induced Apoptosis

Having found increased apoptotic death of germ cells after hCG treatment of cryptorchidism, we assessed the possible long-term consequences of the hCG treatment-induced germ cell death in prepuberty on reproductive functions in adult life.

We studied 21 adult men with a history of surgically treated cryptorchidism, 15 of whom had received unsuccessful hCG therapy before orchidopexy. Apoptosis detected in testis biopsies taken in prepubertal life was correlated with various parameters of testicular function in adult life. At orchidopexy, the age of the patients varied from 1 to 13 years, and at follow-up examination from 16 to 30 years. Testicular volumes were measured with a ruler, using the formula, $0.71 \times$ length $\times$ width^2 (38). Sperm was collected after a minimum of 2 d of abstinence. Semen analysis was done within 2 h and interpreted according to the World Health Organization (WHO) 1992 criteria (39). The results of the semen analysis were considered normal, when the sperm concentration was at least 20×10^6 spermatozoa/mL, the percentage of spermatozoa with progressive motility was at least 50%, and at least 30% of the spermatozoa had a normal morphology. From each patient, the baseline serum FSH, LH, testosterone and prolactin levels were quantified using conventional radioimmunoassays.

Consistent with previous observations, a few scattered apoptotic spermatogonia were seen after DNA end-labeling of the biopsies from patients not treated with hCG, whereas in the samples taken after hCG treatment the labeling of spermatogonia was more extensive. In testis biopsy specimens from boys who had not received hCG treatment for cryptorchidism, no internucleosomal DNA cleavage was detected, even with the highly sensitive method we used to analyzed DNA integrity (Fig. 16.5A). In contrast, the ladder pattern showing the internucleosomal DNA cleavage typical of apoptosis was pronounced in samples obtained from testes after unsuccessful treatment of cryptorchidism with hCG. Quantification of low molecular weight DNA (< 10 kb) showed that testicular DNA from boys who had not received hCG treatment for cryptorchidism showed very low levels of DNA fragmentation. In contrast, the amount of low molecular weight DNA in samples taken after hCG treatment was 2.9-fold higher than the samples without prior hCG treatment ($p < 0.001$, Fig. 16.5B). The amount of DNA fragmentation was not dependent on age at treatment or the location of the cryptorchid testis (data not shown).

About 20 years after the biopsy, the volume of the cryptorchid testis was smaller in the hCG-treated group than in the nontreated group (12.7 ± 0.8 mL versus 9.4 ± 0.9 mL; mean $\pm$ SEM, $p < 0.01$). The sperm densities were also lower, and the FSH levels were higher, in the patients with excessive apoptosis in spermatogonia following treatment with hCG. Among the patients who had

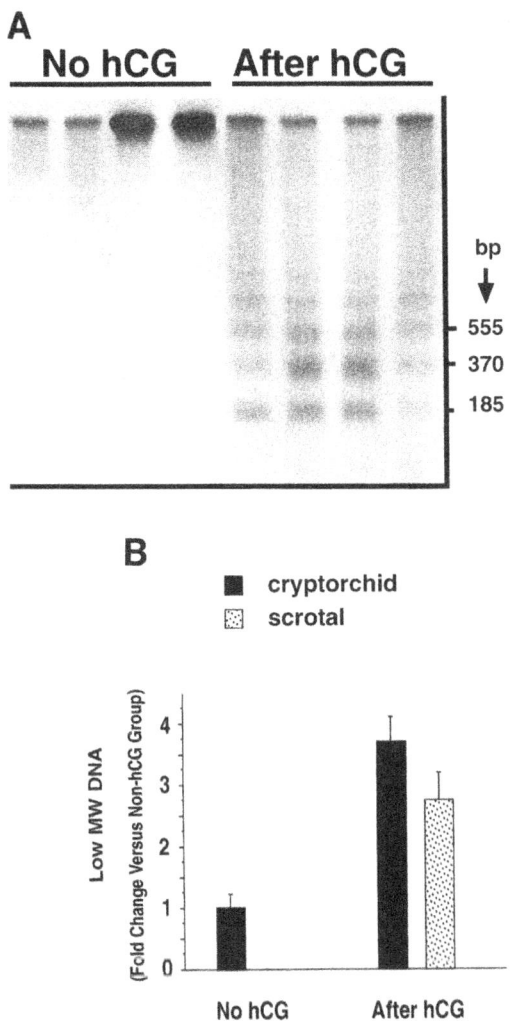

FIGURE 16.5. Analysis of DNA integrity in a human testicular biopsy specimen, indicating the effect of hCG treatment. (A) Autoradiography showing eight representative DNA ladder patterns of testes from patients without and with prior unsuccessful treatment with hCG. For this detection method, one microgram of DNA from each biopsy sample was isolated before labeling its 3′ ends with (α-^{32}P)-dideoxy-ATP. The DNA samples were then loaded on agarose gels and separated by electrophoresis. Evidence of apoptosis, indicated by the occurrence of internucleosomal DNA breakdown into 185-basepair (bp) multiples, is observed in the testis after hCG treatment (B) Quantitative analysis of DNA integrity in the testis biopsy specimen assessed by β-counting of the low molecular weight (MW, <10 kb) DNA fractions following electrophoresis. These data demonstrate that increased apoptotic DNA breakdown occurs in the cryptorchid testis after treatment with hCG.

Testis volume

Sperm density

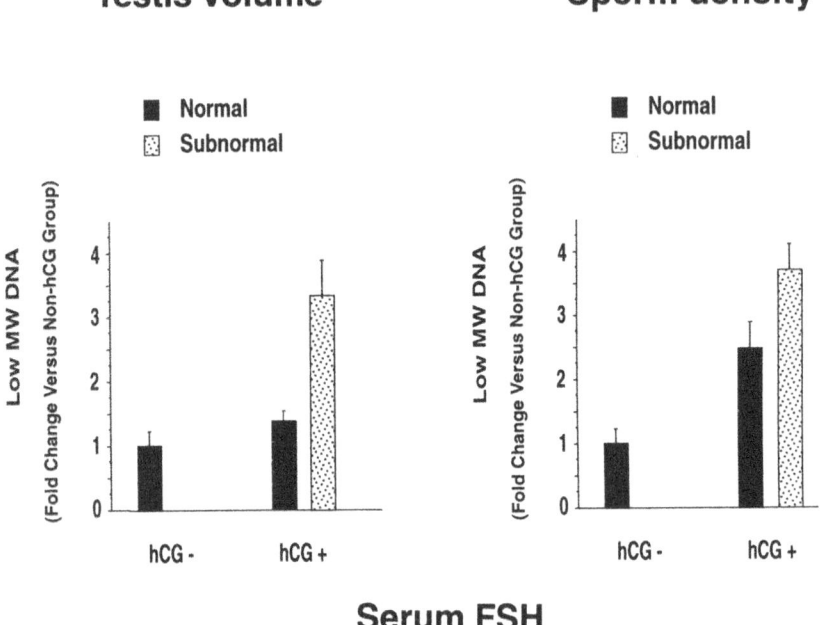

Serum FSH

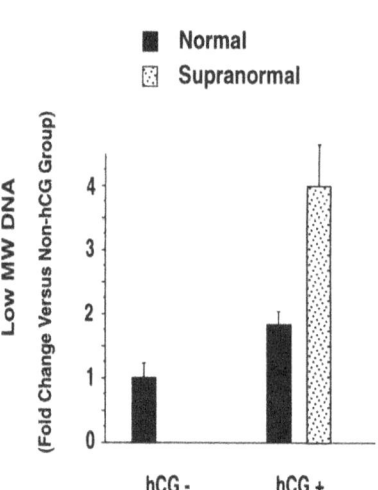

FIGURE 16.6. The outcome of hCG treatment-induced apoptotic DNA fragmentation. The relative levels of low molecular weight DNA fragmentation are shown for patients who were treated for cryptorchidism either operatively only (hCG⁻) or operatively after preceding unsuccessful treatment with hCG (hCG⁺) in prepuberty. The hCG-treated patients were subdivided according to normal or pathological values for testis volume, sperm density or serum FSH levels in adulthood. The clear differences in low molecular weight DNA cleavage between the two groups indicates that the level of DNA fragmentation in prepuberty predicts testis function in adulthood.

received hCG treatment for cryptorchidism in prepuberty, those who had subnormal testis volumes, subnormal sperm density, and/or pathologically elevated serum FSH levels had exhibited significantly more apoptotic DNA fragmentation after hCG treatment (Fig. 16.6). These data suggest that the development of the testis may be severely disrupted by hCG treatment-induced germ cell apoptosis. Furthermore, the degree of apoptotic cell death following hCG treatment in prepuberty seems to predict the outcome of testis function in adulthood.

Conclusion

Our findings lead to the conclusion that apoptosis is a normal, hormonally controlled phenomenon in the human testis in prepuberty. Cryptorchidism decreases its occurrence by reducing the number of germ cells capable of undergoing apoptosis. Apoptotic loss of spermatogonia after hCG treatment of cryptorchidism has dramatic long-term consequences. Because of the poor success rates to induce testicular descent, and the negative effects on subsequent reproductive function, the potential long-term hazards of hCG treatment of the testis should be critically reevaluated.

References

1. Clermont Y. Quantitative analysis of spermatogenesis of the rat: a revised model for the renewal of spermatogonia. Am J Anat 1962;111:111–29.
2. Huckins C. The morphology and kinetics of spermatogonial degeneration in normal adult rats: an analysis using a simplified classification of germinal epithelium. Anat Rec 1978;190:905–26.
3. Wing T-Y, Christiansen AK. Morphometric studies on rat seminiferous tubule. Am J Anat 1982;165:13–25.
4. Oakland E. A description of spermatogenesis in the mouse and its use in analysis of the cycle of seminiferous epithelium and germ cell renewal. Am J Anat 1956;99:391–413.
5. De Rooij DG, Lok D. Regulation of the density of spermatogonia in the seminiferous epithelium of the Chinese hamster: II. Differentiating spermatogonia. Anat Rec 1987;217:131–6.
6. Allan DJ, Harmon BV, Roberts SA. Spermatogonial apoptosis has three morphologically recognizable phases and shows no circadian rhythm during normal spermatogenesis in the rat. Cell Prolif 1992;25:241–50.
7. Kerr JB. Spontaneous degeneration of germ cells in normal rat testis: assessment of cell types and frequency during the spermatogenic cycle. J Reprod Fertil 1992;95:825–30.
8. Johnson L, Lebovitz RM, Samson WK. Germ cell degeneration in normal and microwave-irradiated rats: potential sperm production rates at different developmental steps in spermatogenesis. Anat Rec 1984;209:501–7.
9. Flemming W. Neue Beiträge zur Kenntniss der Zelle. In: Archiv für Mikroskopishe Anatomie, v la Valette St. G, Waldeyer W, eds. Bonn: Verlag von Max Cohen & Sohn (Fr. Cohen), 1887:389–463.

10. Billig H, Furuta I, Rivier C, Tapanainen J, Parvinen M, Hsueh AJW. Apoptosis in testis germ cells: developmental changes in gonadotropin dependence and localization to selective tubule stages. Endocrinology 1995;136:5–12.
11. Sinha Hikim AP, Wang C, Leung A, Swerdloff R. Involvement of apoptosis in the induction of germ cell degeneration in adult rats after gonadotropin-releasing hormone antagonist treatment. Endocrinology 1995;136:2770–75.
12. Henriksen K, Hakovirta H, Parvinen M. Testosterone inhibits and induces apoptosis in rat seminiferous tubules in a stage-specific manner: in situ quantification in squash preparations after administration of ethane dimethane sulfonate. Endocrinology 1995;136:3285–91.
13. Tapanainen J, Tilly JL, Vihko K, Hsueh AJW. Hormonal control of apoptotic cell death in the testis: gonadotropins and androgens as testicular cell survival factors. Mol Endocrinol 1993;7:643–50.
14. Russel LD, Clermont Y. Degeneration of germ cells in normal, hypophysectomized and hormone treated hypophysectomized rats. Anat Rec 1977;187:347–66.
15. Sharpe RM. Regulation of spermatogenesis. In: Knobil E, Neill JD, eds. The physiology of reproduction, 2 ed. New York: Raven Press, 1994:1363–434.
16. Winter JSD, Fairman C, Hobson WC, Prasad AV, Reyes FI. Pituitary-gonadal relations in infancy. I. Patterns of serum gonadotropin concentrations from birth to four years of age in man and chimpanzee. J Clin Endocrinol Metab 1975; 40:545–51.
17. Forest M, Cathiard AM, Bertrand JA. Evidence of testicular activity in early infancy. J Clin Endocrinol Metab 1973;37:148–51.
18. Muller J. Skakkebaek NE. Fluctuations in the number of germ cells during the late foetal and early postnatal periods in boys. Acta Endocrinol (Copenh) 1984;105:271–4.
19. Hadziselimovic F, Herzog B, Buser M. Development of cryptorchid testes. Eur J. Pediatr 1987;146(Suppl 2):8–12.
20. Hadziselimovic F, Thommen L, Girard J, Herzog B. The significance of postnatal gonadotropin surge for testicular development in normal and cryptorchid testes. J Urol 1986.
21. Huff DS, Hadziselimovic F, Snyder HM, III, Blyth B, Duckett JW. Early postnatal testicular maldevelopment in cryptorchidism. J Urol 1991;146:624–6.
22. Clermont Y. Renewal of spermatogonia in man. Am J Anat 1966;118:509–24.
23. Huff DS, Hadziselimovic F, Snyder HM, III, Duckett JW, Keating MA. Postnatal testicular maldevelopment in unilateral cryptorchidism. J Urol 1989;142:546–8.
24. John Radcliffe Hospital Cryptorchidism Study Group. Cryptorchidism: a prospective study of 7500 consecutive male births, 1984-8. Arch Dis Childh 1992;67:892–9.
25. Martin DC. Malignancy in the cryptorchid testis. Urol Clin North Am 1982; 9:371–6.
26. Kogan SJ. Fertility in cryptorchidism. An overview in 1987. Eur J Pediatr 1987; 146:21–4.
27. Hedinger E. Histopathology of undescended testes. Eur J. Pediatr 1982; 139:266–71.
28. Schindler AM, Diaz P, Cuendet A, Sizonenko PC. Cryptorchidism: a morphological study of 670 biopsies. Helv Paediatr Acta 1987;42:145–58.
29. Hadziselimovic F, Herzog B, Hocht B, Hecker E, Miescher E, Buser M. Screening for cryptorchid boys risking sterility and results of long-term buserelin treatment after successful orchidopexy. Eur J Pediatr 1987;146(Suppl):59–62.

30. Nelson WO. Mammalian spermatogenesis: effects of experimental cryptorchidism in the rat and non-descent of the testis in man. Recent Prog Horm Res 1951;6:29–62.

31. Shikone T, Billig H, Hsueh AJW. Experimentally-induced cryptorchidism increases apoptotic cell death in rat testis. Biol Reprod 1994;51:865–72.

32. Rajfer J, Handelsman DJ, Swerdloff RS, et al. Hormonal therapy of cryptorchidism. A randomized double-blind study comparing human chorionic gonadotropin and gonadotropin-releasing hormone. New Engl J Med 1986;314:466–70.

33. Christiansen P, Muller J, Buhl S, et al. Hormonal treatment of cryptorchidism-hCG or GnRH-a multicenter study. Acta Paediatr 1992;81:605–8.

34. Hagberg S, Westphal O. Results of combined hormonal and surgical treatment for underscended testis in boys under 3 years of age. Eur J Pediatr 1987;146 (Suppl):38–9.

35. Karpe B, Eneroth P, Ritzen EM. LHRH treatment in unilateral cryptorchidism: effect on testicular descent and hormonal response. J Pediatr 1983;103:892–7.

36. Hjertkvist M, Lackgren G, Ploen L, Bergh A. Does hCG treatment induce inflammation-like changes in undescended testes in boys? J Pediatr Surg 1993;28:254–8.

37. Dunkel L, Perheentupa P, Apter D. Kinetics of the steroidogenic response to single versus repeated doses of human chorionic gonadotropin in boys in prepuberty and early puberty. Pediatr Res 1985;19:1–4.

38. Lambert B. The frequency of mumps and of mumps orchitis and the consequences for sexuality and fertility. Acta Genet Stat Med 1951;2:1–166.

39. WHO laboratory manual for the examination of human semen and semen-cervical mucus interaction. Cambridge: Cambridge University Press, 1992.

40. Heiskanen P, Billig H, Toppari J, Kaleva M, Rapola J, Arsalo A, Dunkel L. Apoptotic cell death in the normal and cryptorchid human testis: the effect of human chorionic gonadotropin on testicular cell survival. Pediatr Res 1996;40:351–6.

17

Apoptosis and Tumor Invasion in Hormone-Dependent Cancers

Martin Tenniswood, Sean Guenette, Colm Morrissey,
Jacintha O'Sullivan, Zhengqi Wang, Ping Zhan, Srikala Sridhar,
Johnathon Lakins, and Hailun Tang

Morphology of Apoptosis in Glandular Tissues

Accumulating evidence demonstrates that apoptosis is not a single phenomenon, but a series of morphologically and biochemically related processes (1, 2). Cell death of lymphocytes and other cells of reticulo-endothelial origin is dominated by changes in nuclear morphology (3), whereas apoptosis of secretory epithelial cells involves profound cytoplasmic changes and alterations in the cell-cell and cell-substratum interactions typical of highly organized tissues (1, 3). Analysis of electron micrographs has demonstrated that the loss of the secretory epithelium after hormone withdrawal is achieved through the induction of apoptosis in both the prostate (4) and mammary gland (5). In glandular epithelia, the process of apoptosis can be broken down into several stages (Fig. 17.1). During the precondensation stage, many of the genes that are necessary for cell death are induced *de novo,* or recruited from other functions in the gland. The length of the precondensation stage appears to vary from cell to cell within the same tissue and probably reflects the micro-heterogeneity in the hormone or growth factor environment. The precondensation phase is followed by cytoplasmic condensation that involves the loss of the interactions between the dying cell and its neighbors as the extracellular matrix (ECM) is degraded and the cytoplasmic volume decreases. During nuclear condensation, activation of one or more endonucleases results in the fragmentation of the DNA and its marginalization to the nuclear periphery, producing the hyperchromatic, pyknotic nucleus characteristic of apoptotic cells. Since the initiation of apoptosis in glandular tissues is stochastic, it is difficult to clearly separate cytoplasmic and nuclear condensation on a temporal basis in vivo. However, in isolated cells cultured in vitro (such as granulosa cells), nuclear condensation is clearly initiated after cytoplasmic

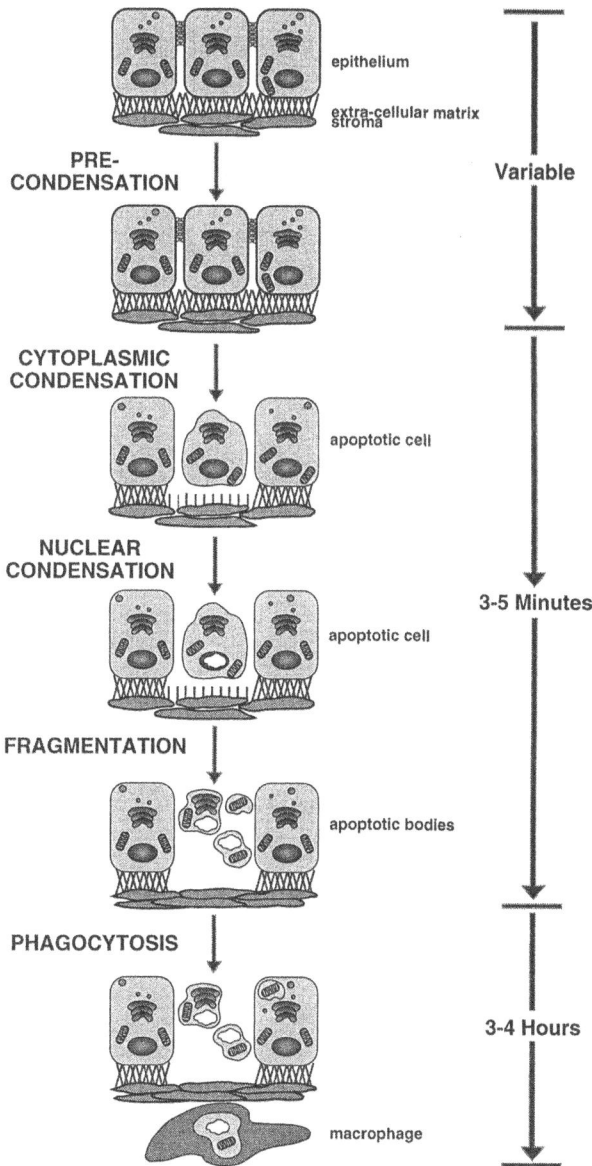

FIGURE 17.1. Schematic representation of the stages of apoptosis in glandular epithelial cells. Approximate time line is shown on right.

condensation (Chapter 9). During the fragmentation phase the apoptotic cell is subdivided into several apoptotic bodies that are subsequently phagocytosed by the neighboring epithelial cells or macrophages and degraded by the lysosomal enzymes either activated in the host cell or in the apoptotic body (6, 7). Remarkably, this latter process occurs without the leakage of the intracellular components into the extracellular space, ensuring that the complement cascade system is not activated, and that there is no inflammatory response. This feature distinguishes apoptosis from necrosis. In the prostate and mammary gland, cytoplasmic and nuclear condensation appear to take between 3 and 5 min and the remaining visible stages of apoptosis are completed within 3 to 4 h. Inhibitors of RNA and protein synthesis delay (but do not completely block) apoptosis in the prostate and other tissues, indicating that both RNA and protein synthesis are required (8, 9).

Control of DNA Fragmentation

One of the most striking characteristics of apoptosis is the formation of hyperchromatic nuclei containing DNA that is then redistributed to the nuclear margins. This marginalization correlates with the activation of Ca^{2+}/Mg^{2+} dependent endonuclease(s) in the dying cell that digests the DNA into oligonucleosome sized fragments that can be detected following agarose gel electrophoresis (10–13). However, internucleosomal DNA cleavage is not seen in all cell types undergoing apoptosis. Indeed, the search for the enzyme(s) responsible for the endonucleolytic cleavage has proved very frustrating. Detailed analysis using pulsed field gel electrophoresis (PFGE) has shown that fragmentation of chromatin probably requires 2 or 3 independent enzyme activities with significantly different cation requirements. There is also evidence that one or more serine proteases may be required, either to proteolytically activate the enzyme or to initiate the degradation of the nuclear matrix that renders the DNA more accessible to the endonucleases. Although several candidate endonucleases have been proposed, including DNase I, DNase II, and nuc 18/cyclophilin A, none of the proposed endonucleases have all of the necessary characteristics to be the endonuclease. Analysis of the fragmentation using PFGE has demonstrated that DNA fragmentation is probably a stepwise process, involving fragmentation at interrosette sites (generating DNA fragments larger than 300 kb), interloop sites (generating fragments of approximately 50 kb) and internucleosomal sites (generating the characteristic 200 bp fragments seen in nucleosome ladders) (14). Furthermore, comparative studies using PFGE have shown that many cells undergoing apoptosis, including PC12, DU145, and MCF-7 cells, do not completely fragment their DNA (15, 16), even though the enzymatic activities necessary to do so are present in isolated nuclei (17). This demonstrates that, although all cells appear to have the enzymatic apparatus necessary for DNA fragmentation, subtle differences in chromatin structure, proteolytic activation, intranuclear pH, or activating divalent cation (Ca^{2+} and Mg^{2+}) or inhibiting ion (Zn^{2+} and K^+) concentrations may significantly alter the extent of DNA fragmentation (13, 18) (Chapter 18).

There is considerable evidence to suggest that members of the interleukin converting enzyme (ICE)-like cysteine protease family participate in an intracellular proteolytic cascade that may lead to the activation of the endonuclease(s). The activation of the endonuclease appears to involve the sequential proteolytic activation of several ICE-like proteases that culminate in the cleavage of poly(ADP)ribose polymerase in the nucleus, and activation of the endonuclease (19). The activation of this pathway has been strongly implicated in the induction of cell death in cells of neuronal and hematopoietic origin (20–23). In most cells, this pathway is inhibited by BCL-2 and is activated by BAX (24–26) and BAK (27). Although the expression of *bcl-2* and its dimerization partners has not been studied in the normal prostate or mammary gland in detail, it is known that *bcl-2* is upregulated in hormone refractory prostate tumors (26). These data suggest that the aberrant expression of this gene may contribute to inappropriate survival of tumor cells (28). Once the DNA has been fragmented, the cell cannot survive for any appreciable length of time and can no longer divide.

The mechanistic role of c-Myc in the apoptosis of epithelial cells in the prostate and mammary gland is not well understood. In fibroblasts elevation of c-Myc is required for the switch from proliferation to apoptosis and appears to require the dimerization with Max (29, 30). The activation of apoptosis can be blocked by a number of growth factors, particularly insulin-like growth factor-I (IGF-I) (31). It is known that c-Myc levels are elevated in the prostate immediately after castration, and it has been suggested that apoptosis in the terminally differentiated epithelial cells requires the re-entry of these cells into a defective cell cycle (32). However, other emerging data suggest that the entry of terminally differentiated secretory cells into the apoptotic pathway is independent of changes in cyclins and other genes involved in the progression through S phase (33, 34). The role of p53 in the death of the glandular epithelium is also unresolved since apoptosis is delayed but not prevented in the p53 knockout mouse (35), and many examples of p53-independent apoptosis can be found in the literature (36). Both of these proteins, and probably other unidentified nuclear factors, play a central role in the initiation and completion of apoptosis, since they regulate the induction of a number of genes that are essential to the process and are themselves regulated by a variety of growth factors.

Survival Signals in Glandular Epithelia

The secretory epithelial cells of both the prostate and mammary gland interact with the basement membrane located between the stroma and the epithelium. Several lines of evidence suggest that the normal function of prostatic and mammary epithelial cells is dependent on a complex interplay between the epithelium, the ECM, and the underlying mesenchyme or stroma (37–41). The mesenchyme probably influences normal epithelial function by two different, but interacting, receptor-mediated systems: growth factors derived from the stromal compartment and components of the ECM.

Role of Soluble Growth Factors in Tissue Homeostasis

There is considerable evidence indicating that the growth and proliferation of epithelial cells in the prostate is influenced by factors such as epidermal growth factor (EGF), transforming growth factor alpha (TGF-α), transforming growth factor beta (TGF-β), nerve growth factor (NGF) and members of the fibroblast growth factor (FGF) families (42–48). Recent evidence has suggested that the interaction of some of these growth factors with their cognate receptors, in particular FGF-7, requires both the specific epithelial receptor and defined components of the ECM, particularly chondroitin sulfate and heparan sulfate (49). As such both may be required, and responsible, for the epithelial differentiation and possibly survival (50, 51). Changes in the relative levels of these growth factors or glycosaminoglycans may influence the induction of apoptosis in glandular epithelial cells and contribute to the microheterogeneity in the ducts of the prostate and breast (40, 52).

Members of the insulin like growth factor family (IGF-I and IGF-II) have recently emerged as key physiological regulators of apoptosis in a number of systems (53–55). For example, it has been demonstrated that IGF-I and its cognate receptor (IGF-RI) can block the induction of apoptosis induced by VM-26, a specific inhibitor of topoisomerase. IGF-I can also prevent apoptosis after the withdrawal of interleukin-3 (IL-3) in IL-3 dependent hematopoietic cells (G. Williams, personal communication). Thus, IGF-I appears to play a central role in tissue homeostasis by repressing the induction of apoptosis, and allowing differentiation or mitosis to occur. As mentioned above, the permissive effect of IGF-I appears to be downstream of c-myc, blocking the entry of the cell into the apoptotic pathway. Elevation of c-myc is necessary but not sufficient for the induction of apoptosis, and down-regulation of IGF-I signaling also appears to be required (30, 31, 56, 57). Other studies suggest that IGF-I is an essential growth factor that is absolutely required for cell survival (58). The prostatic stroma synthesizes and secretes IGF-I (59) and IGF-II (44, 60). The stroma of the normal mammary epithelium secretes IGF-I (53, 61, 62), while the stroma of malignant mammary tumors appears to secrete predominantly IGF-II (63–65). The secretory epithelium of both tissues respond to IGF-I (and to a lesser extent IGF-II and insulin) through the high affinity interaction of these growth factors with the type 1 IGF receptor (IGF-RI) (44, 66, 67). After hormone ablation, insulin-like growth factor binding protein-5 (IGFBP-5) is induced in the epithelial cells of the prostate and mammary gland. The induction of IGFBP-5 is thought to attenuate the cellular response to IGF-I through the high affinity binding of IGF-I to IGFBP-5, which itself associates with several components of the ECM, including types III and IV collagen, laminin, fibronectin, and tenascin (68, 69). These associations probably sequester IGF-I away from its receptor, disrupting the normal homeostatic second messenger signaling downstream of the receptor. Although the mechanisms involved have yet to be elucidated, this appears to result in the induction of several of the genes implicated in the apoptotic process, particularly those involved in the process of cytoplasmic condensation.

Role of ECM in Homeostasis

The ECM contains a number of components, including fibronectin, collagen, laminin, and vitronectin, (70, 71), that are known to interact with their cognate receptors, such as members of the integrin superfamily (72–74). The integrins are localized on the basal surface of epithelial cells and ensure that the cells have the structural underpinnings needed for polarization, vectorial transport and secretion (75–79). The ECM components and their respective membrane receptors are expressed in hormone replete animals, and their synthesis is under tight regulation to ensure that the glandular morphology and differentiated (secretory) phenotype is maintained (37, 38, 79–81).

Tissue Regression Following Hormone Ablation

The rodent prostate is a very useful model for the study of apoptosis induced by hormone withdrawal. The gland is an arborized network of ducts, with a number of cell types that display anatomical and biochemical heterogeneity, and substantially different sensitivities to androgen ablation. After puberty, the rat ventral prostate continues to grow slowly throughout the life span of the animal, and the cells in the replicative zone of the ducts (at the distal tips) continue to divide slowly. After division these cells differentiate and become androgen-dependent epithelial cells that synthesize and secrete prostate steroid binding protein (PSBP) and acid phosphatase (SAP) into the lumen of the ducts. These secretory cells are localized in the distal and intermediate regions of the ducts (52, 82–85) and appear to be critically dependent on androgens for survival (86). In the proximal region of the ducts the luminal epithelial cells do not display any secretory activity and do not express SAP or PSBP (84, 87). Neither these cells, nor the basal cells that are also localized to the proximal region, appear to require androgens for survival.

Following castration there is a gradual depletion of the active androgen, 5α-dihydrotestosterone (5α-DHT), in the prostate over a 24 h period. When the 5α-DHT level falls below that needed to inhibit involution (approximately 8-16 h after castration), the gland starts to regress (88). The striking reduction in prostate size that occurs over the next 3 to 6 d is primarily due to the selective loss of the secretory luminal epithelial cells in the distal and intermediate regions, resulting in the complete obliteration of many of the ducts while maintaining the proximal segments of the ducts following castration.

Identification of Clusterin in Regressing Prostate

Between d 3 and d 7 after castration, there is a well documented reduction in the androgen-dependent activities of the prostate, including the synthesis of SAP, PSBP, and kallikreins, and an alteration in the metabolism of androgens (89–93). Involution of the prostate results in the elimination of 80% of the secretory epithelial cells of the glands, and requires RNA and protein synthesis (8, 86, 94, 95). Results from R_0t analysis and two dimensional gel electrophoresis suggest that between 20 to 30 mRNA species are up-regulated in the regressing prostate (9, 96). A number of different strategies have been used to identify these genes. The first gene to be cloned and characterized in this context, TRPM-2 (testosterone repressed prostate message, now referred to by convention as clusterin), was cloned by differential hybridization (97). This protein is the most abundant protein induced in the prostate and mammary gland after hormone ablation (85, 96–100). It has subsequently been shown that clusterin is expressed in a wide variety of tissues undergoing apoptosis (95, 101). Although the expression of the gene is not exclusively confined to dying cells in other systems (particularly the testis and liver) (102–105), it clearly plays an integral part in the death of secretory epithelial cells in the prostate and mammary gland. Although the role of clusterin in cell death is still not firmly established, the protein appears to be involved in controlling cholesterol efflux from the membranes of the dying cells, facilitating the transfer of cholesterol to ApoA-I/HDL, and protecting the membrane from complement fixation (106). It thus serves to facilitate membrane remodelling and the removal of redundant membranes while protecting the membranes from recognition by the immune system. All of these are critical processes that must occur as the apoptotic cell shrinks prior to fragmentation into apoptotic bodies. It has also been shown that over-expression of clusterin in prostate cancer cells protects them from apoptosis induced by TNFα (107), suggesting that the absolute level of clusterin expression, or the timing of the expression, may be critical in determining the sensitivity of the cells to apoptotic stimuli.

Induction of Cathepsin B During Tissue Regression

Since the induction of apoptosis in individual cells in the rat ventral prostate is asynchronous, the use of differential hybridization has been limited to the identification of the more abundant sequences that are induced during prostate regression. To identify other, less abundant sequences, we have used a screening strategy that was developed to cross-screen recombinant libraries cloned into different lambda-based vectors (108). We have used this lateral cross-screening methodology to clone a number of sequences that are expressed in the ventral prostate after castration and lactating mammary gland after weaning. One of these sequences, originally called RSG-2, is the rat homolog of cathepsin B (109). The steady state levels of cathepsin B mRNA rise significantly in the ventral prostate, reaching a maximum on day 4 after castration (Fig. 17.2). In situ hybridization demonstrates that the mRNA is expressed almost exclusively in the luminal, secretory epithelial cells that undergo apoptosis, and immunofluorescence shows that the protein is clearly associated with apoptotic cells (highlighted by arrowheads). We have demonstrated a similar pattern of induction and cellular distribution of cathepsin B in the regressing mammary gland (109). Furthermore, the synthesis and processing of this enzyme is dramatically increased in both the regressing

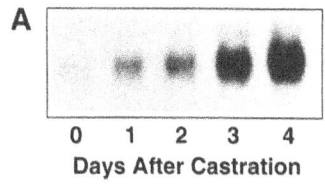

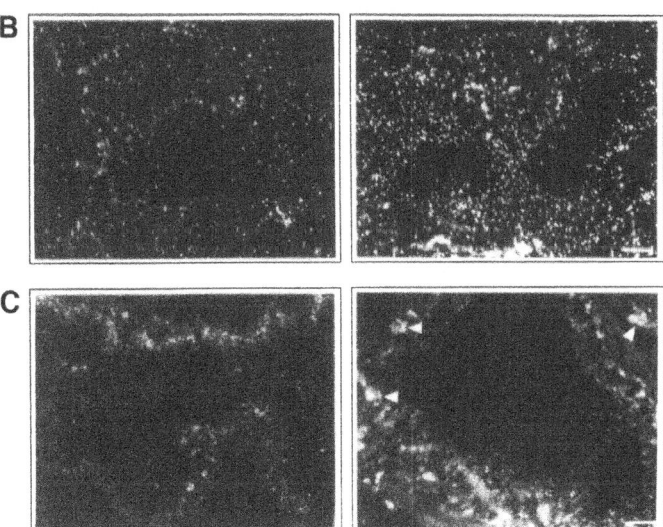

FIGURE 17.2. Expression of cathepsin B during regression of the rat prostate. (A) Northern blot analysis of RNA isolated from the rat ventral prostate on different days after castration. (B) In situ hybridization of cathepsin B mRNA in the normal rat prostate before castration (left) and in the prostate 4 d after castration (right). (C) Immunofluorescent localization of cathepsin B in the normal rat prostate (left) and in the prostate 4 d after castration (right). Arrowheads identify apoptotic cells.

prostate and regressing mammary gland (Fig. 17.3). Cathepsin B has been implicated in the degradation of specific components of the ECM and basement membrane.

Although the composition of the basement membrane is not static, drastic changes in its composition may adversely influence gene expression, cellular function and motility (41, 78, 110). For this reason the synthesis and proteolysis of the components of the basement membrane must be tightly regulated. The ECM contains a number of specific protease inhibitors, including members of the TIMP (tissue inhibitors of metalloprotease) (111), PAI (plasminogen activator inhibitor) (112–115), cystatin and stefin families (116, 117). The balance between the relative level of proteases and their specific inhibitors dictates the degree of protease activity, and essentially ensures that uncontrolled proteolysis of the components of the basement membrane does not occur. Since the degradation of the basement membrane is a prerequisite event for the cytoplasmic condensation of the dying cell, induction and activation of secreted proteases is a necessary part of apoptosis of secretory epithelial cells. The expression and proteolytic activation of

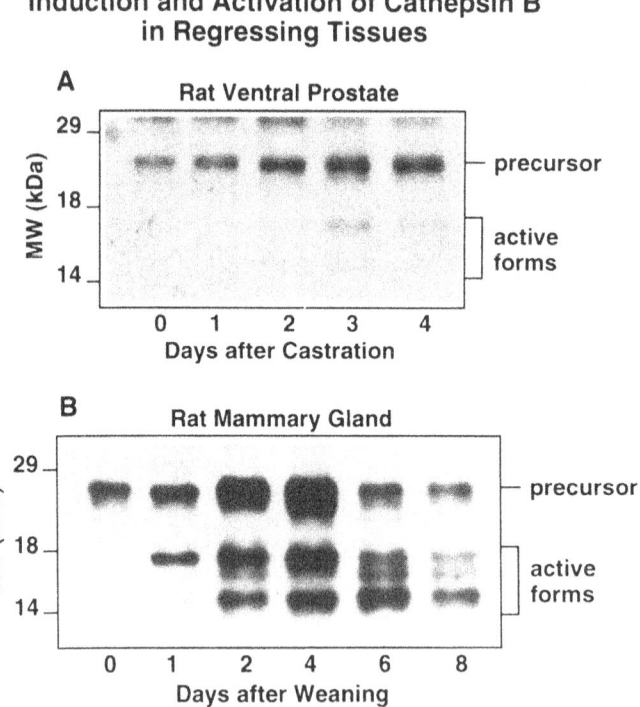

FIGURE 17.3. Western blot analysis of induction and activation of cathepsin B in the regressing prostate and mammary glands after hormone ablation.

cathepsin B correlates temporally with the expression and activation of a number of other extracellular proteases, including matrix metalloproteinases 2 and 9, collagenase, tissue type and urokinase type plasminogen activators, and cathepsin D (81, 111, 118–124). These proteases appear to form a proteolytic cascade that initially degrades the protease inhibitors, abolishing their effects, and subsequently degrades many components of the basement membrane. For example, cathepsin D actively degrades cystatin C, the major inhibitor of cathepsin B, resulting in the enhancement of the enzymatic activity of cathepsin B (116). In turn, cathepsin B activates urokinase type plasminogen activator (125) and degrades collagen type IV, laminin and fibronectin (126). Urokinase type plasminogen activator, in turn, activates plasmin that degrades other components of the ECM and also proteolytically converts procollagenase to the active enzyme. The induction of these ECM-proteases and a number of other lysosomal enzymes, notably β-glucuronidase, in the regressing prostate and mammary gland has led to the suggestion that the degradation of the ECM during apoptosis requires the coordinate activation of several lysosomal proteins in the luminal epithelial cells, and their release from the basal aspect of the cells. This process was first described by Gullino and his colleagues, although its significance was not appreciated at that time (127). As outlined in Figure 17.4, this activation cascade of the ECM-proteases appears to result in the release of the dying cell from the ECM, leading to the membrane mobility that has been described by others as "anoikis."

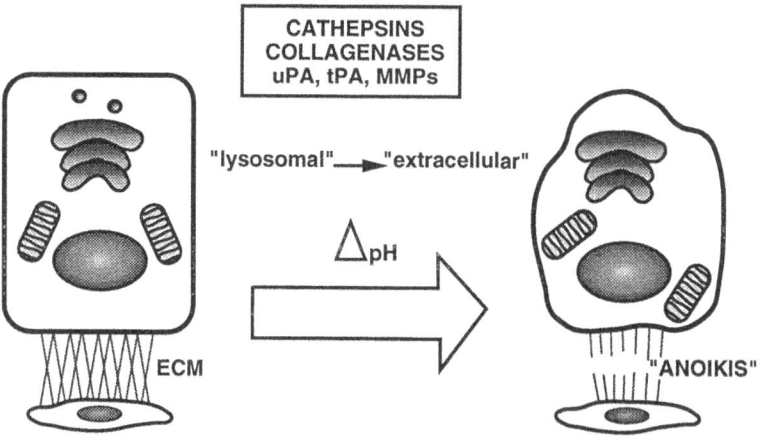

FIGURE 17.4. Proposed role of ECM protease activation during apoptotic cell death in glandular epithelia.

Cloning of p190: A Possible Intracellular Signaling Cascade Responsible for Anoikis

Recently, we have used differential display RT-PCR (dd-RT-PCR) to identify other genes that may be induced during apoptosis in the regressing prostate. We have characterized and sequenced approximately 30 cDNAs that have attributes of expressed sequences, including the AATAAA polyadenylation signal and/or homology to expression tagged sequences already present in the database. One of these sequence is of particular interest since it has striking homology to p190-mRNA. p190 is a GAP (GTPase activating protein)-associated protein that has been implicated in the regulation of the cytoskeletal and focal contacts through the interaction with the rasGAP and rho family of small guanine nucleotide binding proteins. Using ribonuclease protection assays we have shown that the steady state mRNA levels for p190 increase significantly after castration, as does the level of the cognate protein (Morrissey and Tenniswood, unpublished observations). Western blot analysis of the same time course has shown that, whereas the relative levels of rasGAP and rho do not change significantly, the steady state level of focal adhesion kinase (p125^{FAK}) may be down-regulated after castration and the phosphorylation state of p125^{FAK} may also be altered. p125^{FAK} is a tyrosine kinase that regulates the interaction of the cytoskeleton with the extracellular matrix through the integrin superfamily of integral membrane receptors. The protein is presumed to regulate cell adhesion, motility and anchorage-independent growth, and down-regulation of either the absolute level of p125^{FAK} or changes its phosphorylation state would render the cells in the primary tumor less able to associate with the ECM, thus promoting anchorage-independent survival.

These data suggest that after hormone ablation the induction of p190 results in the formation of a p190-rasGAP-rho complex, leading to an attenuation of rho-dependent phosphorylation of p125^{FAK} (128). In the dying cell, this probably results in the loss of actin stress fibers, the loss of the interaction between vinculin, paxilin, and tensin and the $\alpha_5\beta_1$ fibronectin receptor. This loss of integrin binding to fibronectin at focal adhesions presumably causes the cell to lose contact with the ECM and round up, a prelude to the condensation of the cell and DNA fragmentation (Fig. 17.5). In concert with the induction of ECM-proteases described above, these changes in intracellular signalling are probably sufficient to initiate the rounding up and shrinkage of cells during the early cytoplasmic condensation phase of apoptosis that has been termed "anoikis." In several cell culture model systems it has been shown that cytoplasmic condensation precedes DNA fragmentation, and that anoikis and DNA fragmentation are not necessarily coupled (Chapters 9 and 18).

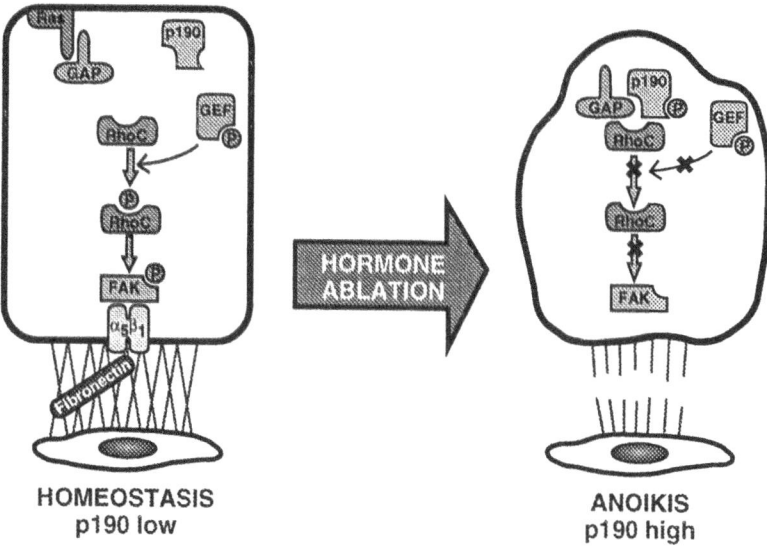

FIGURE 17.5. Proposed role of p190 in the attenuation of rasGAP/rho signaling during "anoikis."

Prostate Disease and Metastatic Progression: Our Hypothesis

The data summarized above suggest that a number of genes are induced in an apparently coordinated manner in the prostate and mammary gland during regression, including a number of extracellular proteases (such as cathepsin B, collagenase, and urokinase and tissue type plasminogen activators), and clusterin. Coupled with the attenuation of focal adhesion kinase activity, the induction of these proteins leads to the disruption of the interaction between the extracellular matrix (ECM) and the cell membrane. These data suggest that in glandular epithelia, the disruption of the ECM is a critical step in the apoptotic pathway. There is considerable evidence indicating that induction of apoptosis occurs in several stages, and that the essential genes are induced *de novo* (or recruited from other cellular functions) *prior to* the initiation of DNA fragmentation, which itself requires the activation of preexisting endonucleases and nuclear proteases (129) (Chapter 18). Remarkably, many of the same genes that are induced during apoptosis are overexpressed in metastatic cells, appear to be required for the invasive phenotype, and are responsible for the extravasation and intravasation of metastatic cells (126, 130, 131). Figure 17.6 highlights the similarity between the apoptotic and invasive phenotypes and emphasizes the point that the differ-

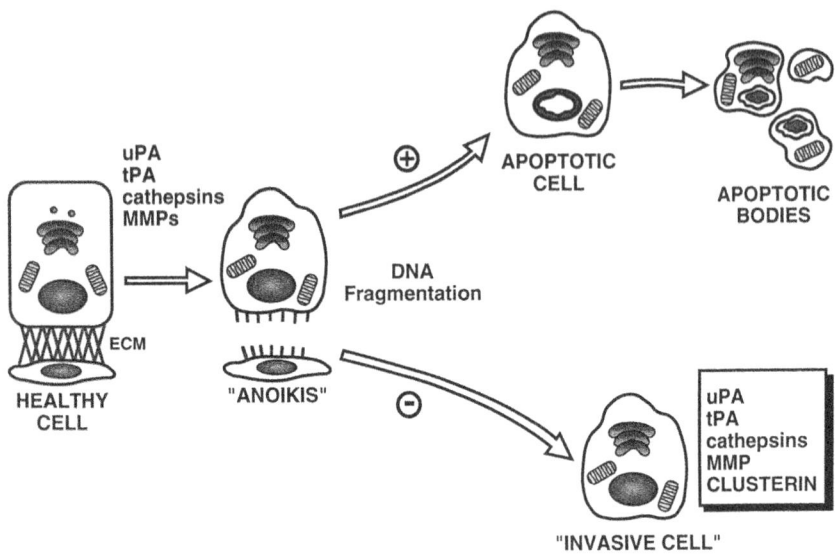

FIGURE 17.6. Hypothetical relationship between attenuated apoptosis and the invasive phenotype.

ence between a preapoptotic cell and a cell with invasive potential is the degradation of the DNA within the apoptotic cell. This suggests that in the natural history of prostate and breast cancer, the progression to an invasive phenotype may be due to abrogated cell death in a few cells that have accumulated mutations in critical pathways essential for the activation of the endonuclease responsible for DNA fragmentation. Abrogation of DNA fragmentation may lead to the progression of prostate and breast cancer, to the development of the invasive phenotype, and ultimately to metastatic progression (Fig. 17.7). Abrogation of DNA fragmentation may result from a number of genetic or epigenetic alterations. For example, failure of the ICE-protease cascade, either due to mutations in the p53 or c-*myc* genes, or due to alterations in the steady state levels of members or the *bcl-2* gene family that disrupt the ratio of BCL-2 and BAX, may block the activation of poly(ADP-ribose) polymerase and DNA fragmentation. It is also possible that stromal cell overexpression of growth factors, such as IGF-I or TGF-β, increased levels of components of the ECM (such as chondroitin sulfate and heparan sulfate), or overexpression of clusterin by the epithelial cells may also block endonuclease activation by as yet undefined intracellular mechanisms. It is also possible that subtle changes in the nuclear concentrations of specific ions, such as K^+ and Ca^{2+}, may also influence the activity of the endonucleases. These observations may have significant implications in the treatment of prostate and breast cancer since nearly all hormone therapies induce apoptosis in hormone-dependent tumor cells. If DNA fragmentation is abrogated in a significant number of cells, effectively uncoupling hormone ablation

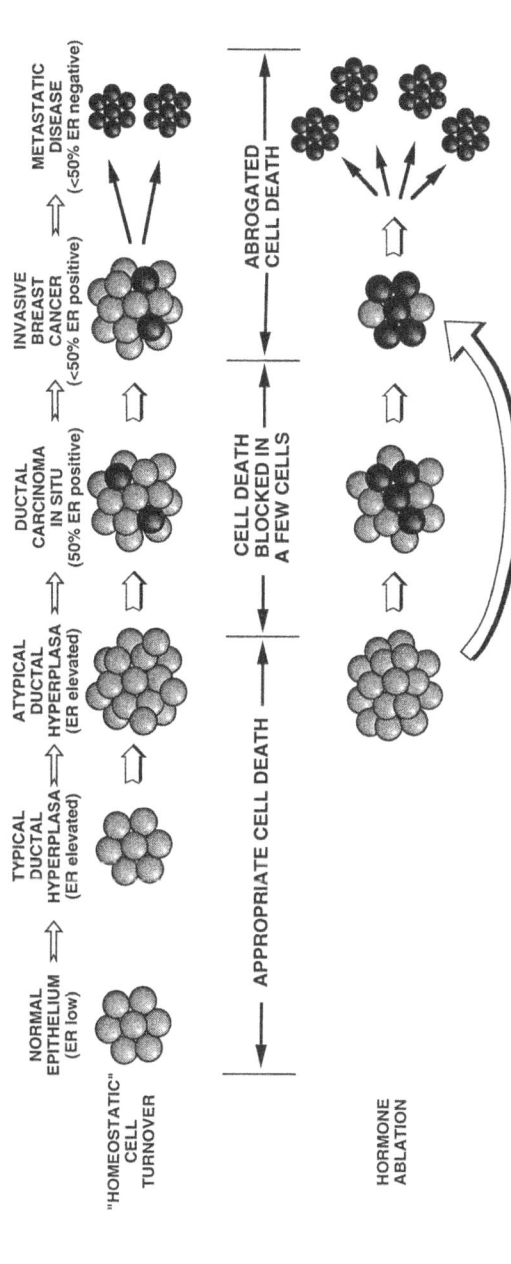

FIGURE 17.7. Proposed involvement of apoptosis in the natural history of breast cancer. The abrogation of apoptosis in a few cells within a localized carcinoma leads to the acquisition of the invasive phenotype and metastatic progression. Early antihormone treatment, which is designed to induce apoptosis in the primary tumor, may lead to an increase in the number of metastases and a poorer long term prognosis. The proposed role of apoptosis in the natural history and treatment of prostatic disease is very similar.

from apoptosis, the surviving cells may acquire a more aggressive phenotype and become a major clinical problem.

If this hypothesis proves to be correct, early hormonal intervention for prostate and breast cancer would be contraindicated, as would neoadjuvant therapy to debulk tumors and to improve positive margins prior to surgery. To improve the treatment of these diseases it will be necessary to determine whether or not all ablative therapies are mechanistically equivalent, and if it is possible to design combination therapies that effectively induce apoptosis without generating cells that acquire an invasive phenotype.

Acknowledgments. The research described in this paper is supported in part by PHS operating grant (CA69233). The authors would like to thank Dr. JoEllen Welsh for reviewing the manuscript and Marina LaDuke and Alice Vera for their expert help with the illustrations.

References

1. Clarke PG. Developmental cell death: morphological diversity and multiple mechanisms. Anat Embryol 1990;181:195–213.
2. Zakeri Z, Bursch W, Tenniswood M, Lockshin R. Cell Death: programmed, apoptosis, necrosis, or other? Cell Death Diff 1995;2:83–92.
3. Cohen JJ. Apoptosis. Immunol Today 1993;14:126–30.
4. Sandford NL, Searle JW, Kerr JFR. Successive waves of apoptosis in the rat prostate after repeated withdrawal of testosterone. Pathol 1984;16:406–10.
5. Guenette RS, Corbeil HB, Leger JG, Wong K, Mezl V, Mooibroek M, et al. Induction of gene expression during involution of the lactating rat mammary gland. J Mol Endocrinol 1994;12:47–60.
6. Bursch W, Paffe S, Putz B, Barthel G, Schulte-Hermann R. Determination of the length of the histological stages of apoptosis in normal liver and in altered hepatic foci of rats. Carcinogen 1990;11:847–53.
7. Bursch W, Kleine L, Tenniswood MP. The biochemistry of cell death by apoptosis. Biochem Cell Biol 1990;68:1071–4.
8. Lee C. Physiology of castration-induced regression of the rat prostate. Proc Clin Biol Res 1981;75A:145–59.
9. Lee C, Sensibar JA. Proteins of the rat prostate. II. synthesis of new proteins in the ventral lobe during castration-induced regression. J Urol 1987;138:903–8.
10. Wyllie AH, Morris RG, Arends MJ, Watt AE. Nuclease activation in programmed cell death. In: Clayton RM, Truman D ed. Co-ordinated regulation of gene expression. New York: Plenum Press;1986:33–41.
11. Arends MJ, Morris RG, Wyllie AH. Apoptosis. The role of the endonuclease. Am J Pathol 1990;136:593–608.
12. Eastman A, Barry MA. The origins of DNA breaks: a consequence of DNA damage, DNA repair, or apoptosis? Cancer Invest 1992;10:229–40.
13. Barry MA, Eastman A. Endonuclease activation during apoptosis: the role of cytosolic Ca^{2+} and pH. Biochem Biophys Res Comm 1992;186:782–9.
14. Walker PR, Sikorska M. Endonuclease activities, chromatin structure, and DNA degradation in apoptosis. Biochem Cell Biol 1994;72:615–23.

15. Dusenbury C, Davis M, Lawrence T, Maybaum J. Induction of megabase DNA fragments by 5-fluorodeoxyuridine in human colorectal tumor (HT29) cells. Mol Pharmacol 1991;39:285–9.
16. Walker PR, Smith C, Youdale T, Leblanc J, Whitfield JF, Sikorska M. Topoisomerase II-reactive chemotherapeutic drugs induce apoptosis in thymocytes. Cancer Res 1991;51:1078–85.
17. Pandey S, Walker PR, Sikorska M. Separate pools of endonuclease activity are responsible for internucleosomal and high molecular mass DNA fragmentation during apoptosis. Biochem Cell Biol 1994;72:625–9.
18. Furuya Y, Lundmo P, Short AD, Gill DL, Isaacs JT. The role of calcium, pH, and cell proliferation in the programmed (apoptotic) death of androgen-independent prostatic cancer cells induced by thapsigargin. Cancer Res 1994;54:6167–75.
19. Lazebnik YA, Kaufmann SH, Desnoyers S, Poirier GG, Earnshaw WC. Cleavage of poly(ADP-ribose) polymerase by a proteinase with properties like ICE. Nature 1994;371:346–7.
20. Jacobson MD, Evan GI. Apoptosis. Breaking the ICE. Curr Biol 1994;4:337–40.
21. Milligan CE, Prevette D, Yaginuma H, Homma S, Cardwell C, Fritz LC, et al. Peptide inhibitors of the ICE protease family arrest programmed cell death of motoneurons in vivo and in vitro. Neuron 1995;15:385–93.
22. Tewari M, Beidler DR, Dixit VM. CrmA-inhibitable cleavage of the 70-kDa protein component of the U1 small nuclear ribonucleoprotein during Fas- and tumor necrosis factor-induced apoptosis. J Biol Chem 1995;270:18738–41.
23. Enari M, Hug H, Nagata S. Involvement of an ICE-like protease in Fas-mediated apoptosis. Nature 1995;375:78–81.
24. Hockenbery D, Zutter M, Hickey W, Nahm M, Korsmeyer S. BCL2 protein is topographically restricted in tissues characterized by apoptotic cell death. Proc Natl Acad Sci USA 1991;88:6961–5.
25. Sentman CL, Shutter JR, Hockenbery D, Kanagawa O, Korsmeyer SJ. bcl-2 inhibits multiple forms of apoptosis but not negative selection in thymocytes. Cell 1991;67:879–88.
26. Colombel M, Symmans F, Gil S, O'Toole KM, Chopin D, Benson M, et al. Detection of the apoptosis-suppressing oncoprotein bcl-2 in hormone-refractory human prostate cancers. Am J Pathol 1993;143:390–400.
27. Chittenden T, Harrington EA, O'Connor R, Flemington C, Lutz RJ, Evan GI, et al. Induction of apoptosis by the Bcl-2 homologue Bak. Nature 1995;374:733–6.
28. Tung PS, Fritz IB. Immunolocalization of clusterin in the ram testis, rete testis, and excurrent ducts. Biol Reprod 1985;33:177–86.
29. Evan G, Harrington E, Fanidi A, Land H, Amati B, Bennett M. Integrated control of cell proliferation and cell death by the c-myc oncogene. Philos Trans R Soc Lond B Biol Sci 1994;345:269–75.
30. Amati B, Littlewood TD, Evan GI, Land H. The c-Myc protein induces cell cycle progression and apoptosis through dimerization with Max. EMBO J 1993;12:5083–7.
31. Harrington EA, Bennett MR, Fanidi A, Evan GI. c-Myc-induced apoptosis in fibroblasts is inhibited by specific cytokines. EMBO J 1994;13:3286–95.
32. Colombel M, Olsson CA, Ng PY, Buttyan R. Hormone-regulated apoptosis results from reentry of differentiated prostate cells onto a defective cell cycle. Cancer Res 1992;52:4313–9.
33. Furuya Y, Isaacs JT. Differential gene regulation during programmed death

(apoptosis) versus proliferation of prostatic glandular cells induced by androgen manipulation. Endocrinology 1993;133:2660–6.

34. Isaacs JT, Furuya Y, Berges R. The role of androgen in the regulation of programmed cell death/apoptosis in normal and malignant prostatic tissue. Semin Cancer Biol 1994;5:391–400.

35. Colombel M, Radvanyi F, Blanche M, Abbou C, Buttyan R, Donehower LA, et al. Androgen suppressed apoptosis is modified in p53 deficient mice. Oncogene 1995;10:1269–74.

36. Bennett MR, Evan GI, Schwartz SM. Apoptosis of rat vascular smooth muscle cells is regulated by p53-dependent and -independent pathways. Circ Res 1995;77:266–73.

37. Bissell MJ, Hall HG, Parry G. How does the extracellular matrix direct gene expression? J Theor Biol 1982;99:31–68.

38. Blum JL, Zeigler ME, Wicha MS. Regulation of rat mammary gene expression by extracellular matrix components. Exp Cell Res 1987;173:322–40.

39. Blum JL, Wicha MS. Role of the cytoskeleton in laminin induced mammary gene expression. J Cell Physiol 1988;135:13–22.

40. Cunha GR, Donjacour AA, Cooke PS, Mee S, Bigsby RM, Higgins SJ, et al. The endocrinology and developmental biology of the prostate. Endocr Rev 1987;8:338–62.

41. Streuli CH, Schmidhauser C, Kobrin M, Bissell MJ, Derynck R. Extracellular matrix regulates expression of TGF-B1 gene. J Cell Biol 1993;120:253–60.

42. Dedhar S, Jewell K, Rojiani M, Gray V. The receptor for the basement membrane glycoprotein entactin is the integrin alpha 3/beta 1. J Biol Chem 1992;267: 18908–14.

43. Chung LW, Li W, Gleave ME, Hsieh JT, Wu HC, Sikes RA, et al. Human prostate cancer model: roles of growth factors and extracellular matrices. J Cell Biochem- Suppl 1992;16H:99–105.

44. Cohen P, Peehl DM, Lamson G, Rosenfeld RG. Insulin-like growth factors (IGFs), IGF receptors, and IGF-binding proteins in primary cultures of prostate epithelial cells. J Clin Endocrinol Metab 1991;73:401–7.

45. Hofer DR, Sherwood ER, Bromberg WD, Mendelsohn J, Lee C, Kozlowski JM. Autonomous growth of androgen-independent human prostatic carcinoma cells: role of transforming growth factor alpha. Cancer Res 1991;51:2780–5.

46. Mori H, Maki M, Oishi K, Jaye M, Igarashi K, Yoshida O, et al. Increased expression of genes for basic fibroblast growth factor and transforming growth factor type beta 2 in human benign prostatic hyperplasia. Prostate 1990;16:71–80.

47. Matuo Y, Nishi N, Takasuka H, Masuda Y, Nishikawa K, Isaacs JT, et al. Production and significance of TGF-β in AT-3 metastatic cell line established from the Dunning rat prostatic adenocarcinoma. Biochem Biophys Res Commun 1990;166:840–7.

48. Matuo Y, McKeehan WL, Yan GC, Nikolaropoulos S, Adams PS, Fukabori Y, et al. Potential role of HBGF (FGF) and TGF-beta on prostate growth. Adv Exp Med Biol 1992;324:107–14.

49. Kan M, Wang F, Xu J, Crabb JW, Hou J, McKeehan WL. An essential heparin-binding domain in the fibroblast growth factor receptor kinase. Science 1993; 259:1918–21.

50. McKeehan WL. Growth factor receptors and prostate cell growth. In Isaacs JT, Franks LM, eds. Prostate Cancer:cell and molecular mechanisms in diagnosis and treatment. Cold Spring Harbor, NY: Cold Spring Harbor Laboratory Press, 1991:165–75.

51. Yan G, Fukabori Y, Nikolaropoulos S, Wang F, McKeehan WL. Heparin-binding keratinocyte growth factor is a candidate stromal to epithelial cell andromedin. Mol Endocrinol 1992;6:2123–8.

52. Sugimura Y, Cunha GR, Donjacour AA. Morphogenesis of ductal networks in the mouse prostate. Biol Reprod 1986;34:961–71.

53. Arteaga CL. Interference of the IGF system as a strategy to inhibit breast cancer growth. Breast Cancer Res Treat 1992;22:101–6.

54. Pietrzkowski Z, Wernicke D, Porcu P, Jameson BA, Baserga R. Inhibition of cellular proliferation by peptide analogues of insulin-like growth factor 1. Cancer Res 1992;52:6447–51.

55. Reeve JG, Morgan J, Schwander J, Bleehen NM. Role for membrane and secreted insulin-like growth factor-binding protein-2 in the regulation of insulin-like growth factor action in lung tumors. Cancer Research 1993;53:4680–5.

56. Evan GI, Littlewood TD. The role of c-myc in cell growth. Curr Opin Genet Dev 1993;3:44–9.

57. Evan GI, Wyllie AH, Gilbert CS, Littlewood TD, Land H, Brooks M, et al. Induction of apoptosis in fibroblasts by c-myc protein. Cell 1992;69:119–28.

58. Guenette RS, Tenniswood M. The role of growth factors in the suppression of active cell death in the prostate: an hypothesis. Biochem Cell Biol 1994;72:553–9.

59. Barni T, Vannelli BG, Sadri R, Pupilli C, Ghiandi P, Rizzo M, et al. Insulin-like growth factor-I (IGF-I) and its binding protein IGFBP-4 in human prostatic hyperplastic tissue: gene expression and its cellular localization. J Clin Endocrinol Metab 1994;78:778–3.

60. Cohen P, Peehl DM, Baker B, Liu F, Hintz RL, Rosenfeld RG. Insulin-like growth factor axis abnormalities in prostatic stromal cells from patients with benign prostatic hyperplasia. J Clin Endocrinol Metab 1994;79:1410–15.

61. Ellis MJ, Singer C, Hornby A, Rasmussen A, Cullen KJ. Insulin-like growth factor mediated stromal-epithelial interactions in human breast cancer. Breast Cancer Res Treat 1994;31:249–61.

62. Singer C, Rasmussen A, Smith HS, Lippman ME, Lynch HT, Cullen KJ. Malignant breast epithelium selects for insulin-like growth factor II expression in breast stroma: evidence for paracrine function. Cancer Res 1995;55:2448–54.

63. Cullen KJ, Allison A, Martire I, Ellis M, Singer C. Insulin-like growth factor expression in breast cancer epithelium and stroma. Breast Cancer Res Treat 1992;22:21–9.

64. Paik S. Expression of IGF-I and IGF-II mRNA in breast tissue. Breast Cancer Res Treat 1992;22:31–8.

65. Brunner N, Moser C, Clarke R, Cullen K. IGF-I and IGF-II expression in human breast cancer xenografts: relationship to hormone independence. Breast Cancer Res Treat 1992;22:39–45.

66. Iwamura M, Sluss PM, Casamento JB, Cockett AT. Insulin-like growth factor I: action and receptor characterization in human prostate cancer cell lines. Prostate 1993;22:243–52.

67. Neuenschwander S, Roberts CT, Jr., LeRoith D. Growth inhibition of MCF-7 breast cancer cells by stable expression of an insulin-like growth factor I receptor antisense ribonucleic acid. Endocrinology 1995;136:4298–303.

68. Clemmons DR. IGF binding proteins and their functions. Mol Reprod Develop 1993;35:368–75.

69. Jones J, Gockerman A, Busby WH, Camacho-Hubner C, Clemmons DR. Extracellular

matrix contains insulin-like growth factor binding protein-5: potentiation of the effects of IGF-1. J Cell Biol 1993;121:679–87.

70. Kofoed JJ, Tumilasci OR, Curbelo HM, Fernandez Lemos SM, Arias NH, Houssay AB. Effects of castration and androgens upon prostatic proteoglycans in rats. Prostate 1990;16:93–102.

71. Paulsson M. Basement membrane proteins: structure, assembly, and cellular inter-actions. Crit Rev Biochem Mol Biol 1992;27:93–127.

72. Hynes RO. Integrins. versatility, modulation, and signalling in cell adhesion. Cell 1992;69:11–25.

73. Damsky CH, Werb Z. Signal transduction by integrin receptors for extracellular matrix: co-operative processing of extracellular information. Cur Opin Cell Biol 1992;4:772–81.

74. Juliano RL, Haskil S. Signal transduction from the extracellular matrix. J Cell Biol 1993;120:577–85.

75. Getzenberg RH, Pienta KJ, Coffey DS. The tissue matrix: cell dynamics and hor-mone action. Endocr Rev 1990;11:399–417.

76. Getzenberg RH, Pienta KJ, Huang EYW, Murphy BC, Coffey DS. Modifications of the intermediate filament and nuclear matrix networks by the extracellular ma-trix. Biochem Biophys Res Commun 1991;179:340–4.

77. Pienta KJ, Partin AW, Coffey DS. Cancer as a disease of DNA organization of dynamic cell structure. Cancer Res 1989;49:2525–32.

78. Pienta KJ, Murphy BC, Getzenberg RH, Coffey DS. The effect of extracellular matrix on morphologic transformation in vitro. Biochem Biophys Res Commun 1991;179:333–9.

79. Petersen OW, Ronnov-Jessen L, Howlett AR, Bissell MJ. Interaction with base-ment membrane serves to rapidly distinguish growth and differentiation pattern of normal and malignant human breast epithelial cells. Proc Natl Acad Sci USA 1992;89:9064–8.

80. Muntzing J. Androgen and collagen as growth regulators of the rat ventral prostate. Prostate 1980;1:71–8.

81. Muntzing J. Collagen synthesis and breakdown in the rat ventral prostate. In: Kerr JP, Sandberg AA, Murphy GP, eds. The prostatic cell: structure and function. New York: Alan R. Liss, Inc, 1981:137–44.

82. Sugimura Y, Cunha GR, Donjacour AA. Morphological and histochemical study of castration-induced degeneration and androgen-induced regeneration in the mouse prostate. Biol Reprod 1986;34:973–83.

83. Sugimura Y, Cunha GR, Donjacour AA. Whole-mount autoradiographic study of DNA synthetic activity during post-natal development and androgen-induced re-generation in the mouse prostate. Biol Reprod 1986;34:985–95.

84. Rouleau M, Léger JG, Tenniswood MP. Ductal heterogeneity of cytokeratins, gene expression and cell death in the rat ventral prostate. Mol Endocrinol 1990;4:2003–13.

85. Sensibar JA, Griswold MD, Sylvester SR, Buttyan R, Bardin CW, Cheng CY, et al. Prostatic ductal system in rats: regional variation in localization of an androgen-repressed gene product, sulfated glycoprotein-2. Endocrinology 1991;128: 2091–102.

86. English HF, Santen RJ, Isaacs JT. Response of glandular versus basal rat ventral prostatic epithelial cells to androgen withdrawal and replacement. Prostate 1987;11:229–42.

87. Lee C, Sensibar JA, Dudek SM, Hiipakka RA, Liao ST. Prostatic ductal system in rats: regional variation in morphological and functional activities. Biol Reprod 1990;43:1079–86.
88. Isaacs JT. Antagonistic effect of androgen on prostatic cell death. Prostate 1984;5:547–57.
89. Clark AF, Tenniswood MP, Bird CE, Flynn TG, Jacobs FA, Abrahams PA. Hormonal control of prostatic biochemical markers. In: Hafez ESE, Springer-Mills E, eds. Clinics in andrology, vol. 6, Prostatic Carcinoma: Biology and Diagnosis. Hingham, MA: Martinus Nijhoff Publishers,1983:98–108.
90. Winderickx J, Swinnen K, van Dijck P, Verhoeven G, Heyns W. Kallikrein-related protease in the rat ventral prostate: cDNA cloning and androgen regulation. Mol Cell Endocrinol 1989;62:217–26.
91. Tenniswood MP, Abrahams PA, Bird CE, Clark AF. Effects of castration and androgen replacement on acid phosphatase activity in the adult rat prostate gland. J Endocrinol 1978;77:301–8.
92. Page MJ, Parker MG. Effect of androgen on the transcription of rat prostatic binding protein genes. Mol Cell Endocrinol 1982;27:343–55.
93. Zhang Y-L, Parker MG. Regulation of prostatic steroid binding protein mRNAs by testosterone. Mol Cell Endocrinol 1985;43:151–4.
94. DeKlerk DP, Coffey DS. Quantitative determination of prostatic epithelial and stromal hyperplasia by a new technique. Invest Urol 1978;16:240–5.
95. Tenniswood MP, Guenette RS, Lakins J, Mooibroek M, Wong P, Welsh JE. Active cell death in hormone-dependent tissues. Cancer Met Rev 1992;11:197–220.
96. Montpetit ML, Lawless KR, Tenniswood M. Androgen-repressed messages in the rat ventral prostate. Prostate 1986;8:25–36.
97. Leger JG, Montpetit ML, Tenniswood MP. Characterization and cloning of androgen-repressed mRNAs from rat ventral prostate. Biochem Biophys Res Commun 1987;147:196–203.
98. Bettuzzi S, Hiipakka RA, Gilna P, Liao S. Identification of an androgen-repressed mRNA in rat ventral prostate as coding for sulphated glycoprotein 2 by cDNA cloning and sequence analysis. Biochem J 1989;257:293–6.
99. Strange R, Li F, Saurer S, Burkhardt A, Friis RR. Apoptotic cell death and tissue remodelling during mouse mammary gland involution. Develop 1992;115:49–58.
100. Wong P, Pineault JM, Lakins J, Taillefer D, Léger JG, Wang C, et al. Genomic organization and expression of the rat TRPM-2 (clusterin) gene, a gene implicated in apoptosis. J Biol Chem 1993;268:5021–31.
101. Jenne DE, Tschopp J. Clusterin the intriguing guises of a widely expressed glycoprotein. Trends Biol Sci 1992;17:154–9.
102. Griswold MD, Roberts K, Bishop P. Purification and characterization of a sulfated glycoprotein secreted by Sertoli cells. Biochem 1986;25:7265–70.
103. Sylvester SR, Morales C, Oko R, Griswold MD. Localization of sulfated glycoprotein-2 (clusterin) on spermatozoa and in the reproductive tract of the male rat. Biol Reprod 1991;45:195–207.
104. Zakeri Z, Curto M, Hoover DM, Wightman K, Engelhardt J, Smith FF, et al. Developmental expression of the S35-S45/SGP-2/TRPM-2 gene in rat testis and epididymis. Mol Reprod Dev 1992;33:373–84.
105. Bursch W, Gleeson TG, Kleine L, Tenniswood MP. Expression of TRPM-2/clusterin mRNA during growth and regression of rat liver. Archiv Toxicol 1995;69:253–8.
106. Wilson MR, Easterbrook-Smith SB, Lakins J, Tenniswood M. Mechanism of in-

duction and function of clusterin at sites of cell death. In: Harmony J, ed. Clusterin: function in vertebrate organ development, function and adaption. Austin, TX: R.G. Landes, 1994:75–100.

107. Sensibar JA, Sutkowski DM, Raffo A, Buttyan R, Griswold MD, Sylvester SR, et al. Prevention of cell death induced by tumor necrosis factor alpha in LNCaP cells by overexpression of sulfated glycoprotein-2 (clusterin). Cancer Res 1995;55:2431–7.

108. Wong P, MacDonald IM, Sood R, Smith C, Pilon R, Tenniswood MP. Identification and partial characterization of a candidate gene for X-linked retinopathies using a lateral approach. Genomics 1993;15:467–71.

109. Guenette RS, Mooibroek M, Wong K, Wong P, Tenniswood M. Cathepsin B, a cysteine protease implicated in metastatic progression, is also expressed during regression of the rat prostate and mammary glands. Eur J Biochem 1994;226:311–21.

110. Lee EY-H, Parry G, Bissell MJ. Modulation of secreted proteins of mouse mammary epithelial cells by the collagenous substrata. J Cell Biol 1984;98:146–55.

111. Lokeshwar BL, Selzer MG, Block NL, Gunja-Smith Z. Secretion of matrix metalloproteinases and their inhibitors (tissue inhibitors of metalloproteinases) by human prostate in explant cultures: reduced tissue inhibitor of metalloproeinase secretion by malignant tissues. Cancer Res 1993;53:4493–8.

112. Stampfer MR, Yaswen P, Alhadeff M, Hosoda J. TGF beta induction of extracellular matrix associated proteins in normal and transformed human mammary epithelial cells in culture is independent of growth effects. J Cell Physiol 1993; 155:210–21.

113. Heegard CE, White JH, Zavizion B, Turner JD, Politis I. Production of various forms of plasminogen activator and plasminogen activator inhibitor by cultured mammary epithelial cells. J Dairy Sci 1994;77:2949–58.

114. Bianchi E, Cohen RL, Dai A, Thor AT, Shuman MA, Smith HS. Immunohistochemical localization of the plasminogen activator inhibitor-1 in breast cancer. Int J Cancer 1995;60:597–603.

115. Heegaard CW, Simonsen AC, Oka K, Kjoller L, Christensen A, Madsen B, et al. Very low density lipoprotein receptor binds and mediates endocytosis of urokinase-type plasminogen activator-type-1 plasminogen activator inhibitor complex. J Biol Chem 1995;270:20855–61.

116. Lenarcic B, Krasovec M, Ritonja A, Olafsson I, Turk V. Inactivation of human cystatin C and kininogen by human cathepsin D. FEBS Lett 1991;280:211–15.

117. Lah TT, Kokalj-Kunovar M, Strukelj B, Pungercar J, Barlic-Maganja D, Drobnic-Kosorok M, et al. Stefins and lysosomal cathepsins B, L and D in human breast carcinoma. Int J Cancer 1992;50:36–44.

118. Sensibar JA, Liu XX, Patai B, Alger B, Lee C. Characterization of castration-induced cell death in the rat prostate by immunohistochemical localization of cathepsin D. Prostate 1990;16:263–76.

119. Freeman SN, Rennie PS, Chao J, Lund LR, Andreasen PA. Urokinase- and tissue-type plasminogen activators are suppressed by cortisol in the involuting prostate of castrated rats. Biochem J 1990;269:189–93.

120. Rennie PS, Bouffard R, Bruchovsky N, Cheng H. Increased activity of plasminogen activators during involution of the rat ventral prostate. Biochem J 1984; 221:171–8.

121. Wilson MJ, Ditmanson JV, Sinha AA, Estensen RD. Plasminogen activator activities in the ventral and dorsolateral prostatic lobes of aging Fischer 344 rats. Prostate 1994;16:147–61.
122. Wilson MJ, Garcia B, Woodson M, Sinha AA. Gelatinolytic and caseinolytic proteinase activities in the secretions of the ventral, lateral, and dorsal lobes of the rat prostate. Biol Reprod 1993;48:1174–84.
123. Wilson MJ, Garcia B, Woodson M, Sinha AA. Metalloproteinase activities expressed during development and maturation of the rat prostatic complex and seminal vesicles. Biol Reprod 1992;47:683–91.
124. Wilson MJ, Strasser M, Vogel MM, Sinha AA. Calcium-dependent and calcium-independent gelatinolytic proteinase activities of the rat ventral prostate and its secretion: characterization and effect of castration and testosterone treatment. Biol Reprod 1991;44:776–85.
125. Kobayashi H, Schmitt M, Goreetzki L, Chucholowski N, Calvete J, Kramer M, et al. Cathepsin B efficiently activates the soluble and the tumor cell receptor bound form of the proenzyme urokinase-type plasminogen activator (pro-uPA). J Biol Chem 1991;266:5147–52.
126. Buck MR, Karustis DG, Day NA, Honn KV, Sloane BF. Degradation of extracellular-matrix proteins by human cathepsin B from normal and tumor tissues. Biochem J 1992;282:273–8.
127. Gullino PM, Grantham FH, Losonczy I, Berghoffer B. Mammary tumor regression. I. Physiopathologic characteristics of hormone dependent tissue. J Natl Cancer Inst 1972;49:1333–48.
128. Burbelo P, Miyamoto S, Utani A, Brill S, Yamada K, Hall A, et al. p-190-B, a new member of the Rho GAP family, and Rho are induced to cluster after integrin cross-linking. J Biol Chem 1995;270:30919–26.
129. Leibovitz D, Koch Y, Fridkin M, Pitzer F, Zwicikl P, Dantes A, et al. Archaebacterial and eukaryotic proteasomes prefer different sites in cleaving gonadotropin-releasing hormone. J Biol Chem 1995;270:11029–32.
130. Liotta L, Stetler-Stevenson WG. Metalloproteinases and cancer invasion. Sem Cancer Biol 1990;1:107–15.
131. Sloane BF. Cathepsin B and cystatins: evidence for a role in cancer progression. Sem Cancer Biol 1990;1:137–52.

18

Cell Volume Regulation and the Movement of Ions During Apoptosis

CARL D. BORTNER AND JOHN A. CIDLOWSKI

Apoptosis is a physiological mode of cell death in which cells are removed or eliminated from the body in response to a given signal or stimulus (1). Apoptosis, also known as programmed cell death, can be distinguished from accidental cell death, referred to as necrosis, by a unique set of characteristics that includes cell shrinkage, nuclear condensation, internucleosomal DNA cleavage, and apoptotic body formation (2). Although many studies have focused on the biochemical (i.e., internucleosomal DNA cleavage) and morphological (i.e., apoptotic body formation) characteristics of apoptosis, relatively few have examined the characteristic cell shrinkage associated with programmed cell death. The distinctive feature of volume loss during this form of cell death was observed from the very first reports of apoptosis (3, 4), and subsequently, cell shrinkage has been observed in all well-defined examples of apoptosis. Thus, the loss of cell volume, along with internucleosomal DNA cleavage, reflects key components of the programmed cell death process.

The loss of cell volume is a unique and distinctive feature of the programmed cell death process. The shrinkage of cells during apoptosis is in marked contrast to that which occurs during necrosis. During necrosis, the cells swell, lose membrane integrity, and eventually lyse eliciting an inflammatory response. This cellular swelling can occur as a result of an early loss of energy in cells, which eliminates all ability to actively move ions in the cell. Especially significant is the loss of sodium/potassium adenosine triphosphate (Na^+/K^+ ATPase) function. The cell becomes flooded with sodium (Na^+) ions, which in turn causes the movement of water into the cell and eventual cell lysis. In contrast, during apoptosis, energy levels remain high until very late in the process, and cells that die by this programmed pathway appear to retain their ability to control the movement of ions. These observations suggest that the loss of volume that occurs during apoptosis must involve an active process. This article examines aspects of cell volume control, the loss of cell volume during apoptosis, and the role volume regulation may play in the programmed cell death process.

Ionic Distributions Across the Cell Membrane: Donnan Equilibrium and the Pump-Leak Hypothesis

An understanding of cell shrinkage during apoptosis must first begin with the recognition of the position and potential (both chemical and electrical) of the various ions that comprise the cells internal and external environment under normal conditions. Three criteria must be met for a cell to maintain homeostasis. First, the intracellular and extracellular contents of the cell must be electrically neutral. Therefore, every positive charge on the outside of the cell must be counter-balanced by a negative charge. A similar condition must exist on the inside of the cell; however, a small unbalanced intracellular charge ratio does exist that accounts for the negative membrane potential of the cell. Second, no osmotic pressure must exist across the cell membrane, or in other words, the total concentration of solute particles inside the cell must equal that on the outside of the cell. Consequently, this absence of osmotic pressure on the cell membrane allows for no net movement of water into or out of the cell.

Since the cell contains impermeant or fixed anions, such as proteins and DNA, balance of charge would lead to passive entry of permeant ions, such as sodium (Na^+) and potassium (K^+). This condition, known as Donnan equilibrium (5), would result in a positive intracellular osmotic pressure. This positive pressure would be balanced by the movement of water into the cell, which in turn would again lead to the passive redistribution of the permeant ions and the cycle would repeat, causing the cell to eventually swell and lyse. Therefore, the cells must actually control ion fluxes to prevent this passive distribution of permeant ions across their membrane to remain viable under normal or isotonic conditions.

Cells compensate for this increase in intracellular osmotic pressure caused by the passive movement of ions into a cell containing impermeant anions by actively transporting ions to the outside of the cell in what has been termed the "pump-leak hypothesis" (6, 7). This hypothesis states that a constant cellular volume is maintained by the extrusion of sodium from the cell by a sodium pump. The exact nature of this sodium pump was not known until Skou (8) described a Na^+/K^+ ATPase. This ATP-dependent pump exchanges three intracellular sodium ions with two extracellular potassium ions and is responsible for the maintenance of a normal cell volume, as well as the high K^+ and low Na^+ intracellular concentrations observed in a most mammalian cells. While the Na^+/K^+ ATPase is used by a majority of mammalian cells to maintain a normal cell volume under isotonic conditions, some exceptions do exist. Renal cortical cells use a Na^+ pump that is not dependent on K^+ and is insensitive to the Na^+/K^+ ATPase inhibitor ouabain (9). Dog erythrocytes use a Ca^{2+} pump in conjunction with a Ca^{2+}/Na^+ exchanger (10). Interestingly, the actions of the Ca^{2+} pump and the Ca^{2+}/Na^+ exchanger in dog red blood cells brings about a high Na^+/low K^+ intracellular concentration, the opposite of what is normally observed in most mammalian cells.

Control of Cell Volume Under Anisotonic Conditions

Cells control their volume under anisotonic conditions by the activation of specific pumps and channels, similar to the manner in which cells control their volume under isotonic conditions (11, 12). Ions can cross the cell membrane by one of four general mechanisms in response to a change in a cell's extracellular environment: electrodiffusion, cotransporter, electroneutral exchange, and electrogenic exchange (i.e., Na^+/K^+ ATPase). These ion transport mechanisms are usually quiescent or are active at only a low level under normal isotonic conditions, but can be stimulated to function at a much higher rate of activity upon a change of cell volume. In general, cells maintain their volume within very narrow limits (approximately 3%). However, certain cell types (liver cells, renal medulla epithelial cells, intestinal mucosa) do encounter extreme conditions of tonicity where their extracellular environment may vary greatly. Additionally, during pathophysiological disturbances, any cell may be exposed to extreme anisotonic conditions.

When cells are exposed to hypotonic conditions they initially swell, but after a period of time they decrease their cell volume by what has been termed a regulatory volume decrease (RVD) response (Fig. 18.1). This decrease in cell volume is mediated by the loss of K^+ and Cl^- from the cell, with the concomitant movement of water out of the cell. Different cell types use different mechanisms during a RVD response, but all achieve the same result of reducing cell volume back to near normal homeostatic levels. Figure 18.2 shows three mechanisms known to extrude K^+ and Cl^- from the cell during RVD. The first mechanism is the bumetanide-sensitive, electroneutral K^+/Cl^- cotransporter. This cotransporter has been observed to be active in the RVD response of red blood cells in several species, including duck (13), sheep (14), and human (15). A second way in which a RVD response is achieved is through the activation of a K^+/H^+ exchanger coupled to a Cl^-/HCO_3^- exchanger as observed in amphibian red blood cells (16). A final mechanism of RVD is achieved through activation of individual K^+ and Cl^- channels. The transport of both K^+ and Cl^- through separate pathways has been shown to occur in human lymphocytes (17–19), Ehrlich ascites tumor cells (20), MDCK cells (21), and in frog urinary bladder (22).

In contrast to hypotonic conditions, when cells are exposed to hypertonic conditions, they shrink and can subsequently increase their cell volume by a regulatory volume increase (RVI) response (Fig. 18.1). During this response, there is an initial influx of Na^+ and Cl^- into the cell, but the Na^+ is then replaced by K^+ by the action of the Na^+/K^+ ATPase giving a net increase of K^+ and Cl^-. Figure 18.2 shows the mechanisms known to be involved in a RVI response. The $Na^+/K^+/2Cl^-$ cotransporter has been shown to be responsible for this response in rat red blood cells (23), human red blood cells (24), and Ehrlich ascites tumor cells (25). A second RVI mechanism, which is observed in lymphocytes (26, 27), liver (28), and gallbladder (29), is the activation of

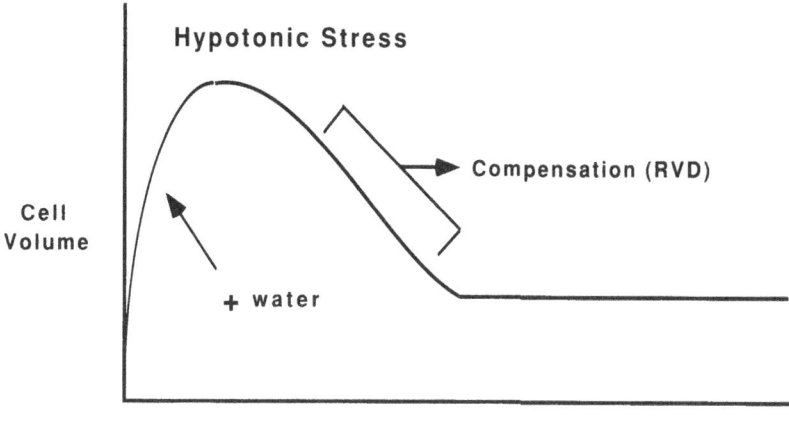

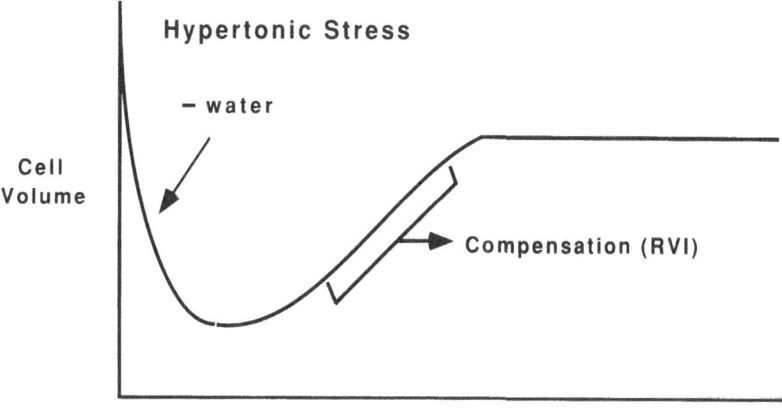

FIGURE 18.1. Regulation of cell volume in mammalian cells. When cells are exposed to hypotonic stress, an immediate increase in cell volume is observed due to the movement of water into the cell. However, most cells can compensate for this increase in volume by the activation of a RVD response which returns the cell to a near normal size. In contrast, cells exposed to hypertonic stress immediately decrease their cell volume, but compensating mechanisms known as a RVI response again return the cell to a near normal size.

a Na^+/H^+ exchanger coupled to a Cl^-/HCO_3^- exchanger. Both of these RVI mechanisms allow the cells to return to a normal cell volume after hypertonic stress.

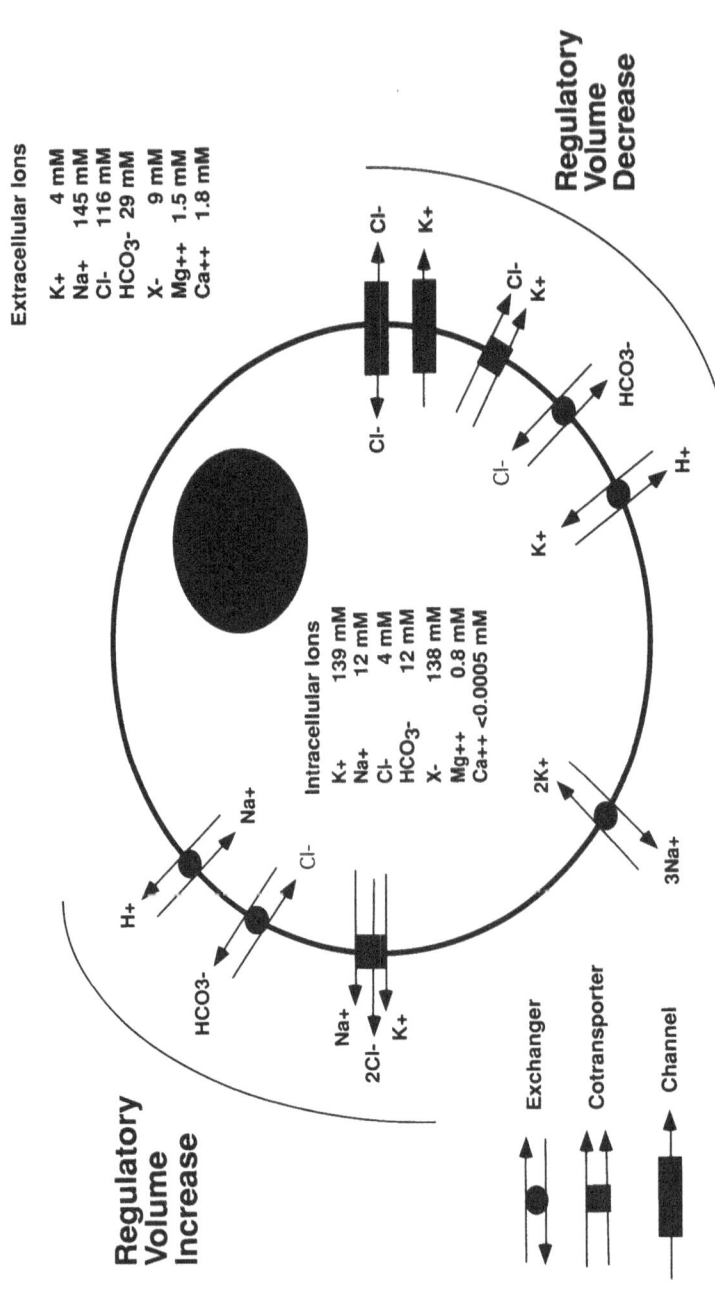

FIGURE 18.2. Volume regulatory mechanisms in mammalian cells. The ionic composition of both the intracellular and extracellular environments of a cell are shown along with the mechanisms responsible for both a regulatory volume increase (RVI) and a regulatory volume decrease (RVD) response.

Volume Regulation in Lymphocytes

Lymphocytes have been one of only a few cell types extensively examined in the content of cell volume regulation. Lymphocytes possess the two important characteristics for accurate and precise cell volume measurements. These are an essentially spherical shape of the cell and the non-adherent nature of substrate interaction. In fact, considerable information exists on volume regulation of the various subtypes of lymphoid cells (30, 31). When T-lymphocytes are exposed to hypotonic conditions, they show an increase in both K^+ and Cl^- efflux from the cell (30). This loss of ions from the cell accounts for the RVD response and returns these swollen cells to a near normal cell volume. In contrast, B-lymphocytes show only a modest increase in K^+ efflux under hypotonic conditions, and therefore, very little RVD (30). This difference in RVD activity has been shown to be a failure of hypotonic conditions to stimulate a loss of K^+ in B lymphocytes, while, surprisingly, the anion component (a loss of Cl^- in these cells) remains fully functional. These data support the idea that the loss of K^+ and Cl^- during RVD in lymphocytes is through independent pathways.

On the other hand, T-lymphocytes exposed to hypertonic conditions shrink and remain in a state of reduced volume (32, 33). This absence of a RVI response in T-cells is not absolute, as Grinstein et al. (34) showed that peripheral blood lymphocytes can undergo a RVI response if first primed by hypotonic conditions. When peripheral blood lymphocytes are allowed to equilibrate under hypotonic conditions, and are then placed back into an isotonic environment (which is now hypertonic to the original condition), the cells shrink and regain their cell volume by a RVI response. This return to a near normal cell volume, observed only after an initial RVD response, has been termed a secondary RVI response and was shown to occur by the activation of a Na^+/H^+ exchanger coupled to a Cl^-/HCO_3^- exchanger. The mechanism for activation of this secondary RVI response in T-cells is unclear. Additionally, the exact mechanisms that trigger many of these exchangers and transporters involved in volume regulation are unknown.

The Loss of Cell Volume During Apoptosis

T-lymphocytes have been an excellent model system for studying programmed cell death because of their marked susceptibility to undergo apoptosis. We have used an immature mouse lymphoma cell line, designated S49 Neo, to study the characteristics of glucocorticoid-induced apoptosis. When S49 Neo cells are treated with the glucocorticoid, dexamethasone, a characteristic apoptotic response is observed (35). This response includes a loss of cell volume, internucleosomal DNA cleavage, and apoptotic body formation.

Glucocorticoid-induced cell size changes in thymocytes have been investigated as early as 1981 by Thomas and Bell (36). Treatment of freshly-isolated rat thymocytes with dexamethasone showed the progressive appearance of apoptotic cells which was a dose-dependent required protein synthesis. No steroid effects were observed during the first hour of treatment, but subsequently a steady increase in a smaller sized population of cells was detected. Electronic sizing techniques showed that these apoptotic cells had a mean diameter of 4.6 mm, compared to a diameter of 5.2 mm for normal cells. Interestingly, no distinct intermediate-sized population of cells was observed. Therefore, cells were either of a normal size or of a markedly reduced size.

Isopyknic centrifugation is a technique that has been used to ascertain differences in the buoyant density between normal and apoptotic cells (37–39). Wyllie and Morris (37) measured the buoyant density of untreated and methylprednisolone-treated rat thymocytes. Untreated thymocytes, after 1 h, had a buoyant density of 1.075 g/ml. However, the population of treated thymocytes had two distinctly different densities, one similar to the untreated population of cells and a second with an increased buoyant density. Similar to the observations of Thomas and Bell (36), the population with an increased buoyant density grew in number over a period of time with no significant intermediate population observed. Percoll separation of these treated thymocytes showed that the more dense cells consisted almost entirely of apoptotic cells, whereas the less dense cells were similar to the untreated or normal cell population. This stage of high density was only transient since during extended incubations, when chromosomal condensation was observed, this increase in density was lost. These investigators concluded that the observed increase in cell density of apoptotic cells must be due to a loss of water from the cell.

A recent study by Benson et al. (39) characterized the loss of cell volume in an established T-cell line, designated CEM-C7A, during dexamethasone-induced apoptosis. Treatment of CEM-C7A cells for 48 h with dexamethasone resulted in a 42% decrease in cell volume compared to untreated cells. However, measurement of the buoyant density of dexamethasone-treated CEM-C7A cells showed that no change had occurred during the first 24 h, and a less dense fraction was observed at the later time. The authors interpreted these results to suggest that a loss of cell volume occurs in two phases. The first phase of volume loss is not solely mediated by a loss of intracellular ions and water, but the export of cytoplasmic contents, decreasing all cellular components, and therefore maintaining a normal cell density. The second phase was suggested to represent a loss of macromolecules from the cell during the fragmentation process associated with apoptosis. These data contradict conclusions made in previous studies (37, 38), where a transient increase in buoyant density was observed in thymocytes treated with glucocorticoids. The manner in which a cell shrinks or loses volume during apoptosis may thus depend on the cell type (primary versus established cell lines) being examined and/or the frequency at which the density measurements are made.

The idea that a loss of cell volume during apoptosis occurs in stages has been previously proposed by other investigators (40–42). Cell shrinkage during radiation-induced apoptosis of rat thymocytes has recently been reported to occur as two distinct stages (42). First, there was a rapid loss of cell volume which decreased the size of these thymocytes to approximately 75% of their original volume. Subsequently, these cells showed a more gradual decrease in cell size to approximately 57% of their original volume over a period of several hours. Using two different sets of density figures, these authors concluded that most of the material lost during both stages of cell shrinkage was water, and not cytoplasmic or cellular contents. Therefore, the way in which cell death is induced, along with the cell type and frequency of observations as described earlier, may account for differences in the manner in which cell shrinkage or volume loss is reported during apoptosis.

Human eosinophils also lose cell volume during apoptosis upon cytokine withdrawal (43). The viability of purified human eosinophils was reduced to 24% after 48 h in the absence of cytokine, with over 70% of these cells showing a decrease in cell size. Electronic sizing of these apoptotic cells showed on average a 63.2% reduction in cell volume, which is even greater than the 43% reported in irradiated rat thymocytes as described above. Again, a clear bimodal size distribution was observed, with very few cells in an intermediate state of cell volume. Therefore, the cells appeared either normal or shrunken, with no intervening cell population. This study also examined volume reduction in neutrophils that spontaneously undergo apoptosis in culture. Interestingly, after 24 h, cells which comprised the smaller-sized population had only a 30% decrease in cell volume, and this population overlapped the volume distribution of the normal or non-apoptotic cells. Thus, although there may be cell type specific pathways to reduce cell volume during the initiation of apoptosis, the loss of cell volume is a common feature of the process in a wide variety of systems.

Analysis of the Relationship Between Cell Volume and Apoptosis by Flow Cytometry

Flow cytometry has become an important tool for studying volume loss during apoptosis, particularly with non-adherent cells. Flow cytometry utilizes the light scattering properties of the cell as a measure of cell size and cell density. A decrease in forward-scattered and side-scattered light directly correlates with a decrease in cell size and cell density, respectively (44). With the use of fluorescent dyes, such as propidium iodide, other parameters, such as DNA content, can be simultaneously ascertained, along with the light-scattering properties of the cell. A decrease in DNA content has been associated with apoptotic fragmentation of the DNA, and this characteristic can therefore be directly related to the size of individual cell populations. Interestingly, examination of dexamethasone-treated S49 Neo cells by flow cytometry showed that only cells that exhibited a loss in cell volume had DNA which was degraded (45). This observation suggests an important relationship between the loss of cell volume and activation of the nucleases associated with apoptosis. Therefore, knowledge on how these cells lose volume and how this reduction in cell size may relate to other characteristics of apoptosis should facilitate our understanding of the entire programmed cell death process. Although, many investigators have utilized flow cytometry to study the various parameters of apoptosis (46), we will limit our discussion here to those studies which have specifically examined cell volume.

Swat et al. (47) have used flow cytometry to examine apoptosis in immature thymocytes. When isolated mouse thymocytes were compared to thymocytes cultured for a period of 24 h, an increased number of cells with a decrease in forward-scattered light (reduced cell volume) and an increase in side-scattered light (increased cell density) was observed. HL-60 cells also showed a decrease in forward-scattered light that occurred as early as 2 h after treatment with the topoisomerase inhibitor, camptothecin (48). However, this decrease in forward-scattered light was not accompanied by an immediate change in side-scattered light. A change in side-scattered light did not occur until 3 h after treatment, when a decrease was observed, in stark contrast to the increase in side-scattered light observed in 24 h-cultured primary thymocytes. Darzynkiewicz hypothesized that the increase in cell density observed after a 24 h culture of primary thymocytes probably reflects chromatin condensation and nuclear fragmentation of the cell. Therefore, it appears that T-cells undergo an initial decrease in cell density followed by an increase in density during apoptosis, with the latter concurrent with other morphological characteristics of apoptosis. The kinetics of the programmed cell death process have also been studied through the use of flow cytometry in mature human T-cells (49). Activation of apoptosis in these cells by either ionomycin, PHA, or anti-T-cell receptor (TCR) monoclonal antibodies showed that cell shrinkage occurred prior to DNA degradation. Although the time of activation varied between apoptotic inducing agents, a loss of cell volume was always observed before DNA fragmentation. These data support the observation from our lab that only cells that shrink contain DNA that has been degraded.

Whether or not a loss of cell volume can trigger apoptosis in immature thymocytes has recently been investigated (45). As described earlier, thymocytes do not initially respond to a loss in cell volume with a RVI response. Interestingly, when S49 Neo cells are cultured under hypertonic conditions (using mannitol, sucrose, or NaCl) they underwent a very rapid loss of viability. The increase in the extracellular osmolality produced not only a loss of viability and a sustained decrease in cell volume, but also triggered internucleosomal DNA cleavage and apoptotic body formation, both recognized characteristics of apoptosis. In contrast to dexamethasone-induced apoptosis, cell death induced by hypertonic conditions does not require protein synthesis, suggesting that these cells contain all the necessary proteins to execute the apoptotic process. When S49 Neo cells were cultured under hypotonic conditions (where the cells swell) for a similar period of time, apoptosis did not occur. This suggests that a specific change in cell volume is required to trigger the apoptotic process. We extended these studies to include other thymic lymphoid cells that also do not demonstrate a RVI response, such as CEM-C7 cells and primary thymocytes, and observed a similar apoptotic response. Interestingly, when nonlymphoid cells (COS, HeLa, GH$_3$, and L-cells) were exposed to similar hypertonic conditions, an initial decrease in cell volume was observed followed by a RVI response with no subsequent apoptosis. Therefore, the presence of a volume regulatory response may act as a first line of defense to protect cells against programmed cell death. These results demonstrate that cell volume plays a crucial role during apoptosis, and suggests that mechanisms that control cell volume must be either inhibited or overridden for cells to undergo programmed cell death (summarized in Fig. 18.3).

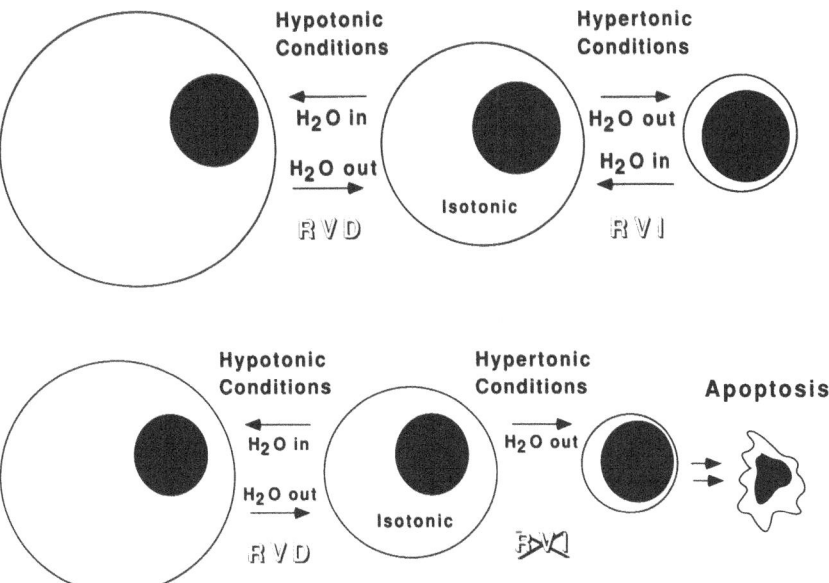

FIGURE 18.3. (A) Regulation of cell volume in most cells compared to volume regulation in thymic lymphocytes. Most cells can control their cell volume in hypotonic and hypertonic conditions by a RVD or RVI response, respectively. (B) However, thymic lymphocytes, which do not elicit a primary RVI response to hypertonic conditions, can subsequently undergo apoptosis. (Reproduced with permission from Bortner and Cidlowski (45).)

The Loss of Ions and Transport Mechanisms Contribute to the Loss of Cell Volume During Apoptosis

The link between a loss of cell volume and the loss of ions from the cell during the apoptotic process has recently become an area of active investigation (39, 43, 45). Since K^+ is the most abundant ion inside the cell, a loss of K^+ from the cell would cause water to follow, allowing the cell to lose volume. In a study of cytokine withdrawal-induced cell death of human eosinophils, 4-aminopyridine (4-AP), a potent K^+ channel blocker, was shown to inhibit apoptotic shrinkage in a dose-dependent manner (43). The loss of cell volume was inhibited by approximately 60% at a concentration of 3 mM 4-AP. Unfortunately, biochemical characteristics of apoptosis were not assessed in the presence of 4-AP. However, the presence of 4-AP did inhibit the previously described bimodal cell volume distibution, as intermediate volumes were observed. These observations were not restricted to 4-AP since similar results were also detected in the presence of two other K^+ channel blockers, sparteine and quinidine (43).

In an effort to understand the various stages of cell shrinkage observed during dexamethasone-induced apoptosis of CEM-C7A cells, Benson et al. (39) also examined the role of K^+. During the first phase of volume loss (which occurs between 12 and 36 h posttreatment), a net decrease in intracellular K^+ content was observed, allowing for the movement of water out of the cell. However, calculations measuring the volume of intracellular water led to the conclusion that no change in K^+ concentration had occurred, as the loss of both K^+ and water were balanced. Interestingly, induction of apoptosis was reversible at this time if the glucocorticoid was removed during this early phase. In contrast to this first phase of cell shrinkage, the second phase of volume reduction that occurred was associated with chromatin condensation and cellular fragmentation, and not the loss of ions from the cell.

Flow cytometry has also been used to examine intracellular ionic variations during apoptosis (50). L-cells treated with either etoposide or excess thymidine underwent cell cycle arrest and apoptosis. Cell death characterized by flow cytometry showed a population of cells with a subdiploid DNA fluorescence and reduced forward and side-light scatter. The size of treated cells was shown to range from 7 to 10 mm in diameter, almost half of what was observed in the untreated population. A slow and steady increase in Ca^{2+} was observed during the first 6 h of treatment; however, after 24 h the level of Ca^{2+} returned to normal. Attempts to inhibit this increase Ca^{2+} caused the cells to undergo necrosis, suggesting a precise orchestration of ion movements during apoptosis. The Na^+ concentration doubled in the treated L-cells to a steady state value of 30 mM, whereas K^+ decreased to approximately half of its initial intracellular concentration. In further support of the idea that a movement of ions plays an important role during apoptosis, Jonas et al. (51) also showed that T-cells treated with low doses of the staphylococcal alpha-toxin became permeable to monovalent ions and underwent internucleosomal DNA cleavage similar to that observed during apoptosis. Higher doses of alpha-toxin allowed for the movement of divalent cations, along with larger molecules such as propidium iodide, and caused a rapid depletion of ATP and no DNA cleavage. Interestingly, when cells were cultured in Na^+-free buffer in the presence of low doses of alpha-toxin, no loss of viability or DNA degradation occurred. These investigators suggested that the movement of Na^+ may also contribute to the cascade of events observed during programmed cell death and supports the notion of a specific and precisely-timed movement of ions during apoptosis.

Recently, several investigators have begun to examine the effects of various transporters and exchangers during apoptosis. The Na^+/H^+ exchanger and the $Na^+/K^+/2Cl^-$ cotransporter have been two of the various cellular transport mechanisms that have been investigated for their role during cell death (52–54). Interestingly, both of these transport mechanisms are also known to be involved during the RVI response. The Na^+/H^+ exchanger is a integral plasma membrane protein that exchanges extracellular Na^+ for intracellular H^+. Multiple forms of this exchanger exist, representing a gene family that are structurally similar, but differ in their tissue distribution, kinetics, and response to external stimuli (55). The Na^+/H^+ exchanger is also known to be involved in pH regulation. At a neutral pH, this exchanger is inactive; however as the pH decreases, this exchanger becomes highly active. The modulation of this exchanger has been shown to occur by phosphorylation, such that activation of protein kinase C (PKC) by phorbol esters can stimulate exchanger activity (56).

Li and Eastman (53) showed that a cytotoxic T-lymphocyte cell line underwent intracellular acidification and apoptosis upon growth factor withdrawal. Using flow cytometry to measure both intracellular H^+ and Ca^{2+}, an increase in both ions was observed, but only an increase in H^+ content was consistently associated with the appearance of degraded DNA. To ascertain the role that the Na^+/H^+ exchanger may play during apoptosis, these investigators showed that normal cells initially controlled their intracellular pH in response to changes in the extracellular pH. In contrast, apoptotic cells had a lower intracellular pH in response to changes in the extracellular pH, except under the most extreme extracellular acidic conditions. However, in the presence of EIPA, an inhibitor of the Na^+/H^+ exchanger, the intracellular pH dropped dramatically under these extreme extracellular acidic conditions, suggesting that the Na^+/H^+ exchanger was responsible for the control of pH only at extreme acidic conditions. Therefore, the Na^+/H^+ exchanger was active in apoptotic cells, but an alteration in the set-point at which this exchanger can control intracellular pH had occurred, which allows for a lower or acidic intracellular pH.

In contrast to the cellular acidification described during apoptosis for cytotoxic T-cells, cellular alkalinization has also been observed during programmed cell death (52). Thapsigargin and the calcium ionophore, A23187, which induce apoptosis in HL-60 cells, were shown to stimulate the Na^+/H^+ exchanger allowing for intracellular alkalinization. Inhibition of this exchanger with dimethyl amiloride blocked intracellular alkalinization and apoptosis. These results suggest that intracellular alkalinization or the stimulation of the Na^+/H^+ exchanger contributes to the induction of apoptosis. However, these investigators also showed that when the Na^+/H^+ exchanger was independently stimulated with the phorbol ester, PMA, neither thapsigargin- nor A23187-induced apoptosis occurred. Additionally, inhibition of the Na^+/H^+ exchanger upon treatment with PMA and either thapsigargin or A23187 did not allow for the induction of cell death. These data suggest that PMA inhibition of apoptosis and activation of the PKC pathway are independent of intracellular alkalinization or stimulation of the Na^+/H^+ exchanger. In support of this conclusion, the presence of PKC inhibitors abolished the PMA-mediated inhibition of apoptosis, and once again A23187 and thapsigargin could induce apoptosis. These results exemplify the complexity of the programmed cell death process, and suggest that multiple pathways can be simultaneously involved, some seeming in direct conflict with others to initiate or inhibit apoptosis.

A second ion transport mechanism, the $Na^+/K^+/2Cl^-$ cotransporter, had been shown earlier to play a potentially important role in apoptosis (54). The nitrogen mustard [bis(2-chloroethyl)methylamine; HN_2], which inhibits the $Na^+/K^+/2Cl^-$ cotransporter, was shown to suppress the growth of murine leukemia cells. Although the mode of cell death was not defined, several features of these cells resembled apoptotic, as opposed to necrotic, death, including the cells ability to exclude the vital dye, trypan blue, the maintenance of control-level ATP concentrations, and a reduction in cell volume. Interestingly, a decrease in K^+ content was observed, but intracellular K^+ concentrations were not affected due to an approximate 35% reduction in cell size.

Can a Loss of Cell Volume Initiate and Control Cell Death?

It is apparent from the studies described in this article that the simple observation of a loss of cell volume during apoptosis probably involves a complex set of physiological mechanisms. These mechanisms could allow for the controlled movement of ions during apoptosis to orchestrate the programmed cell death process. The activation of specific ion transport pathways, along with the inhibition of volume regulatory mechanisms, may also facilitate the occurrence of other characteristics of apoptosis. Alternatively, the loss of cell volume may physically disrupt the cell structure, allowing for the release of enzymes that, in turn, could degrade inhibitory apoptotic proteins. In contrast, a decrease in cell volume may result in the concentration of a positive apoptotic factor to initiate cell death. Cell shrinkage may also alter gene expression to induce cell death. Short-term modulation of cell volume has been shown to modify cellular metabolism and gene expression in liver cells (57). Cellular swelling and cell shrinkage lead to opposite patterns of cellular metabolism in liver cells and may act as second or third messengers during the regulation of cellular processes. Recently, the regulation of glutamine and glucose metabolism by cell volume has been observed in lymphocytes and macrophages (58). Under hypotonic conditions, the rates of glutamine metabolism and glycolysis were shown to increase, suggesting that changes in cell volume may be an unrecognized mechanism for regulating metabolism in these cells. Therefore, the movement of ions would play a more indirect role in apoptosis and only induce the loss of cell volume in these above mentioned mechanisms.

However, many studies, including our own, suggests the movement of ions plays a direct role in the loss of cell volume during apoptosis along with the activation of other apoptotic mechanisms responsible for the various characteristics observed during programmed cell death. In Figure 18.4 we propose the following hypothesis to explain the events that occur during apoptosis. When a cell is exposed to an apoptotic stimulus, various events are initiated that set the stage for cell death. These events may include reorganization of cellular contents and structure, gene transcription, and protein synthesis. Interestingly, various apoptotic stimuli require different lengths of time, as shown by the various arrows in Figure 18.4, to initiate cell death. This suggests that specific apoptotic agents use selective pathways to program the cell for apoptosis. For example, we have shown that S49 Neo cells treated with dexamethasone undergo apoptosis, and this process requires a period of several hours and is dependent on protein synthesis. In contrast, when S49 Neo cells are treated under hypertonic conditions, apoptosis occurs

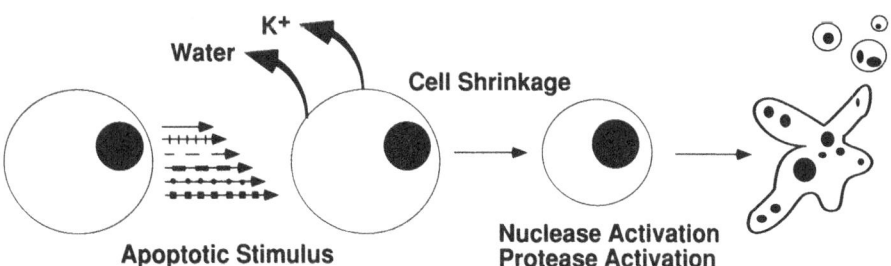

FIGURE 18.4. Model for the role that ions play during apoptosis. Individual pathways leading to cell death may be activated depending on the specific apoptotic stimulus applied, as shown by the different arrows. However, once the cell has been programmed to undergo apoptosis, there is a rapid loss of K^+ followed by the movement of water out of the cell, which accounts for the characteristic cell shrinkage observed during this mode of cell death. This rapid loss of ions and water permits the immediate activation of both nucleases and proteases associated with apoptosis. At this time, the cell is at a point of no return and apoptotic body formation ensues.

very rapidly and is independent of protein synthesis. Therefore, the extended period of time required for dexamethasone-induced apoptosis may be to transcribe and translate various components essential for this pathway of cell death. However, treatment of cells under hypertonic conditions allows apoptosis to occur by an existing cellular pathway. This idea of different apoptotic pathways to program the cell death process would account for the disparity in time between treatment of cells with various apoptotic agents and the occurrence of any physical evidence related to cell death.

It is clear that a loss of cell volume, once triggered during apoptosis, is a rapid event and appears to occur in distinct stages. Whether or not an initial loss of cell volume, as described by several investigators and reviewed in this article, occurs during the programming phase of cell death is not clear. However, once the stage for cell death is set, there appears to be a substantial loss of cell volume (a second phase of volume loss?) in which the cell goes from a near normal volume to one of a much reduced size. This idea is supported by the observations of several laboratories, in which no intermediate-sized populations of cells are detected during apoptosis. The loss of ions, most likely K^+, from the cell would allow for cell shrinkage along with the activation of various nucleases and proteases associated with apoptosis. Several recent lines of evidence support this conclusion. Apoptotic nuclease activity has been shown to be inhibited in dexamethasone-treated thymocytes cultured in high K^+ (Hughes et al., manuscript in preparation). The high K^+ concentration on the outside of these cells eliminates the normal K^+ concentration gradient, therefore inhibiting the loss of K^+ from the cell and protecting the cells from apoptosis. Additionally, low extracellular K^+ concentrations were shown to induce apoptosis in immature cerebellar granule neurons of the central nervous system (59), whereas higher K^+ concentrations inhibited cell death. Inhibitors of transcription and translation suppressed this low K^+-induced apoptosis, suggesting that protein synthesis is needed for apoptotic induction in these cells. Additionally, interleukin-1b-converting enzyme (ICE), a cellular protease functionally similar to the nematode cell death gene product CED-3, has been shown to be involved in apoptosis (reviewed in ref. 60). Interestingly, agents that depleted monocytes of K^+ stimulated the cleavage of prointerleukin(IL)-1b to mature IL-1b, suggesting that a loss of K^+ stimulates ICE activity (61). Therefore, a loss of K^+ from a cell during apoptosis may allow for the reduction of cell volume along with the activation of apoptotic nucleases and proteases to trigger cell death. At this stage, the cell is at a point of no return in the apoptotic process, and apoptotic body formation will follow.

In conclusion, the manner and degree of cell shrinkage observed during cell death is likely to be cell-type specific. Additionally, the method in which cell death is initiated may also alter the way in which cell shrinkage is viewed. The results reviewed in this article support the idea that the loss of cell volume during apoptosis may occur via different pathways depending on the mode of cell death induction, and that the movement of ions, along with the inhibition of volume regulatory mechanisms, probably plays a critical role. Flow cytometry has become an indispensable tool for studying volume loss during apoptosis, and future studies to examine the movement of ions and activity of ion transporters will undoubtedly further our understanding of the entire programmed cell death process.

References

1. Schwartzman RA, Cidlowski JA. Apoptosis: the biochemistry and molecular biology of programmed cell death. Endocr Rev 1993;14:133–51.
2. Martin SJ, Green DR, Cotter TG. Dicing with death: dissecting the components of the apoptotic machinery. Trends Biochem Sci 1994;19:26–30.
3. Kerr JFR, Wyllie AH, Currie AR. Apoptosis: a basic biological phenomenon with wide-ranging implications in tissue kinetics. Br J Cancer 1972;26:239-57.
4. Wyllie AH. Glucocorticoid-induced thymocyte apoptosis is associated with endogenous endonuclease activation. Nature 1980;284:555–6.
5. Sten-Knudson O. Passive transport processes. In: Giebisch G, Tosteson DC, Ussing HH, eds. Membrane transport in biology. Vol. I. Concepts and Models. Berlin and New York: Springer-Verlag, 1978:5–113.
6. Wilson TH. Ionic permeability and osmotic swelling of cells. Science 1954; 120:104–5.
7. Leaf A. On the mechanism of fluid exchange of tissues in vitro. Biochem J 1956; 62:241–8.
8. Skou JC. The influence of some cations on an adenosine triphosphatase from peripheral nerves. Biochem Biophys Acta 1957;23:294–401.
9. Proverbio F, Duque JA, Proverbio T, Marin R. Cell volume-sensitive Na^+-ATPase activity in rat kidney cortex cell membranes. Biochem Biophys Acta 1988;941:107–10.
10. Parker JC. Dog red blood cells: adjustment of density in vivo. J Gen Physiol 1973;62:147–56.
11. Al-Habori M. Cell volume and ion transport regulation. Int J Biochem 1994;26:319–34.
12. Hoffmann EK. Volume regulation in cultured cells. Curr Top Membr Transport 1987;30:125–80.
13. McManus TJ, Haas M, Starke C, Lytle CW. The duck red cell model of volume-sensitive chloride-dependent cation transport. Anno NY Acad Sci 1985;456:183–6.
14. Lauf PK. K^+:Cl^- cotransport: sulfhydryls, divalent cations and the mechanism of volume regulation. J Membr Biol 1985;88:1–13.
15. Hall AC, Ellory JC. Effects of high hydrostatic pressure on "passive" monovalent cation transport in human red cells. J Membr Biol 1986;94:1–17.
16. Cala PM. Volume regulation by amphiuma red blood cells: characteristics of volume-sensitive K/H and Na/H exchange. Mol Physiol 1985;8:199–214.
17. Grinstein S, Rothstein A, Sarkadi B, Gelfand EW. Responses of lymphocytes to anisotonic media: volume-regulating behavior. Am J Physiol 1984;246:C204–15.
18. Sarkadi B, Mack E, Rothstein A. Ionic events during the volume response of human peripheral blood lymphocytes to hypotonic media. I. Distinctions between volume-activated Cl^- and K^+ conductance pathways. J Gen Physiol 1984;83:497–512.

19. Sarkadi B, Mack E, Rothstein A. Ionic events during the volume response of human peripheral blood lymphocytes to hypotonic media. II. Volume- and time-dependent activation and inactivation of ion transport pathways. J Gen Physiol 1984;83:513–27.

20. Hoffmann EK, Simonsen LO, Lambert IH. Volume-induced increase of K^+ and Cl^- permeabilities in Ehrlich ascites tumour cells: role of internal Ca^{2+}. J Membr Biol 1984;78:211–22.

21. Banderali U, Roy G. Activation of K^+ and Cl^- channels in MDCK cells during volume regulation in hypotonic media. J Membr Biol 1992;126:219–34.

22. Davis CW, Finn AL. Interactions of sodium transport, cell volume, and calcium in frog urinary bladder. J Gen Physiol 1987;89:687–702.

23. Duhm J, Gobel BO. Na^+-K^+ transport and volume of rat erythrocytes under dietary K^+ deficiency. Am J Physiol 1984;246:C20–9.

24. O'Neil WC. Volume-sensitive Cl^--dependent K transport in human erythrocytes. Am J Physiol 1987;258:F1657–65.

25. Levinson C. Regulatory volume increase in Ehrlich ascites tumour cells is mediated by the 1Na:1K:2Cl cotransport system. J Membr Biol 1992;126:277–84.

26. Grinstein S, Clarke CA, Dupre A, Rothstein A. Volume-induce increase of anion permeability in human lymphocytes. J Gen Physiol 1982;80:801–23.

27. Grinstein S, Goetz JD, Cohen S, Rothstein A, Gelfand EW. Regulation of Na^+/H^+ exchange in lymphocytes. Ann NY Acad Sci 1985;456:207–19.

28. Lang F, Stehle T, Haussinger D. Water, K^+, H^+, lactate and glucose fluxes during cell volume regulation in perfused rat liver. Pflugers Arch 1989;413:209–16.

29. Ericson AC, Spring KR. Volume regulation by necturus gallbladder: apical Na^+-H^+ and Cl^--HCO_3^- exchange. Am J Physiol 1982;243:C146–50.

30. Cheung RK, Grinstein S, Gelfand EW. Volume regulation by human lymphocytes: identification of differences between the two major lymphocyte subpopulations. J Clin Invest 1992;70:632–38.

31. Grinstein S, Clarke CA, Rothstein A, Gelfand EW. Volume-induced anion conductance in human B lymphocytes is cation independent. Am J Physiol 1983;245: C160–3.

32. Roti-Roti LW, Rothstein A. Adaptation of mouse leukemic cells (L5178Y) to anisotonic media. I. Cell volume regulation. Exp Cell Res 1973;79:295–310.

33. Hempling HG, Thompson S, Dupre A. Osmotic properties of human lymphocytes. J Cell Physiol 1977;93:293–302.

34. Grinstein S, Clarke CA, Rothstein A. Activation of Na^+/H^+ exchange in lymphocytes by osmotically-induced volume changes and by cytoplasmic acidification. J Gen Physiol 1983;82:619–38.

35. Hughes Jr. FM, Cidlowski JA. Regulation of apoptosis in S49 cells. J Steroid Biochem Molec Biol 1994;49:303–10.

36. Thomas N, Bell PA. Glucocorticoid-induced cell-size changes and nuclear fragility in rat thymocytes. Mol Cell Endocrinol 1981;22:71–84.

37. Wyllie AH, Morris RG. Hormone-induced cell death. Purification and properties of thymocytes undergoing apoptosis after glucocorticoid treatment. Am J Path 1982;109:78–87.

38. Morris RG, Hargreaves AD, Duvall E, Wyllie AH. Hormone-induced cell death. 2. Surface changes in thymocytes undergoing apoptosis. Am J Path 1984; 115:426–36.

39. Benson RSP, Heer S, Dive C, Watson AJM. Characteristics of cell volume loss in

CEM-C7A cells during dexamethasone-induced apoptosis. Am J Physiol 1996;270:C1190–1203.

40. Ohyama H, Yamada T, Watanabe I. Cell volume reduction associated with inter-phase death in rat thymocytes. Radiat Res 1981;85:333–9.

41. Ohyama H, Yamada T, Ohkawa A, Watanabe I. Radiation-induced formation of apoptotic bodies in rat thymus. Radiat Res 1985;101:123–30.

42. Klassen NV, Walker PR, Ross CK, Cygler J, Lach B. Two-stage cell shrinkage and the OER for radiation-induced apoptosis of rat thymocytes. Int J Radiat Biol 1993;64:571–81.

43. Beauvais F, Michel L, Dubertret L. Human eosinophils in culture undergo a strik-ing and rapid shrinkage during apoptosis. Role of K⁺ channels. J Leukoc Biol 1995;57:851–5.

44. Willman, CL, Stewart CC. General principles of multiparameter flow cytometric analysis: applications of flow cytometry in the diagnostic pathology laboratory. Semin Diagnostic Pathol 1989;6:3–12.

45. Bortner CD, Cidlowski JA. The absence of volume regulatory mechanisms con-tributes to the rapid activation of apoptosis in thymocytes. Am J Physiol 1996;271:C950–61.

46. Cidlowski JA, King KL, Evans-Storms RB, Montague JW, Bortner CD, Hughes FM Jr. The biochemistry and molecular biology of glucocorticoid-induced apoptosis in the immune system. Recent Horm Res 1996;52:1–35.

47. Swat W, Ignatowicz L, Kisielow P. Detection of apoptosis of immature CD4⁺8⁺ thymocytes by flow cytometry. J Immunol Meth 1991;137:79–87.

48. Darzynkiewicz Z, Bruno S, Del Bino G, Gorczyca W, Hotz MA, Lassota P, Traganos F. Features of aoptotic cells measured by flow cytometry. Cytometry 1992;13:795–808.

49. Wesselborg S, Kabelitz D. Activation-driven death of human T cell clones: time course kinetics of the induction of cell shrinkage, DNA fragmentation, and cell death. Cell Immunol 1993;148:234–41.

50. Barbiero G, Duranti F, Bonelli G, Amenta JS, Baccino FM. Intracellular ionic variations in the apoptotic death of L cells by inhibitors of cell cycle progression. Expt Cell Res 1995;217:410–8.

51. Jonas D, Walev I, Berger T, Liebetrau M, Palmer M, Bhakdi S. Novel path to apoptosis: small transmembrane pores created by Staphylococcal alpha-toxin in T lymphocytes evokes internucleosomal DNA degradation. Infect Immun 1994;62:1304–12.

52. Zhu WH, Loh TT. Effects of Na⁺/H⁺ antiport and intracellular pH in the regulation of HL-60 cell apoptosis. Biochim Biophys Acta 1995;1269:122–8.

53. Li J, Eastman A. Apoptosis in an interleukin-2-dependent cytotixic T-lymphocyte cell line is associated with intracellular acidification. J Biol Chem 1995;270:3203–11.

54. Wilcock C, Chahwala SB, Hickman JA. Selective inhibition by bis(2-chloroethyl)methylamine (nitrogen mustard) of the Na⁺/K⁺/Cl⁻ cotransporter of murine L1210 leukemia cells. Biochim Biophys Acta 1988;946:368–78.

55. Yun CHC, Tse CM, Nath SK, Levine SA, Brant SR, Donowitz M. Mammalian Na⁺/H⁺ exchanger gene family: structure and function studies. Am J Physiol 1995, 269:G1–11.

56. Sardet, C, Counillon, L, Franchi, A, Pouyssegur, J. Growth factors induce phos-phorylation of the Na⁺/H⁺ antiporter, a glycoprotein of 110 kD. Science 1990;247:723–6.

57. Haussinger D, Lang F, Gerok W. Regulation of cell function by the cellular hydration state. Am J Physiol 1994;267:E343–55.
58. Wu G, Flynn NE. Regulation of glutamine and glucose metabolism by cell volume in lymphocytes and macrophages. Biochem Biophys Acta 1995;1243:343–50.
59. D'Mello SR, Galli C, Ciotti T, Calissano P. Induction of apoptosis in cerebellar granule neurons by low potassium: inhibition of death by insulin-like growth factor I and cAMP. Proc Natl Acad Sci USA 1993;90:10989–93.
60. Kumar S. ICE-like proteases in apoptosis. Trends Biochem Sci 1995;20:198–202.
61. Walev I, Reske K, Pamer M, Valeva A, Bhakdi S. Potassium-inhibited processing of IL-1b in human monocytes. EMBO 1995;14:1607–14.

19

Apoptosis in the Ovary:
The Role of DNase I

DAVID L. BOONE AND BENJAMIN K. TSANG

The essential function of the ovary, to produce oocytes and hormones, involves follicular development, ovulation, and luteinization. The predominant event in the ovary however, is the loss of follicles through atresia. Luteal regression depends on the reproductive state of the female and must occur before the next wave of follicular development can proceed. Follicular atresia and luteal regression account for the removal of most steroidogenic cells from the ovary, but the physiological mechanisms that underlie these two predominant and important aspects of ovarian function are not fully understood.

Apoptosis is a physiological form of cell death characterized by endonucleolytic degradation of DNA into oligonucleosomal fragments (1). Apoptotic DNA degradation occurs during ovarian follicular atresia (2–4) and luteal regression (5). Hormones and intraovarian factors previously reported to induce or suppress these processes have been demonstrated to activate or inhibit DNA degradation in ovarian cells (6–9). Identification of the endonuclease(s) responsible for DNA degradation in the ovary may be essential for a complete understanding of the role and regulation of apoptosis in follicular atresia and luteal regression. This chapter summarizes the current understanding of endonucleases as they relate to apoptosis, documents the identification of DNase I in the rat ovary, the potential role and regulation of DNase I during apoptosis, and presents a model for the role of DNase I in apoptosis during follicular atresia or luteal regression

Role of Endonucleases and DNA Degradation in Apoptosis

Although DNA degradation remains the most consistent and striking feature of apoptosis, the role of Ca^{2+}/Mg^{2+}-dependent endonuclease activity in the process remains unresolved. Initially, the Ca^{2+} sensitivity of the endonuclease suggested that increased intracellular calcium could be the activating mechanism of apoptosis (10–13). Diverse apoptosis-inducing agents, such as hormones, toxins, and radiation, can increase intracellular Ca^{2+} concentration or disrupt calcium regulation (11–15) leading to endonuclease activation and cell death. Calcium ionophores are known to induce death of cells (16), whereas elimination of calcium can prevent cell death (12, 16). The role of Ca^{2+} in apoptosis is unclear, however, since an increase in intracellular Ca^{2+} concentration is not always associated with cell death and can sometimes result in proliferation or protection from apoptosis (17, 18). The response of a cell to elevated Ca^{2+} likely depends on a number of factors, including the extent and duration of the rise in intracellular Ca^{2+}, the differentiated state of the cell, and the combined presence of other mitogenic or apoptotic signals.

Oncogenes, such as bcl-2 and ras, have been shown to modulate apoptosis (19–22). Transfection of fibroblasts with Ha-ras correlated with the appearance and enhanced expression of Ca^{2+}/Mg^{2+}-dependent endonuclease activity (21, 22), whereas cotransfection with bcl-2 suppressed the endonuclease in isolated nuclei, suggesting that some effects of the products of these genes on apoptosis may be exerted at the level of the endonuclease. Follicle stimulating hormone (FSH) suppresses apoptotic DNA production and regulates expression of members of the bcl-2 gene family in ovarian follicles (23). A direct interaction of bcl-2 oncogene products and endonucleases has not been demonstrated in the ovary, but this may be one mechanism whereby FSH suppresses apoptotic DNA degradation in follicles.

Interleukin-1β-converting enzyme (ICE), or ICE-like proteases, induce apoptosis following transfection (24) and ICE activation causes the degradation of another enzyme, poly-ADP-ribose polymerase (PARP) (25). In addition to its role in DNA repair (26), PARP may serve to tonically suppress endonuclease activity (27). Inhibition of PARP by ICE could, therefore, derepress the endonuclease, leading to the internucleosomal cleavage of DNA and apoptosis. Apoptotic DNA degradation in isolated nuclei was inhibited by treatment with the PARP substrate, NAD (27). The PARP inhibitors potentiated liver damage by hepatotoxic agents (28) and induced apoptosis in leukemic and other human malignant cells (29). PARP inhibited the 35 to 37 kDa, Ca^{2+}/Mg^{2+}-dependent endonuclease in bull seminal plasma (30). Although the identity of this endonuclease was not known, DNase I has been identified as an endonuclease activity in human (31, 32) and rabbit (33) seminal plasma. Thus, the induction of apoptosis by ICE and ICE-related proteases may involve inactivation of PARP and the subsequent derepression (activation) of DNase I. Tilly and coworkers recently reported that mRNA for ICE-related proteases decreased in ovaries of gonadotropin-treated rats (34). Inhibition of ICE-related proteases did not prevent morphological signs of apoptosis but did prevent internucleosomal DNA cleavage. This suggests that ICE or related proteases in ovarian follicles are linked to the activity of endonucleases but not to the entire process of apoptosis. In addition to its effects on endonuclease activity via bcl-2 gene family members, FSH may serve to prevent apoptotic DNA degradation by suppression of ICE-related proteases in the ovary.

The importance of DNA degradation in apoptosis is demonstrated by the fact that zinc (35) or aurintricarboxyllic acid (36), which can prevent apoptosis, inhibit the Ca^{2+}/Mg^{2+}-dependent endonuclease activity in most cells examined to date. These two compounds also inhibit activity of ovarian DNase I (Fig. 19.1C). Conversely, cell death with the morphological features of apoptosis can proceed in the absence of the generation of any apparent low molecular weight DNA cleavage products (37–39). Enucleated cells retained the ability to undergo a "cytosolic" apoptosis characterized by the distinct morphology of apoptosis and disruption of mitochondrial activity (40). Futhermore, Bcl-2, a protein known to block apoptotic cell death, inhibited this cytosolic apoptosis (40). These findings demonstrated that the degradation of DNA and destruction of the nucleus may not be an essential component of the apoptotic process, but is instead perhaps one branch of a cascade of events controlled within the cytosol, similar to the control of nuclear events during the cell cycle by elements in the cytoplasm.

Ca^{2+}/Mg^{2+}-Dependent Endonuclease Activity in the Rat Ovary

Zeleznik et al. (41) first reported the presence of Ca^{2+}/Mg^{2+}-dependent endonuclease activity in the rat ovary, and showed that the activity was developmentally expressed in granulosa and luteal cell nuclei such that the nuclei of differentiated granulosa and luteal cells, but not undifferentiated granulosa cells, contained Ca^{2+}/Mg^{2+}-dependent endonuclease activity. The identity of the endonuclease was not determined in that study. We, therefore, isolated nuclei from undifferentiated (diethylstilbestrol or DES-treated rat) and differentiated (equine chorionic gonadotropin or eCG-treated rat) granulosa cells, as well as from luteal cells (human chorionic gonadotropin or hCG-treated rat), and examined the identity and characteristics of the endonuclease(s) involved in DNA degradation (42). Ovarian cell nuclei degraded their DNA in an apoptotic fashion when exposed to Ca^{2+} and Mg^{2+} in vitro (Fig. 19.2). As reported by Zeleznik et al. (41), the endogenous endonuclease was present in ovarian cell nuclei in a developmental pattern. Atresia is seldomly observed in small undifferentiated follicles, and is most likely to occur during early differentiation as an antrum begins to form (4, 43–44). The developmental pattern of endonuclease activity demonstrated in Figure 19.2 is consistent with the developmental pattern of follicular atresia. The ability of follicles to undergo atresia may be a reflection of the ability of their cells to degrade DNA. Small immature follicles do not possess an endonuclease and, therefore, are not capable of DNA degradation and atresia. However, larger mature follicles contain an endonuclease and thus possess the machinery to die by apoptosis. The presence or absence of an endonuclease is not

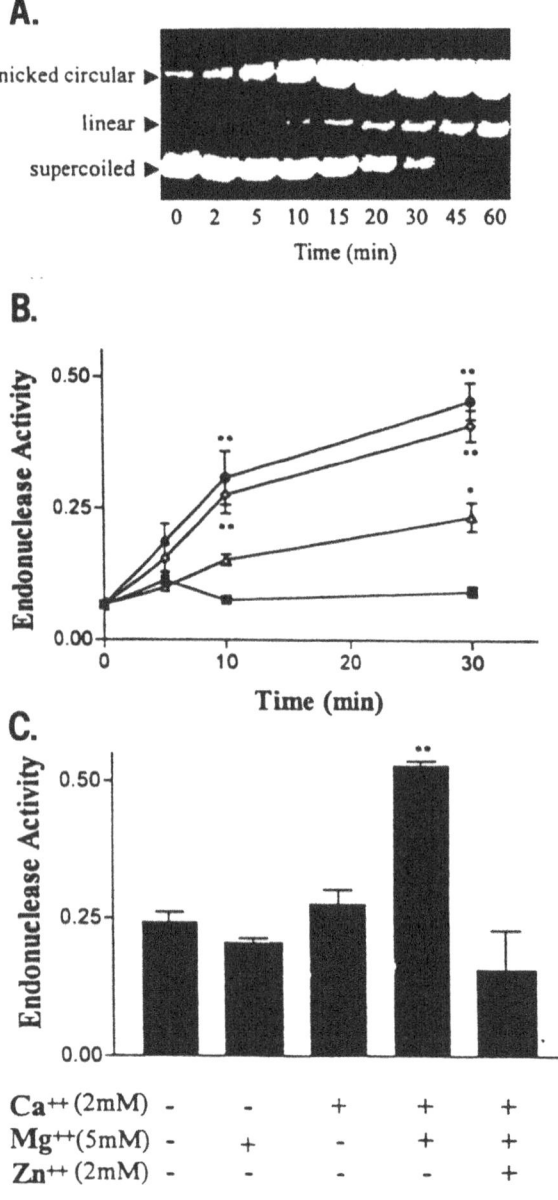

FIGURE 19.1. Developmental pattern and cation dependency of endonuclease activity in ovarian nuclear protein extracts. (A) Measurement of endonuclease activity in nuclear extracts. The degradation of plasmid DNA incubated (37°C) with 2.0 µg of nuclear protein extract from eCG-treated rat ovaries is resolved by 1% agarose gel electrophoresis was visualized by ethidium bromide staining and photographed on a UV transilluminator. Endonuclease activity (the production of nicked circular DNA) is measured by densitometrically scanning the photographic negative. (B) 0.4 µg of protein extracts from nuclei of DES- (triangle), eCG- (diamond) and hCG- (circle) treated rat ovaries

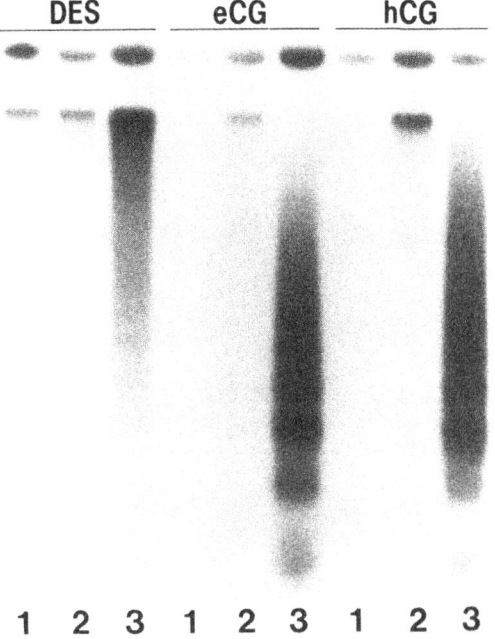

FIGURE 19.2. Developmental pattern of endonuclease activity in isolated rat ovarian nuclei. Nuclei isolated from DES-, ECG-, and hCG- treated rat ovaries were incubated for 30 min at 37°C in the absence (lane 2) or presence (lane 3) of Ca^{2+}(2 mM) and Mg^{2+}(5 mM). Lane 1 depicts activity at time = 0. DNA was extracted, 3'-end labeled with ^{32}P-ddATP (250 ng/reaction), resolved by agarose (2%) gel electrophoresis and analyzed by autoradiography. (Reproduced with permission from Boone et al. (42).)

likely to be the only factor in determining the susceptibility of cells to undergo apoptosis. The ability to die is likely determined by the extent of cytodifferentiation and acquisition of a number of components of the cellular machinery necessary for cell death.

The model in Figure 19.3 incorporates what is known about the expression of Ca^{2+}/Mg^{2+}-dependent endonuclease activity into our current understanding of follicular development and atresia as well as luteal regression. Preantral follicles contain granulosa cells that are undifferentiated except for the presence of FSH receptors (41, 44). Experimentally, estrogen in vivo is sufficient to produce follicles of this type. These cells have little or no endonuclease and these follicles do not frequently undergo atresia. When atresia does occur

←——

were incubated (37°C) with plasmid DNA in the presence of Ca^{2+}(2 mM) and Mg^{2+}(5 mM) (square: control, no protein extract). (C) Protein extract from eCG-treated rat ovaries incubated (37°C, 15 min) with plasmid DNA in the presence of Ca^{2+}(2 mM), Mg^{2+}(5 mM) and/or Zn^{2+}(2 mM).(** $p < 0.001$ compared to control; * $p < 0.05$ compared to control). (Reproduced with permission from Boone et al. (42).)

254   D.L. Boone and B.K. Tsang

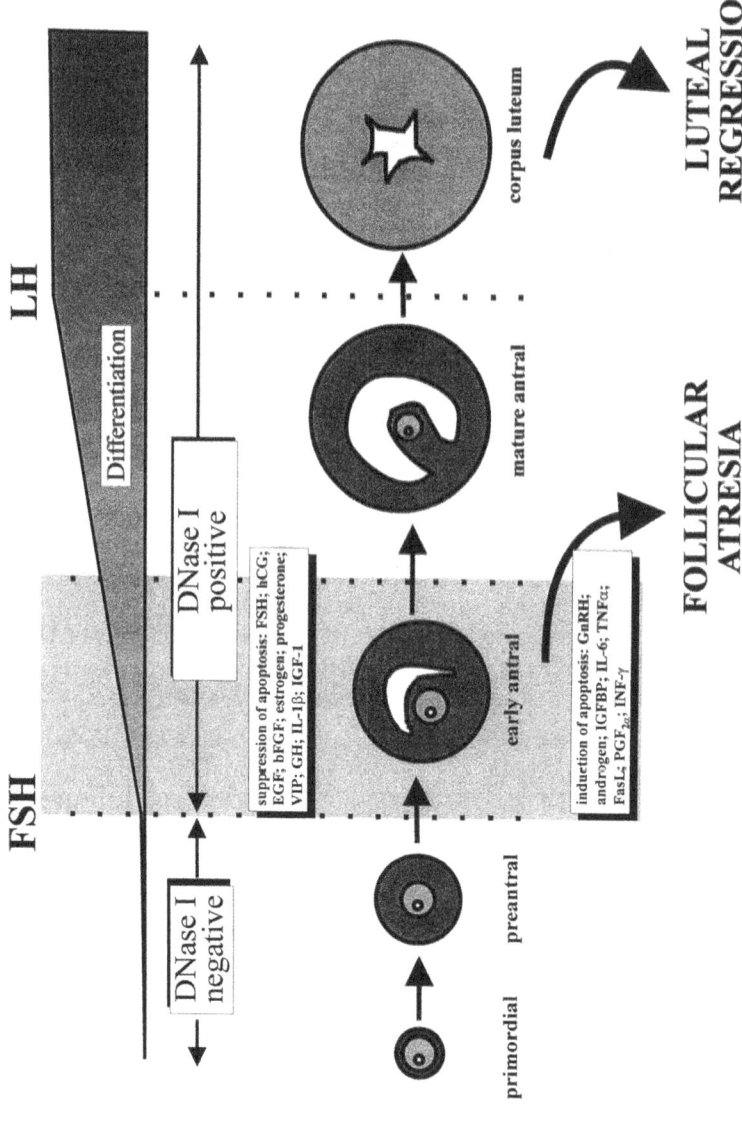

FIGURE 19.3. A model illustrating the developmental pattern of ovarian DNase I expression during follicular development, atresia, and luteal regression.

TABLE 19.1. Activators and suppressors of ovarian DNA degradation.

Activators	References	Suppressors	References
GnRH	7	FSH	7,8,23,103
androgen	6	hCG	9
IGFBP	8	estrogen	6
IL-6	74	progesterone	102
TNFα	108	bFGF	78
Fas antibody	107	EGF	78,102
PGF$_{2\alpha}$	9	VIP	79
		IGF-I	8
		GH	105
		IL-1β	106

in such follicles it is initiated in the oocyte and the granulosa cells die subsequent to this oocyte atresia (44). FSH stimulates ovarian granulosa differentiation and expression of a Ca^{2+}/Mg^{2+}-dependent endonuclease. The follicle can now undergo atresia unless it also acquires the ability to keep the endonuclease in an inactive state. Potential mediators of ovarian endonuclease activity are listed in Table 19.1. Activation of the endonuclease results from an absence of suppressing signals or presence of activating signals (or a combination of both) and is dependent on the differentiated state of the cell. The corpus luteum also possesses an endonuclease and the ability to suppress its activity. Prostaglandin $F_{2\alpha}$ initiates luteolysis, which is associated with apoptosis and endonuclease activation (5, 9). This may be mediated directly by increasing luteal cell intracellular Ca^{2+} levels or indirectly by suppression of luteal cell progesterone production.

Identification of the Endonuclease in Ovarian Cell Nuclei

To identify the endonuclease responsible for apoptotic DNA degradation in the ovary, nuclear proteins were extracted from DES-, eCG-, and hCG-treated rats and examined by plasmid degradation and zymographic assays. Consistent with the ability of intact nuclei to degrade DNA, the extracts were found to possess a developmentally regulated Ca^{2+}/Mg^{2+}-dependent endonuclease (Fig. 19.1). The endonuclease cleaves plasmid DNA in a manner consistent with apoptotic DNA production, in that it first cuts one strand of the DNA to produce open circular DNA and then the other strand to produce linear DNA (Fig. 19.1, inset). Consistent with the developmental pattern of endonuclease activity in intact nuclei, we found a developmental pattern of nuclease activity corresponding to molecular weights of 32 and

34 kDa, as determined by zymography (45) following SDS-PAGE (Fig. 19.4). In addition, a nuclease activity corresponding to 27 kDa was also present in the nuclear protein extracts. Since the intact nuclei and plasmid degradation assays demonstrated that the apoptotic endonuclease activity was Ca^{2+}/Mg^{2+}-dependent, the cation requirements of the 32, 34, and 27 kDa bands of nuclease activity were examined. The 32/34 kDa activity was found to be Ca^{2+}/Mg^{2+}-dependent whereas the 27 kDa activity required only Mg^{2+} for full activity. In contrast to the developmental pattern of expression consistently observed for the 32/34 kDa nuclease activity, this characteristic was not evident for the 27 kDa activity. The 27 kDa activity was often found in equal amounts in ovaries of DES-, eCG-, and hCG-treated rats, active only on single-stranded DNA in the zymographic assay and observed only after completely denaturing electrophoretic separation (presence of dithiothreitol in gel loading buffer). The 32/34 kDa activity was more active on double-stranded than single-stranded DNA and after semi-denaturing electrophoresis (no dithiothreitol in gel loading buffer). Based on the developmental pattern of expression, cation requirement and activity on double- or single-stranded DNA, we concluded that the 32/34 kDa, but not the 27 kDa, nuclease activity was responsible for the apoptotic DNA degradation in rat ovarian nuclei (42).

It is possible that the 32/34-kDa bands of activity in the zymographic gel could represent a false positive signal due to histone H1 interference since histones are known to bind and prevent intercalation of ethidium bromide to DNA (46, 47). To address this possibility, we extracted nuclei with a high salt buffer (1 M NaCl), instead of the usual 0.35 M NaCl extraction, and demonstrated that the extracts contained much more histone and indeed exhibited false positive staining (Fig. 19.5). This false positive staining, however, could be distinguished from actual nuclease activity, by silver staining the DNA in the gel. In this assay, nuclease activity, but not false positive histones, appeared as clear spots in a brown-stained gel. The silver staining was slightly more sensitive than ethidium bromide staining for the detection of nuclease activity and was subsequently used to confirm nuclease activity in the 27, 32, and 34-kDa bands.

As shown in Figure 19.4, bovine pancreatic DNase I exhibited the same cation dependency, DNA substrate preference and optimal electrophoresis conditions, as well as a similar pH optimum and molecular weight, as the 32/34 kDa activity, suggesting that the 32/34 kDa might be DNase I. In this context, we also noted that nuclear protein extracts analyzed by plasmid degradation assays also shared the same biochemical characteristics as DNase I, including dependence on Ca^{2+}/Mg^{2+} or active in Mn^{2+} alone, and inhibition by zinc, aurintricarboxylic acid or dithiothreitol. In addition, the degraded DNA in apoptotic nuclei contained 3'-OH ends available for labeling by terminal transferase, consistent with the action of DNase I on DNA. The endonuclease was also immunologically indistinguishable from DNase I since an

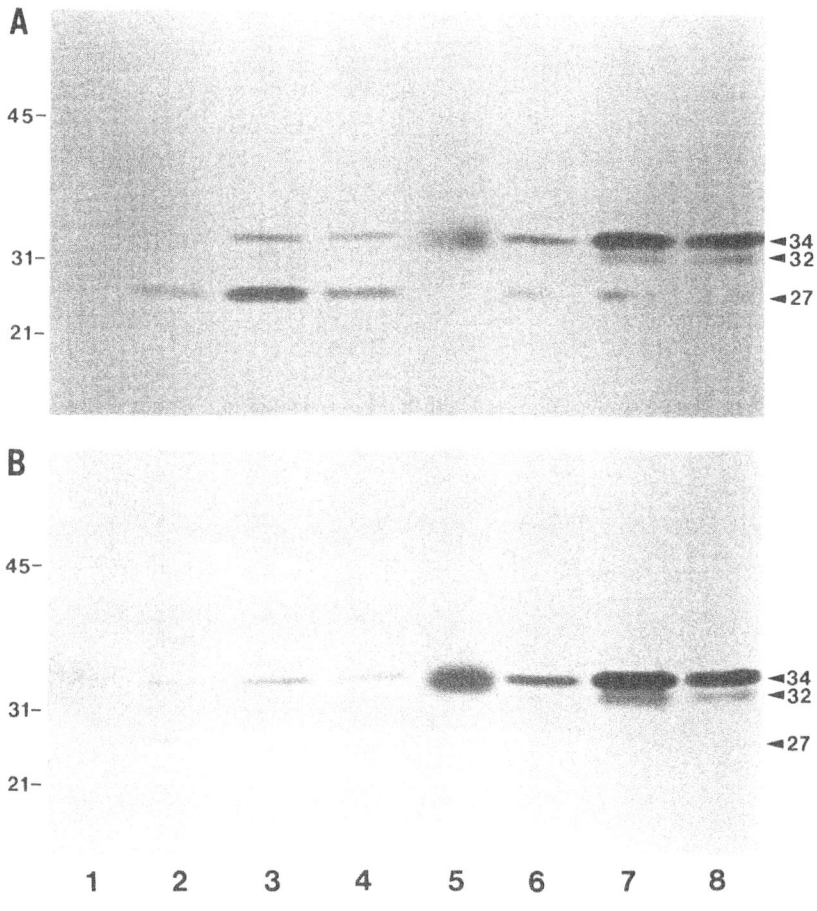

FIGURE 19.4. Zymographic analysis of nuclease activity in ovarian nuclear extracts. Ten ng of bovine pancreatic DNase I (lanes 1,5) or 40 µg of nuclear protein extracts from DES- (lanes 2,6), eCG- (lanes 3,7) and hCG- (lanes 4,8) treated rat ovaries were subjected to electrophoresis in SDS-polyacrylamide (12%) gels containing 250 µg/mL of either single stranded DNA (panel A) or double stranded DNA (panel B). Electrophoresis was carried out under complete denaturing (lane 1–4) or semi-denaturing (lane 5–8) conditions. After incubation (37°C, 72 h) in the presence of Ca^{2+}(2 mM) and Mg^{2+}(5 mM) nucleases were identified by staining the DNA remaining in the gel with ethidium bromide and photography under UV transillumination. All nuclease activities were confirmed as indicated in Figure 19.5. Molecular weight markers are indicated on the left, in kilodaltons. The arrows on the right indicate the positions of the 27, 32, and 34 kDa nuclease activities in the nuclear extracts. (Reproduced with permission from Boone et al. (42).)

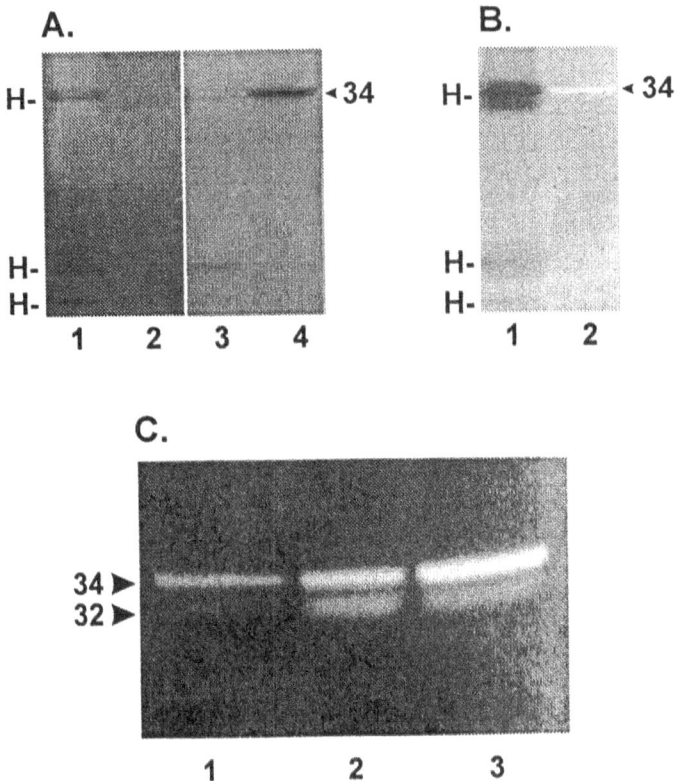

FIGURE 19.5. Nuclease activity is distinct from false positive histone staining. Zy-mograms using electrophoresis conditions as described in Figure 19.4. High salt (1.0 M NaCl) nuclear protein extract (40 μg) was used to demonstrate false positive histone (H) staining (Panel A, lane 1,3; Panel B, lane 1) as compared to true nu-clease activity (indicated by 34) in eCG nuclear extract (10 μg; Panel A, lanes 2,4; Panel B, lane 2). Ethidium bromide stained gels were photographed under UV trans-illumination before (Panel A, lanes 1,2) and after (Panel A, lanes 3,4) incubation [37°C, 24 h; Ca^{2+}(2 mM) and Mg^{2+}(5 mM)]. Gels were then washed in SDS (5%, 1 h) and DNA in the gel was identified by silver staining to distinguish false positive histone (H; Panel B, lane 1) from true nuclease activity (Panel B, lane 2). Panel C shows the silver staining of 32/34 kDa nuclease activities in DES- (lane 1) eCG- (lane 2), and hCG- (lane 3) nuclear protein extracts after zymography as described in Figure 19.4.

anti-DNase I antibody was successfully used for Western blot analysis and immunoprecipitation of the 32/34-kDa activity. Finally, zymographic analy-sis with gels containing both DNA and G-actin have shown that the 32/34 kDa activity is inhibited by G-actin, a highly diagnostic feature of DNase I (Table 19.2) (48–51).

DNase I and Apoptosis

DNase I as an Endonuclease

DNase I is an endonuclease that produces free 3'-OH-ends of DNA, has no apparent sequence specificity and hydrolyzes both native and denatured DNA, albeit at different rates (50). The presence of DNase I in pancreatic and parotid secretions reflects a digestive function for this enzyme. DNase I has been demonstrated in tissues with no digestive function, including the heart, lung, kidney, prostate, and thymus, suggesting a role for the enzyme in cellular functions (51, 52). Mannherz and coworkers have implicated DNase I as the endonuclease involved in apoptosis (52–54, 56). Nuclei from thymocytes and lymph node cells were found to contain a Ca^{2+}/Mg^{2+}-dependent endonuclease that was precipitated by antibodies to DNase I and inhibited by G-actin (52). Transfection of a DNase I cDNA into COS cells was sufficient to induce the morphological and biochemical changes typical of apoptosis (53). DNase I mRNA has now been identified in rat thymus, lymph node, kidney, brain, stomach, parotid, small intestine, seminal vesicle, and testis (Table 19.2) (52, 54). Furthermore, DNase I has been immunolocalized in stratified epithelium and in enterocytes of the small intestine in a pattern consistent with the differentiation of these cells, and often in the perinuclear region of cells destined to be apoptotic as well as in apoptotic cells (54). Finally, androgen ablation is known to induce apoptosis (55) and DNase I mRNA upregulation (56) in the rat prostate, and DNase I-like endonuclease activity has also been implicated in apoptosis in leukemic HL-60 cells (57).

DNase I as an Actin Binding Protein

DNase I may have roles in apoptosis independent of its DNA degrading activity. As stated earlier, a characteristic feature of DNase I is its ability to bind and be inhibited by G-actin (48–51). Indeed the crystallization and atomic modelling of G-actin required that it be complexed to DNase I (58). The polymerization of G-actin (45 kDa) to F-actin is inhibited by DNase I and F-actin can be depolymerized by DNase I (59). The actin-DNase I interaction occurs naturally in cells (60), although isoforms of the enzyme differ in their affinity for actin (50, 51, 54). Clearly, cytoskeletal changes occur during apoptosis and are reflected in the shape changes of the cell. DNase I may participate in the morphology of apoptosis by its interaction with G-actin or F-actin. The cytoskeleton is more than just a structural element as it participates in the regulation of diverse cellular functions, including steroidogenesis and responses to growth factors (61). Agents that disrupt the cytoskeleton also interfere with steroidogenesis in ovarian granulosa (62, 63) and luteal (64) cells. DNase I delivered to adrenal cells in liposomes disrupted steroid production (65), which was attributed to cytoskeletal changes induced by DNase I. The induction of cytosolic apoptosis

TABLE 19.2. Endonucleases implicated in apoptosis.

Endonuclease	Tissue/cell type	Molecular weight (kDa)	Unique features	References
DNase I	ovarian follicle or corpus luteum; prostate; distal tubule and collecting ducts of kidney; stratified epithelium; enterocytes of gut; embryonic lens cells; thymocytes; HL-60 cells; lymph node cells; fibroblasts.	31–36	inhibited by G-actin; depolymerises F-actin; optimal activity at neutral pH; Ca^{2+}/Mg^{2+}-dependent	42, 52, 54, 56, 57, 91
Nuc18	thymocytes; germinal centre B-cells	18–22	cyclophilin; Ca^{2+}/Mg^{2+}-dependent	81–83,
104				
DNase II	CHO cells; embryonic lens cells; HL-60 cells	27	optimal activity at acidic pH; Ca^{2+}/Mg^{2+}-independent	87, 90–92
unidentified	spleen	27	Ca^{2+}/Mg^{2+}-dependent	96
unidentified	proximal tubule of kidney	15		95
unidentified	CTLL2 (cytotoxic T) cell line	40, 58	Ca^{2+}/Mg^{2+}-dependent (58kDa); Mg^{2+}-dependent (40 kDa)	93
unidentified	hepatoma, pheochromocytoma, MCF7 breast and prostatic carcinoma cells	97		94

in enucleated cells (40) might therefore not exclude a central role of the endonuclease in apoptosis, but point to a primary function of the endonuclease (DNase I) in the disruption of cellular activity at the level of the cytoskeleton while secondarily degrading the genome.

The DNase I Gene and Its Potential Regulation

The cDNA sequences for human (66, 69), rat (67), and mouse (68) DNase I are now known. GC content and CpG/GpC dinucleotide ratios in the 5' regions of the human DNase I gene suggest that DNase I is not a housekeeping gene but is subject to potential regulation (69). Basic transcriptional elements in the 5' region of the human DNase I gene include putative CAAT and TATA box-like sequences and consensus binding elements for the transcription factor Sp1. Besides these elements, there are other putative regulatory elements including three Pan1 motifs (transcription activators in pancreatic acinar cells; 70), a consensus sequence for TRE (thyroid hormone response element), two sequences similar to GRE (glucocorticoid response element), a sequence for AP-2 (cyclic adenonine monophosphate [cAMP] or phorbol ester signals for growth control), two GATA-1 motifs, an interleukin-6 (IL-6) response element and a putative AP-1 recognition sequence (69). The pan1 motifs probably reflect the high level of expression of DNase I in the human pancreas where the enzyme is synthesized for use in alimentary digestion. The GREs and AP-2 sites are interesting, as endonuclease expression and apoptosis are known to be regulated by glucocorticoids, cAMP and phorbol ester in different cell types (71–73). FSH stimulates cAMP production in the granulosa cell and initiates its differentiation and expression of DNase I (42). Stimulation of granulosa cells with high doses of cAMP has been shown to induce apoptosis (73), but it is not known if the endonucleases are upregulated. The presence of an IL-6 response element may account for the observation that IL-6 induces apoptosis in granulosa cells (74). The regulation of DNase I may involve a variety of transcription factors, some of which can be implicated in apoptosis in ovarian cells. It must be emphasized, however, that the 5' region of the rat ovarian DNase I gene may differ from that of the human leukocyte DNase I gene, and that the mere presence of consensus motifs for transcription factors in the 5' region of the gene is suggestive, but not proof, that such factors are involved in the regulation of the gene in any given cell type.

There are no studies to date on the role of new gene activity or protein synthesis in ovarian cell apoptosis. The requirement of new protein synthesis for apoptosis appears to be dependent on the cell type or apoptogenic signal (13, 15, 75–77). Hsueh and coworkers reported (6) that DES withdrawal induced apoptosis in the ovarian follicles of DES-treated rats and that this was enhanced by administration of androgen after DES withdrawal. Since we did not observe DNase I in the nuclei of DES-treated rats it is possible that estrogen suppresses the expression of DNase I. Estrogen withdrawal would allow the expression of DNase I to proceed in the absence of gonado-tropic support, resulting in apoptosis in the follicle. It is not known whether or not estrogen and androgen regulate the expression of DNase I in the ovary, but androgen ablation does upregulate DNase I mRNA in the prostate (56). Further studies are needed to determine the regulation of DNase I expression by steroids in the ovary, and to determine the need for new gene or protein synthesis in ovarian cell apoptosis.

Role of DNase I in Ovarian Follicular Atresia and Luteal Regression

The following hypothetical model incorporates the potential roles for DNase I in ovarian follicular atresia or luteal regression: upon the initiation of granu-losa differentiation the gene for DNase I is activated by AP-2 transcription factor binding sites in the 5' region. As the cells differentiate they produce DNase I, the activity of which is suppressed by its interaction with G-actin and/or ribosylation by PARP. An apoptogenic signal from GnRH (7), andro-gen (6), cytokines (74), or perhaps lack of sufficient survival signals from FSH (7, 8, 23, 103), estrogen (6), or growth factors (78, 79), disrupts the suppression of DNase I in the cell (Fig. 19.3). The signals or mechanisms that regulate apoptosis may include elevation of intracellular Ca^{2+}, ICE inhibi-tion of PARP, or changes in cytoskeletal dynamics. DNase I derepression in the nucleus leads to DNA degradation, particularly in areas of active tran-scription, as has been shown for the genes for gonadotropin receptor and P450-aromatase (80). DNase I changes in the cytosol lead to disruption of cytoskeletal regulation, causing rapid changes in steroidogenesis, disrup-tion of the endoplasmic reticulum and release of calcium, altered mitochon-drial function, membrane blebbing and cell shrinkage (Fig. 19.6).

Other Endonucleases

Endonucleases have been identified and implicated in apoptosis in various tissue and cell types (Table 19.2). Nuc18 was first demonstrated as an en-dogenous endonuclease in thymocytes and increased levels of this enzyme were found in nuclear extracts during glucocorticoid-induced apoptosis (81). Although suspected to be histone 2B (46, 47), nuc18 was subsequently found to be a legitimate nuclease (82) with sequence similarity to cyclophilins (83), a group of proteins with high affinity binding for the immunosuppres-sive drug cyclosporin A. Interestingly, recombinant cyclophilins also have nuclease activity that is activated by cyclosporin A. It is possible that the immunosuppressive effects of cyclosporin A may be mediated by the endo-nucleolytic activity of cyclophilins in thymocytes (83). Cyclophilins have been found in a wide range of organisms and in every mammalian tissue studied to date, and are present in the cytoplasm, nucleus, ER, or secretory

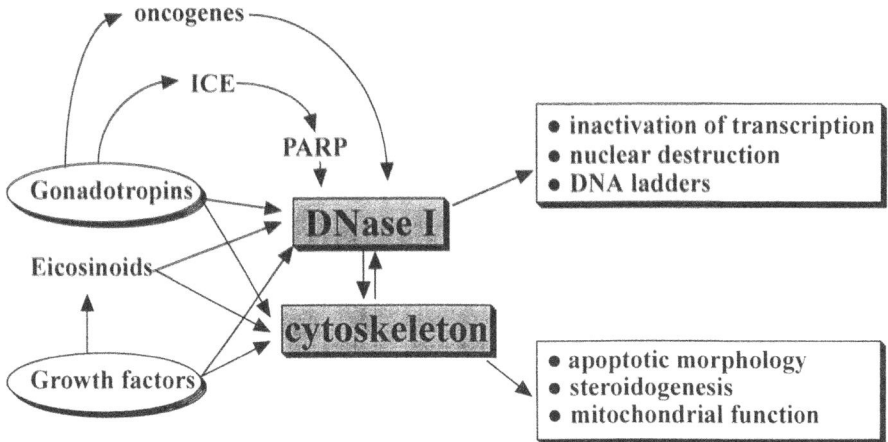

FIGURE 19.6. The potential role of DNase I in follicular atresia or luteal regression. This model illustrates the possible regulation of DNase I and the cytoskeleton by gonadotropins, growth factors, and eicosinoids, as well as the interaction of DNase I with the cytoskeleton, to elicit apoptosis in ovarian cells.

pathway of cells (83). Cyclophilins are likely present in the ovary since cyclosporins have been shown to increase estradiol levels in rats (84) and decrease progesterone levels and numbers of corpora lutea in gonadotropin-primed mice (85). Cyclophilin mRNA is routinely used as a housekeeping gene for Northern blot analysis in many tissue types, including the ovary (86). Although there is no evidence for nuc18 in ovarian nuclear extracts, cyclophilins may be primarily localized in other cellular compartments in ovarian cells.

The possible role of DNase II in apoptosis has also been examined. DNase II is a 27 kDa, Ca^{2+}/Mg^{2+}-independent nuclease. It is active at pH below 6 but does not produce 3'-OH ends that are characteristic of apoptotic DNA. Although unable to detect significant Ca^{2+}/Mg^{2+}-dependent endonuclease activity in Chinese hamster ovary cells, Barry and Eastman (87) have successfully identified a cation-independent, acidic endonuclease believed to be DNase II. Similar observations have been reported in bladder cancer T24 cells (88). Polymorphonuclear leukocytes and differentiated CD34+ cells are believed to contain both Ca^{2+}/Mg^{2+}-dependent endonuclease and DNase II-like activities (89). DNase II has been immunolocalized and implicated in apoptosis during chicken lens cell differentiation (90, 91). This chicken embryo DNase II-like nuclease exists in three forms (18, 23, and 60 kDa) and can produce the DNA ladder pattern of apoptosis. Decreased pH has been shown to be associated with apoptotic signals and to be sufficient to induce apoptosis in bladder cancer and HL-60 cells (88, 92). It is of interest that although DNase II may be involved in apoptosis in certain cell types, we were unable to detect any cation-independent, acidic endonuclease activity in ovarian cell nuclear extracts, suggesting that DNase II may not be important for apoptotic DNA degradation in the ovary.

Various unidentified and apparently unique endonuclease activities with apparent molecular weights of 40, 58, and 97 kDa (93, 94) have been demonstrated in different cell lines. A 15 kDa endonuclease activity is increased in rat renal proximal tubules after hypoxia/reoxygenation injury (95). A 27 kDa activity has been purified from the nuclei of human spleen and shown to produce apoptotic DNA degradation in isolated nuclei in vitro (96). There are several reports of high molecular weight DNA fragmentation in the absence of low molecular weight DNA cleavage, suggesting the presence of functionally-different pools of endonuclease activity in the nucleus (94, 97–99). These studies often use intercalating dyes, such as ethidium bromide, for comparison of the abundance of low (100 to 1000 base pairs) and high (50 to 300 kilobase pairs) molecular weight DNA species. Since such dyes will stain the larger DNA with 50 to 3000 times more intensity than oligonucleosomes, the relative production of high and low molecular weight DNA fragments during apoptosis must be studied with caution. The differential sensitivity of endonucleases to various inhibitors, as well as their differences in cation requirements, supports the theory that distinct nucleases are responsible for the high and low molecular weight DNA cleavage observed during apoptosis (94, 100). The high molecular weight DNA is apparently cleaved by an activity that requires either Ca^{2+} or Mg^{2+}, is insensitive to zinc, and may involve the action of proteases and/or topoisomerases.

Conclusion

DNA degradation is a consistent feature of ovarian follicular atresia and luteal regression. The role and regulation of endonucleases and DNA degradation in apoptosis remains unresolved. Potential regulators of Ca^{2+}/Mg^{2+}-dependent endonuclease activity include increased intracellular Ca^{2+}, bcl-2 or other oncogenes, and the enzyme PARP, a known substrate for ICE. A developmental pattern of endogenous Ca^{2+}/Mg^{2+}-dependent endonuclease activity is present in the rat ovary. Expression of this endonuclease correlates with the formation of antral follicles, which is associated with granulosa cell differentiation. The presence Ca^{2+}/Mg^{2+}-dependent endonucleases may determine whether or not activating or suppressing signals cause an ovarian cell to undergo apoptosis and therefore determine the fate of the follicle (i.e., ovulation versus atresia; Fig. 19.3). Three nuclease activities with molecular weights of 27, 32, and 34 kDa have been identified in rat ovarian cell nuclear extracts. The 32/34 kDa activity exhibits characteristics consistent with apoptotic DNA degradation in ovarian cells and is indistinguishable from DNase I, the endonuclease implicated in apoptosis of a variety of other cells and tissues.

DNase I may be regulated by G-actin or by PARP, an enzyme known to be inhibited by ICE. In addition to degrading DNA, DNase I could potentially function as a regulator of cytoskeletal dynamics and thus play a role in the morphological changes associated with apoptosis (Fig. 19.6). The sequence of the human gene for DNase I suggests that its expression may be subject to regulation by a variety of transcription factors, such as those activated by the cAMP pathway and the IL-6 response element, both of which have been implicated as mediators of apoptosis in granulosa cells. The regulation of DNase I expression and the role of new gene or protein synthesis in ovarian apoptosis remains to be determined. Although other endonucleases, such as nuc18 or DNase II, are present in some cells, there is no evidence to date implicating these nucleases in ovarian apoptosis. We have presented a hypothetical model for the role of DNase I in ovarian follicular atresia or luteal regression that incorporates the developmental pattern of susceptibility to apoptosis, possible regulatory mechanisms for DNase I expression, and the consequences of DNase I activation on DNA degradation and cytoskeletal changes during apoptosis.

Acknowledgments. This work was supported by a grant from the Medical Research Council of Canada (MT-10368; B.K.T.) and a Genesis Research Foundation Graduate Studentship and Ontario Graduate Scholarship (D.L.B.). We thank Professor K. Kishi and Dr. T. Yasuda (Department of Legal Medicine, Fukui Medical School, Fukui, Japan) for generously providing the anti-rat DNase I antibody.

References

1. Wyllie AII. Glucocorticoid-induced thymocyte apoptosis is associated with endogenous endonuclease activation. Nature 1980;284:555–56.
2. Hughes FM Jr, Gorospe WC. Biochemical identification of apoptosis (programmed cell death) in granulosa cells: evidence for a potential mechanism underlying follicular atresia. Endocrinology 1991;129:2415–22.
3. Tilly JL, Kowalski KI, Johnson AL, Hsueh AJW. Involvement of apoptosis in ovarian follicular atresia and postovulatory regression. Endocrinology 1991;129:2799–801.
4. Hsueh AJW, Billig H, Tsafriri A. Ovarian follicular atresia: a hormonally controlled process. Endocr Rev 1994;15:707–24.
5. Juengal JL, Garverick HA, Johnson AL, Youngquist RS, Smith MF. Apoptosis during luteal regression in cattle. Endocrinology 1993;132:249–54.
6. Billig H, Futura I, Hsueh AJW. Estrogens inhibit and androgens enhance ovarian granulosa cell apoptosis. Endocrinology 1993;133:2204–12.
7. Billig H, Futura I, Hsueh AJW. Gonadotropin-releasing hormone directly induces apoptotic cell death in the rat ovary: biochemical and in situ detection of

deoxyribonucleic acid fragmentation in granulosa cells. Endocrinology 1994; 134:245–52.

8. Chun SY, Billig H, Tilly J, Furuta I, Tsafriri A, Hsueh AJW. Gonadotropin suppression of apoptosis in cultured preovulatory follicles: mediatory role of endogenous insulin like growth factor-1. Endocrinology 1994;135:1845–53.

9. Dharmarajan AM, Goodman SB, Tilly KI, Tilly JL. Apoptosis during functional corpus luteum regression: evidence of a role for chorionic gonadotropin in promoting luteal cell survival. Endocr J 1994;2:295–303.

10. Jones DP, McConkey DJ, Nicotera P, Orrenius S. Calcium-activated DNA fragmentation in rat liver nuclei. J Biol Chem 1989;264:6398–403.

11. McConkey DJ, Hartzell P, Nicotera P, Orrenius S. Calcium-activated DNA fragmentation kills immature thymocytes. FASEB J 1989;3:1843–9.

12. McConkey DJ, Nicotera P, Hartzell P, Bolloma G, Wyllie AH, Orrenius S. Glucocorticoids activate a suicide process in thymocytes through elevation of cytosolic Ca^{2+} concentration. Arch Biochem Biophys 1989;269:365–70.

13. Cohen JJ, Duke RC. Glucocorticoid activation of a calcium-dependent endonuclease in thymocyte nuclei leads to cell death. J Immunol 1984;132:38–42.

14. McConkey DJ, Hartzell P, Duddy SK, Hakansson H, Orrenius S. 2378-Teterachlorodibenzo-p-dioxin kills immature thymocytes by Ca^{2+}-mediated endonuclease activation. Science 1988;242:256–9.

15. Schwartzman RA, Cidlowski JA. Apoptosis: the biochemistry and molecular biology of programmed cell death. Endocr Rev 1993;14:133–51.

16. Kizaki H, Tadakuma T, Odaka C, Muramatsu J, Ishamura Y. Activation of a suicide process of thymocytes through DNA fragmentation by calcium ionophores and phorbol esters. J Immunol 1989;143:1843–9.

17. Thompson EB. Apoptosis and steroid hormones. Mol Endocrinol 1994;8:665–73.

18. Rodriguez-Tarduchey G, Collin M, Lopez-Rivas S. Regulation of apoptosis in interleukin 3-dependent hemopoeitic cells by interleukin-3 and calcium ionophores. EMBO J 1990;9:2997–3002.

19. Hockenbury DM, Nunez G, Milliman C, Schreiber RD, Korsmeyer SJ. Bcl-2 is an inner mitochondrial membrane protein that blocks programmed cell death. Nature 1990;348:334–6.

20. Strasser A, Harris AW, Cory S. Bcl-2 transgene inhibits T cell death and perturbs thymic self-censorship. Cell 1991;67:889–99.

21. Arends MJ, McGregor AH, Toft NJ, Brown EJ, Wyllie AH. Susceptibility to apoptosis is differentially regulated by c-myc and mutated Ha-ras oncogenes and is associated with endonuclease availability. Br J Cancer 1993;68:1127–33.

22. Fernandez A, Fosdick LJ, Marin MC, Diaz C, McDonnell TJ, Ananthaswamy HN, McConkey DJ. Differential regulation of endogenous endonuclease activation in activated murine fibroblast nuclei by ras and bcl-2. Oncogene 1995;10:769–74.

23. Tilly JL, Tilly KI, Kenton ML, Johnson AL. Expression of the *bcl-2* gene family in the immature rat ovary: equine chorionic gonadotropin-mediated inhibition of apoptosis is associated with decreased *bax* and constitutive *bcl-2* and *bcl-x*$_{LONG}$ messenger ribonucleic acid levels. Endocrinology 1995;136:232–41.

24. Miura M, Zhu H, Rotello R, Hartweig EA, Yuan J. Induction of apoptosis in fibroblasts by IL-1β-converting enzyme, a mammalian homologue of the *C. elegans* death gene *ced-3*. Cell 1993;75:653–60.

25. Lazebnik YA, Kaufmann SH, Desnoyers S, Poirier GG, Earnshaw WC. Cleavage of poly(ADP-ribose) polymerase by a protease with properties like ICE. Nature 1994;371:346–7.

26. Carson DA, Seto S, Wasson DB, Carrera CJ. DNA strand breaks, NAD metabolism, and programmed cell death. Exp Cell Res 1986;164:273–81.

27. Nelipovich PA, Nikonova LV, Umansky SR. Inhibition of poly (ADP-ribose) polymerase as a possible reason for activation of Ca^{2+}/Mg^{2+}-dependent endonuclease in thymocytes of irradiated rats. Int J Radiat Biol 1988;53:749–65.

28. Ray SD, Sorge CL, Kamendulis LM, Corcoran GB. Ca^{++}-activated DNA fragmentation and dimethylnitrosamine-induced hepatic necrosis: effects of Ca^{++}-endonuclease and poly(ADP-ribose) polymerase inhibitors in mice. J Pharmacol Exp Ther 1992;263:387–94.

29. Rice WG, Hillyer CD, Harten B, Schaeffer CA, Dorminy M, Lackey DA III, Kirsten E, Mendeleyev J, Buki KG, Hakam A, Kun E. Induction of endonuclease-mediated apoptosis in tumor cells by C-nitroso-substituted ligands of poly(ADP-ribose) polymerase. Proc Natl Acad Sci USA 1992;89:7703–7.

30. Tanaka Y, Yoshihara K, Itaya A, Kamiya T, Koide SS. Mechanism of the inhibition of Ca^{2+}, Mg^{2+}-dependent endonuclease of bull seminal plasma induced by ADP-ribosylation. J Biol Chem 1984;259:6579–85.

31. Sawazaki K, Yasuda T, Nadano D, Tenjo E, Iida R, Takeshita H, Kishi K. A new individualization marker of human semen: deoxyribonuclease I (DNase I) polymorphism. Forensic Sci Int 1992;57:39–44.

32. Yasuda T, Sawazaki K, Nadano D, Takeshita H, Nakanaga M, Kishi K. Human seminal deoxyribonuclease I (DNase I): purification, enzymological and immunological characterization and origin. Clin Chim Acta 193;218:5–16.

33. Takeshita H, Yasuda T, Nadano D, Tenjo E, Sawazaki K, Iida R, Kishi K. Detection of deoxyribonucleases I and II (DNases I and II) activities in reproductive organs of male rabbits. In J Biochem 1994;26:1025–31.

34. Flaws JA, Kugu K, Trbovich AM, DeSanti A, Tilly KI, Hirshfield AN, Tilly JL. Interleukin-1β-converting enzyme-related proteases (IRPs) and mammalian cell death: dissociation of IRP-induced oligonucleosomal endonuclease activity from morphological apoptosis in granulosa cells of the ovarian follicle. Endocrinology 1995;136:5042–53.

35. Fleiger D, Reithmuller G, Zeigler-Heitbrock HWL. Zinc^{++} inhibits both tumor necro sis factor-mediated DNA fragmentation and cytolysis. Int J Cancer 1989;44:315–9.

36. Batistatou A, Greene LA. Aurintricarboxylic acid rescues PC12 calls and sympathetic neurons from cell death caused by nerve growth factor deprivation: correlation with suppression of endonuclease activity. J Cell Biol 1991;115:461–71.

37. Ormerod MG, O'Neill CF, Robertson D, Harrap KR. Cisplatin induces apoptosis in a human ovarian carcinoma cell line without concomitant internucleosomal degradation of DNA. Exp Cell Res 1994;211:231–7.

38. Oberhammer F, Fritsch G, Schmied M, Pavelka M, Printz D, Purchio T, Lassman H, Schulte-Hermann R. Condensation of the chromatin at the membrane of an apoptotic nucleus is not associated with activation of an endonuclease. J Cell Sci 1993;104:317–26.

39. Ucker DS, Obermiller PS, Eckhart W, Apgar JR, Berger NA, Meyers J. Genome digestion is a dispensable consequence of physiological cell death mediated by toxic T lymphocytes. Mol Cell Biol 1992;12:3060–9.

40. Jacobson MD, Burne JF, Raff MC. Programmed cell death and bcl-2 protection in the absence of a nucleus. EMBO J 1994;13:1899–910.

41. Zeleznik AJ, Ihrig LI, Bassett SG. Developmental expression of Ca^{++}/Mg^{++}-dependent endonuclease activity in rat granulosa and luteal cells. Endocrinology 1989;125:2218–20.

42. Boone DL, Yan W, Tsang BK. Identification of a deoxyribonuclease I-like endonuclease in rat granulosa and luteal cell nuclei. Biol Reprod 1995;53:1057–65.
43. Hirshfield AN. Development of follicles in the mammalian ovary. Int Rev Cytol 1991;124:43–101.
44. Greenwald GS, Terranova PF. Follicular selection and its control. In Knobil E, Neill J, eds. The physiology of reproduction. New York: Raven Press, 1988:387–435.
45. Rosenthal AL, Lacks SA. Nuclease activity detection in polyacrylamide gels. Anal Biochem 1977;80:76–90.
46. Alnemri ES, Litwack G. Glucocorticoid-induced lymphocytolysis is not mediated by an induced endonuclease. J Biol Chem 1989;264:4104–11.
47. Baxter GD, Smith PJ, Lavin MF. Molecular changes associated with induction of cell death in a human T-cell leukaemia line: putative nucleases identified as histones. Biochem Biophys Res Comm 1989;162:30–2.
48. Hitchcock SE. Actin: deoxyribonuclease I interaction. J Biol Chem 1980;255:5668–73.
49. Lazarides E, Lindberg U. Actin is the naturally occurring inhibitor of deoxyribonuclease I. Proc Natl Acad Sci USA 1974;71:4742–6.
50. Kreuder V, Diekhoff J, Sittig M, Mannherz HG. Isolation, characterisation and crystallization of deoxyribonuclease I from bovine and rat parotid gland and its interaction with rabbit skeletal muscle actin. Eur J Biochem 1984;139:389–400.
51. Lacks SA. Deoxyribonuclease I in mammalian tissues. Specificity of inhibition by actin. J Biol Chem 1981;256:2644–8.
52. Peitsch MC, Polzar B, Stephan H, Crompton T, MacDonald HR, Mannherz HG, Tschoop J. Characterization of the endogenous deoxyribonuclease involved in nuclear DNA degradation during apoptosis (programmed cell death). EMBO J 1993;12:371–7.
53. Polzar B, Peitsch MC, Loos R, Tschopp J, Mannherz HG. Overexpression of deoxyribonuclease I (DNase I) transfected into COS-cells: its distribution during apoptotic cell death. Eur J Cell Biol 1993;62:397–405.
54. Polzar B, Zanotti S, Stephan H, Rauch F, Peitsch MC, Irlmer M, Tschoop J, Mannherz HG. Distribution of deoxyribonuclease I in rat tissues and its correlation to cellular turnover and apoptosis (programmed cell death). Eur J Cell Biol 1994;64:200–10.
55. Kyprianou N, Isaacs JT. Activation of programmed cell death in the rat ventral prostrate after castration. Endocrinology 1988;122:552–62.
56. Bacher M, Rausch U, Goebel HW, Polzar B, Mannherz HW, Aumuller G. Stromal and epithelial cells from rat ventral prostate during androgen deprivation and estrogen treatment-regulation of transcription. Exp Clin Endocrinol 1993;101:78–86.
57. Lipskaia LA. Calcium and magnesium ion-dependent endonuclease 37 kDa is activated during colchicine-induced apoptosis in HL-60 cells. Tsitologiia 1994;36:303–9.
58. Mannherz HG. Crystallization of actin in complex with actin-binding proteins. J Biol Chem 1992;267:11661–4.
59. Mannherz HG, Barrington Leigh J, Leberman R, Pfrang H. A specific 1:1 actin:DNase I complex formed by the action of DNase I on F-actin. FEBS Lett 1975;60:34–8.
60. Malika-Blaszkiewicz M, Roth JS. Evidence for the presence of DNase-actin complex in L1210 leukemia cells. FEBS Lett 1993;153:235–9.

61. Peppelenbosch MP, Tertoolen LGJ, Hage WJ, de Laat SW. Epidermal growth factor-induced actin remodelling is regulated by 5-lipoxygenase and cyclooxygenase products. Cell 1993;74:565–75.
62. Carnegie JA, Tsang BK. Microtubules and the calcium-dependent regulation of rat granulosa cell steroidogenesis. Biol Reprod 1987;36:1007–15.
63. Carnegie JA, Tsang BK. The cytoskeleton and rat granulosa cell steroidogenesis: possible involvement of microtubules and microfilaments. Biol Reprod 1988; 38:100–8.
64. Gwynne A, Condon WA. Effects of cytochalasin B, colchicine, and vinblastine on progesterone synthesis and secretion by bovine luteal tissue in vitro. J Reprod Fertil 1982;65:151–6.
65. Osawa S, Betz G, Hall PF. Role of actin in the responses of adrenal cells to ACTH and cyclic AMP: inhibition by DNase I. J Cell Biol 1984;99:1335–42.
66. Shak S, Capon DJ, Hellmiss R, Marsters SA, Baker CL. Recombinant human DNase I reduces the viscosity of cystic fibrosis sputum. Proc Natl Acad Sci USA 1990;87:9188–92.
67. Polzar B, Mannherz HG. Nucleotide sequence of a full length cDNA clone encoding the deoxyribonuclease I from rat parotid gland. Nuc Acid Res 1990;18:7151.
68. Peitsch MC, Irmler M, French LE, Tschopp J. Genomic organisation and expression of mouse deoxyribonuclease I. Biochem Biophys Res Comm 1995;207:62–8.
69. Yasuda T, Kishi K, Yanagawa Y, Yoshida A. Structure of the human deoxyribonuclease I (DNase I) gene: identification of the nucleotide substitution that generates its classical genetic polymorphism. Ann Hum Genet 1995;59:1–15.
70. Meister A, Weinrich SL, Nelson C, Rutter WJ. The chymotrypsin enhancer core. Specific factor binding and biological activity. J Biol Chem 1989;264:20744–51.
71. McConkey DJ, Orrenius S, Jondal M. Agents that elevate cAMP stimulate DNA fragmentation in thymocytes. J Immunol 1990;145:1227–30.
72. McConkey DJ, Hartzell P, Jondal M, Orrenius S. Inhibition of DNA fragmentation in thymocytes and isolated thymocyte nuclei by agents that stimulate protein kinase C. J Biol Chem 1989;264:13399–402.
73. Aharoni D, Dantes A, Oren M, Amsterdam A. cAMP-mediated signals as determinants for apoptosis in primary granulosa cells. Exp Cell Res 1995;218:271–82.
74. Gorospe WC, Spangelo BL. Interleukin-6: potential roles in neuroendocrine and endocrine function. Endocr J 1993;1:3–9.
75. Duke RC, Chervenak R, Cohen JJ. Endogenous endonuclease-induced DNA fragmentation: an early event in cell mediated cytolysis. Proc Natl Acad Sci USA 1983;80:6361–5.
76. Waring P. DNA fragmentation induced in macrophages by gliotoxin does not require protein synthesis and is preceded by raised inositol phosphate levels. J Biol chem 1990;265:14476–80.
77. Martin DP, Schmidt RE, Distefano PS, Lowry OH, Carter JG, Johnson EM Jr. Inhibitors of protein synthesis and RNA synthesis prevent neuronal death caused nerve growth factor deprivation. J Cell Biol 1988;106:829–44.
78. Tilly JL, Billig H, Kowalski KI, Hsueh AJW. Epidermal growth factor and basic fibroblast growth factor suppress the spontaneous onset of apoptosis in cultured rat ovarian granulosa cells and follicles by a tyrosine kinase dependent mechanism. Mol Endocrinol 1992;6:1942–50.
79. Flaws JA, DeSanti A, Tilly K, Javid RO, Kugu K, Johnson AL, Hirshfield AN, Tilly JL. Vasoactive intestinal peptide-mediated suppression of apoptosis in the ovary:

potential mechanisms of action and evidence of a conserved antiatretogenic role through evolution. Endocrinology 1995;136:4351–9.

80. Tilly JL, Kowalski KI, Schomberg DW, Hsueh AJW. Apoptosis in atretic ovarian follicles is associated with selective decreases in messenger ribonucleic acid transcripts for gonadotropin receptors and cytochrome P450 aromatase. Endocrinology 1992;131:1670–6.

81. Compton MM, Cidlowski JA. Identification of a glucocorticoid-induced nuclease in thymocytes. J Biol Chem 1987;262;8288–92.

82. Gaido ML, Cidlowski JA. Identification purification and characterization of a calcium dependent endonuclease (NUC18) from apoptotic rat thymocytes. J Biol Chem 1991;266:18580–5.

83. Montague JW, Gaido ML, Frye C, Cidlowski JA. A calcium-dependent nuclease from apoptotic rat thymocytes is homologous with cyclophilin. J Biol Chem 1994;269:18877–80.

84. Esquifino AI, Moreno ML, Agrasal C, Villanua MA. Effects of cyclosporin on sham-operated and pituitary-grafted young female rats. Proc Soc Exp Biol Med 1995;208:397–403.

85. Husein M, Pingle S. Effect of cyclosporin A at therapeutic and toxic doses on the superluteinized ovaries in BALB/c mice. Transplant Proc 1992;24:1663–8.

86. Michel U, McMaster JW, Findlay JK. Regulation of steady-state follistatin mRNA levels in rat granulosa cells in vitro. J Mol Endocrinol 1992;9:147–56.

87. Barry MA, Eastman A. Identification of deoxyribonuclease II as an endonuclease involved in apoptosis. Arch Biochem Biophys 1993;300:440–50.

88. Shemtov MM, Cheng DL, Kong L, Shu WP, Sassaroli MA, Droller MJ, Liu BC. LAK cell mediated apoptosis of human bladder cancer cells involves a pH-dependent endonuclease system in the cancer cell: possible mechanism of BCG therapy. J Urology 1995;154:269–74.

89. Anzai N, Kawabata H, Hirama T, Masutani H, Ueda Y, Yoshida Y, Okuma M. Types of endonuclease activity capable of inducing internucleosomal DNA fragmentation are completely different between human CD34+ cells and their granulocytic descendants. Blood 1995;86:917–23.

90. Aarruti C, Chaudin E, De Maria A, Courtois Y, Counis MF. Characterisation of eye-lens DNases: long term persistence of activity in postapoptotic lens fibre cells. Cell Death Diff 1995;2:47–56.

91. Torriglia A, Chaudin E, Chany-Fournier F, Jeanny JC, Courtois Y, Counis MF. Involvement of DNase II in nuclear degeneration during lens cell differentiation. J Biol Chem 1995;270:28579–85.

92. Perez-Sala D, Collado-Escobar D, Mollinedo F. Intracellular alkinization suppresses lovastatin-induced apoptosis in HL-60 cells through the inactivation of a pH-dependent endonuclease. J Biol Chem 1995;270:6235–45.

93. Deng GE, Podack ER. Deoxyribonuclease induction in apoptotic cytotoxic T lymphocytes. FASEB J 1995;9:665–9.

94. Pandey S, Walker PR, Sikorska M. Separate pools of endonuclease activity are responsible for internucleosomal and high molecular mass DNA fragmentation during apoptosis. Biochem Cell Biol 1994;72:625–9.

95. Ueda N, Walker PD, Hsu SM, Shah SV. Activation of a 15-kDa endonuclease in hypoxia/reoxygenation injury without morphological features of apoptosis. Proc Natl Acad Sci USA 1995;92:7202–6.

96. Ribeiro JM, Carson DA. Ca^{2+}/Mg^{2+}-dependent endonuclease from human spleen: purification properties and role in apoptosis. Biochemistry 1993;32:9129–36.
97. Walker PR, Weaver VM, Lach B, LeBlanc J, Sikorska M. Endonuclease activities associated with high molecular weight and internucleosomal DNA fragmentation in apoptosis. Exp Cell Res 1994;213:100–6.
98. Earnshaw WC. Nuclear changes in apoptosis. Curr Opin Cell Biol 1995;7:337–43.
99. Cohen GM, Sun X-M, Fearnhead H, MacFarlane M, Brown DG, Snowden RT, Dinsdale D. Formation of large molecular weight fragments of DNA is a key committed step of apoptosis in thymocytes. J Immunol 1994;153:507–16.
100. Weaver VM, Lach B, Walker PR, Sikorska M. Role of proteolysis in apoptosis: involvement of serine proteases in internucleosomal DNA fragmentation in immature thymocytes. Biochem Cell Biol 1993;71:488–500.
101. Walker PR, Sikorska M. Endonuclease activities, chromatin structure, and DNA degradation in apoptosis. Biochem Cell Biol 1994;72:615–23.
102. Luciano AM, Pappalardo A, Ray C, Peluso JJ. Epidermal growth factor inhibits large granulosa cell apoptosis by stimulating progesterone synthesis and regulating the distribution of intracellular free calcium. Biol Reprod 1994;51:646–54.
103. Tilly JL, Tilly KI. Inhibitors of oxidative stress mimic the ability of follicle-stimulating hormone to suppress apoptosis in cultured rat ovarian follicles. Endocrinology 1995;136:242–52.
104. Lindhout E, Lakeman A, de Groot C. Follicular dendritic cells inhibit apoptosis in human B lymphocytes by a rapid and irreversible blockade of preexisting endonuclease. J Exp Med 1995;181:1985–95.
105. Eisenhauer KM, Chun SY, Billig H, Hsueh AJ. Growth hormone suppression of apoptosis in preovulatory rat follicles and partial neutralization by insulin-like growth factor binding protein. Biol Reprod 1995;53:13–20.
106. Chun SY, Eisenhauer KM, Kubo M, Hsueh AJ. Interleukin-1 beta suppresses apoptosis in rat ovarian follicles by increasing nitric oxide production. Endocrinology 1995;136:3120–7.
107. Quirk SM, Cowan RG, Joshi SG, Henrikson KP. Fas antigen-mediated apoptosis in human granulosa/luteal cells. Biol Reprod 1995;52:279–87.
108. Jo T, Tomiyama T, Ohashi K, Saji F, Tanizawa O, Ozaki M, Yamamoto R, Yamamoto T, Nishizawa Y, Terada N. Apoptosis of cultured mouse luteal cells induced by tumor necrosis factor-alpha and interferon-gamma. Anat Record 1995;241:70–6.

20

Synthetic Estrogen-Mediated Alterations in Uterine Cell Fate

WILLIAM J. HENDRY III, XINGLONG ZHENG, WENDELL W. LEAVITT, WILLIAM S. BRANHAM, AND DANIEL M. SHEEHAN

Our experimental system is uniquely suited to the pursuit of two complementary objectives: first, delineation of the mechanisms whereby estrogens
regulate uterine growth and morphogenesis, and second, identification of
mechanistic alterations that cause degeneration of the normal growth process to the unregulated neoplastic state. These are biomedically important
issues because: first, successful conception and gestation demands normal
uterine form and function, and second, estrogen-dependent uterine neoplasms
are responsible for considerable morbidity and mortality. Since estrogens
normally elicit a striking, yet ultimately limited, growth response in the
mature mammalian uterus, a well-integrated interplay of positive and negative regulatory pathways must be involved. To probe this topic, we have
exploited a model of atypical estrogen responsiveness that reflects lesions in
such regulatory pathways.

History of the Experimental System

From the 1940s to the 1960s, the synthetic estrogen, diethylstilbestrol (DES),
was prescribed primarily for the support of high-risk pregnancies (1). Consequently, an estimated 1 to 4 million fetuses in the USA were exposed to
this agent (1). In 1971, Herbst et al. (2) described the surprising development of vaginal clear cell adenocarcinoma in eight young women that had
been exposed to DES *in utero*. Since then, numerous reports have shown that
perinatal DES exposure induces teratogenic and neoplastic effects in the
male and female reproductive tracts of both humans (3) and experimental
animals (4). We chose the hamster as an experimental model because its
gestation period is very predictable and short (16 d) compared to other rodents. We predicted that the reproductive tract in newborn hamsters would
be at the same developmental stage as in humans exposed to DES *in utero*,

and treating neonates rather than fetuses with DES would avoid uncertainties associated with maternal metabolism and placental transfer of the drug. For the studies described below, detailed descriptions of materials and methods can be found in previous publications (5–9).

Gross Reproductive Tract Morphology

Within 6 h of birth, animals received a single subcutaneous injection of 50 µl of corn oil vehicle either alone (control) or containing 100 µg DES (≈40 mg/kg body weight). This dosage level, which is within the range administered to pregnant women (1), had profound consequences throughout the female reproductive tract (5). As shown in Figure 20.1, we observed massive enlargement of both uterine horns plus inflammatory lesions of the oviduct/ovarian bursa (salpingitis). The latter phenomenon was also reported in neonatally DES-exposed mice (10). We soon discovered that the oviduct/ovarian lesions accelerated with the onset of puberty and ultimately proved fatal. We were also concerned that the neonatally DES-exposed hamster would enter a

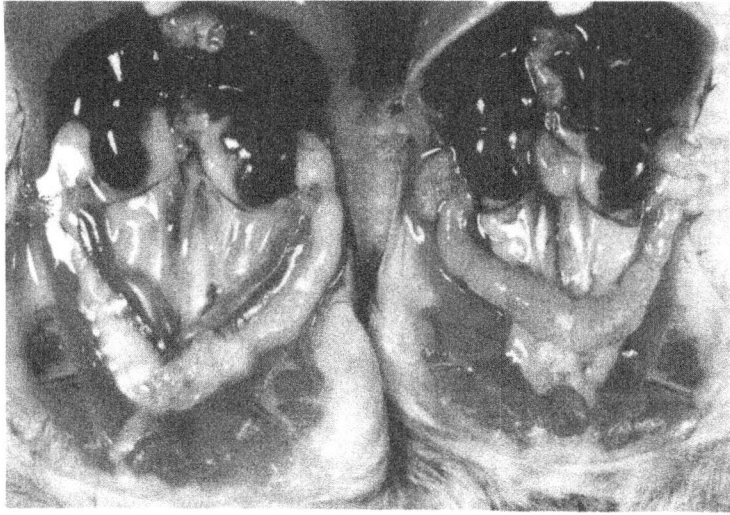

FIGURE 20.1. Effect of neonatal DES exposure on gross morphology of the female reproductive tract. Both the control (CON) and neonatally DES-exposed animal were 4 months of age. Note the greatly enlarged uterine horns and the cystic oviductal/ovarian masses (lateral to the kidneys) in the DES-exposed animal.

hyperestrogenic "persistent estrous" state as do other rodents that are perinatally exposed to estrogens (11). These considerations prompted us to perform prepubertal ovariectomies plus reestablish sustained circulating levels of the natural estrogen, estradiol-17β (E$_2$).

Serum Steroids

The effects of neonatal DES treatment, ovariectomy alone, and ovariectomy plus estrogen replacement (E$_2$ implant) on the endocrine status of mature hamsters are shown in Figure 20.2. In intact animals, E$_2$ levels were increased and progesterone (P) levels were decreased in the DES-exposed group compared to the control group. This indicates that the DES-exposed animals did enter a "persistent estrous" state in which anovulatory and cystic ovaries continuously secrete high levels of E$_2$ but little or no P (11). In both the control and DES-exposed groups, ovariectomy reduced serum P

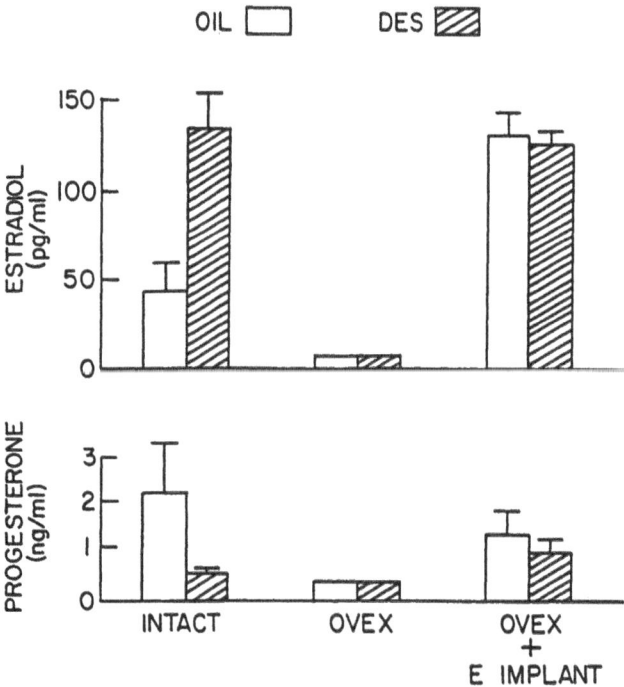

FIGURE 20.2. Serum steroid levels in mature hamsters. Blood was collected from animals that were subjected to sham surgery (INTACT), were ovariectomized alone (OVEX) or were ovariectomized and also received a Silastic implant filled with E$_2$ (OVEX + E IMPLANT). Serum levels of estradiol and progesterone were measured by specific radioimmunoassay. Animals were those described in Table 20.2. (Reproduced with permission from Leavitt et al. (5).)

and E_2 to undetectable levels, whereas the E_2 implants produced serum levels that were comparable to those measured in the intact, DES-exposed group.

Uterine Morphology

Regardless of endocrine status, DES-exposed uteri were always heavier than the corresponding controls (Table 20.1). In intact animals, uterine weight was three-fold greater in the DES-exposed group compared to controls. Although prepubertal ovariectomy reduced uterine growth in both groups, DES-exposed uteri remained heavier than controls (initial indication that neonatal DES exposure induced an early, ovary-independent, and permanent alteration in the hamster uterus). After comparable E_2 stimulation, uterine weight was still three-fold greater in DES-exposed animals compared to controls (initial indication that neonatal DES exposure altered subsequent estrogen responsiveness in the adult hamster uterus). The DES-induced effects on overall dimensions and general endometrial morphology of the adult hamster uterus are illustrated in Figure 20.3. Under the same levels of postpubertal E_2 stimulation, DES-exposed uterine horns were more massive than the controls (upper panel). Internally, the endometrium of DES-exposed uteri was extremely hyperplastic and populated with numerous polyps and cystic structures (lower panel).

Endometrial Tumors

Tumor masses also developed in DES-exposed uteri both from intact animals and from ovariectomized plus E_2-implanted animals. A representative example of the endometrial adenocarcinomas that developed in DES-exposed

TABLE 20.1. Uterine weight in adult hamsters.

	Uterine weight (mg)	
Surgical procedure	CON	DES
Sham Ovex.	558 ± 14 (13)	1812 ± 185 (13)
Ovex.	23 ± 1 (3)	42 ± 1 (3)
Ovex. + E_2 Implant	584 ± 41 (26)	1705 ± 182 (20)

Neonates were injected within 6 h of birth with 50 µL corn oil vehicle either alone (control, CON) or containing 100 µg of DES. At three weeks of age, some animals underwent sham surgery (Sham Ovex.) whereas others were ovariectomized only (Ovex.) or also received an E_2-filled Silastic implant (Ovex + E_2 Implant). Animals were sacrificed between 3 and 10 months of age. Results are expressed as means ± SEM (n). Uterine weights for all three DES-treated groups were significantly different from treatment-matched controls ($p < 0.01$) according to Students t test. (Reproduced with permission from Leavitt et al. (5).)

DES   **CON**

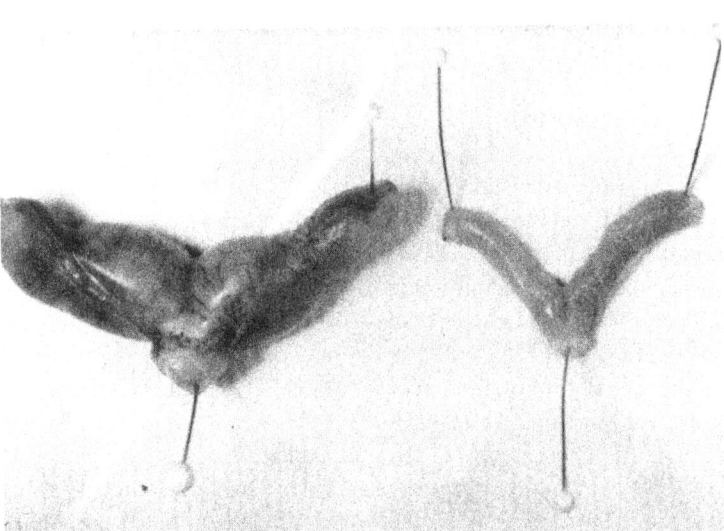

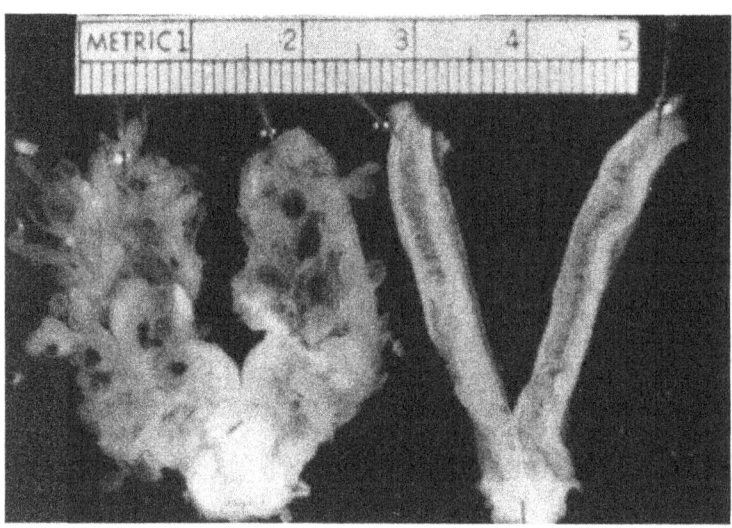

FIGURE 20.3. Uterine morphology in adult hamsters. Uteri were excised from con-trol (CON) and neonatally DES-exposed animals. Those in the upper panel were from 7-month old animals that were ovariectomized and E$_2$ implanted on d 21 of life. Those in the lower panel were from 4-month old intact animals (control animal sacrificed at proestrous stage). The uterine horns in the lower panel were opened longitudinally and immersed in fluid to reveal the abnormal endometrial lining of the DES-exposed organ. (Reproduced with permission from Leavitt et al. (5).)

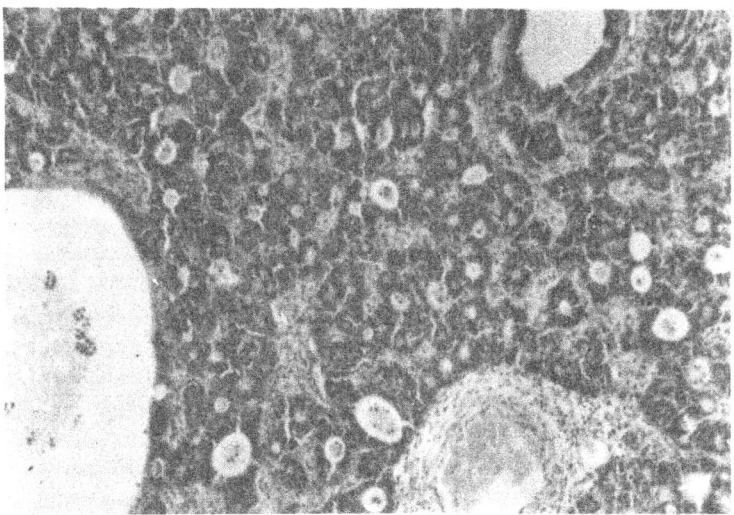

FIGURE 20.4. Histology of endometrial adenocarcinoma. The photomicrograph illustrates a neoplastic region in the uterus of an intact hamster that was neonatally exposed to DES and sacrificed at 7 months of age. Tissue was fixed in formalin, embedded in paraffin, sectioned at 6 μm thickness and stained with hematoxylin and eosin.

uteri is shown in Figure 20.4. In some areas of the tumors, neoplastic cells were quite distinct from the stromal connective tissue and often formed closely packed glandular structures with either small or distended lumina (Fig. 20.4). In other cases, the distinction between neoplastic and connective tissue cells was less clear such that glandular structures were often interspersed with solid nests of neoplastic cells (not shown).

The representative tumor specimens discussed above were derived from the large tumorigenesis study presented in Table 20.2. The relationships between neonatal DES treatment, postpubertal estrogen stimulation, and tumor induction become evident when the tumorigenesis data (Table 20.2) are correlated with endocrine status (Fig. 20.2) of the experimental groups. Tumor induction accompanied the hyperestrogenic environment in intact DES-exposed animals and was completely prevented by E_2 withdrawal. However, generation of comparable hyperestrogenic environments in control and DES-exposed animals produced uterine tumors only in the DES-exposed group. Therefore, tumor induction was dependent upon neonatal DES treatment plus chronic postpubertal E_2 stimulation. These data provide further evidence that neonatal DES exposure alters subsequent estrogen responsiveness in the adult hamster uterus.

TABLE 20.2. Induction of endometrial tumors in adult hamsters.

Surgical procedure	Tumor incidence	
	CON	DES
Sham Ovex.	0/35 (0%)	11/29 (38%)
Ovex.	0/34 (0%)	0/18 (0%)
Ovex. + E_2 Implant	0/28 (0%)	30/35 (86%)

Neonatal treatment and range of animal age at time of necropsy was as described for Table 20.1. Sham surgery and ovariectomy was performed on day 3 of life and animals received E_2 implants on day 21 of life. (Reproduced with permission from Leavitt et al. (5).)

Early Effects of Neonatal DES Treatment

To investigate the consequences of DES exposure on earlier stages of uterine morphogenesis, we analyzed uterine tissue morphometry and determined cell labeling indices following in vivo pulse-labeling with [^{3}H]thymidine in 3 to 21-d old animals (9). The latter approach provided an estimate of proliferative activity in individual uterine tissue compartments. Between 9 and 21 d of age, cell labeling was detectable in the epithelial and stromal tissue compartments (Fig. 20.5). Neonatal DES treatment did not alter proliferative activity in luminal epithelial tissue on d 9 (Fig. 20.5A), but did result in a surge of activity in this tissue between d 15 and 21. There was precocious development of endometrial glands on d 9 in DES-exposed uteri, and there were sites of intense proliferative activity (Fig. 205B). Glandular proliferation subsequently declined to the levels measured in control uteri on d 15 and 21. In the uterine stroma (Fig. 20.5C), the percentage of cells engaged in proliferative activity declined between d 9 and 21 in both groups, but the levels on d 9 and 15 were lower in DES-exposed uteri compared to controls. Observations at earlier ages (not shown) revealed that neonatal DES treatment also induced:

1. A precocious (d 3) burst of cellular proliferation throughout the uterus.
2. An extended period (d 3 to 9) of hypertrophy and cell crowding in the luminal epithelium.
3. An extreme acceleration of uterine growth resulting in a persistent increase in total uterine mass (>three-fold enhancement on d 5 to 21). In summary, the morphogenesis of the prepubertal hamster uterus was profoundly altered by a single neonatal exposure to DES. Both acute and persistent changes in mitotic activity, organization, and dimensions of individual uterine tissue compartments were observed.

Specificity of the Neonatal DES Effects

To determine if the consequences of neonatal DES treatment reflect a specific attribute of that chemical or is a function only of its estrogenic activity, female hamsters were treated or not on the day of birth with the same dose

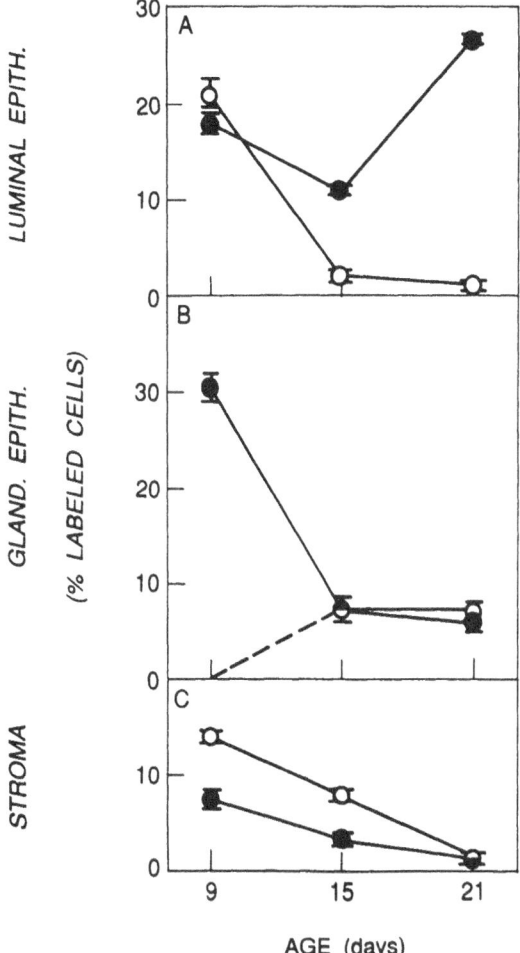

FIGURE 20.5. Proliferative activity in uterine tissues of immature hamsters. At the indicated ages, control (open symbols) and neonatally DES-treated animals (closed symbols) were sacrificed 1 h after an injection of (^{3}H]thymidine. Labeling indices [% labeled cells] were determined in luminal epithelium (A), glandular epithelium (B), and stromal tissue (C). Points represent the mean ± SEM from three uteri of the accumulated counts in nine representative areas/uterus. (Reproduced with permission from Hendry and Leavitt (9).) EPITH, epithelium.

(100 µg) of either DES or E$_2$ and their reproductive tracts were studied at various ages. We found that the two agents exerted drastically different consequences on the female hamster reproductive tract. During the prepubertal period, accelerated reproductive tract growth (primarily the uterus) was much greater in DES-exposed animals than in E$_2$-exposed animals. The left half of Figure 20.6 demonstrates the extent of this phenomenon in 15-d old animals. After puberty, this DES-dominant trend continued (not shown) due to

a greater increase in uterine mass and to an accumulation of pyoid inflammatory products in the oviduct/ovarian bursa (salpingitis). Under conditions of equivalent postpubertal E_2 stimulation, uterine growth remained accelerated in the neonatally DES-exposed group compared to the neonatally E_2-exposed group. The right half of Figure 20.6 demonstrates the extent of this phenomenon in 4-month-old animals. Thus when administered neonatally to hamsters, DES is much more effective than E_2 in inducing:

1. Accelerated growth of the immature uterus.
2. Oviductal and ovarian lesions.
3. A general uterotropic response to post-pubertal E_2 stimulation.

   Histologically, differential treatment effects were most evident in the endometrial epithelial cell compartment. The micrographs in Figure 20.7 are representative of the responses seen throughout most of the prepubertal period when ovarian estrogen secretion is minimal (5). Low magnification (left panels), reveals the dominant effect of DES exposure on uterine size and

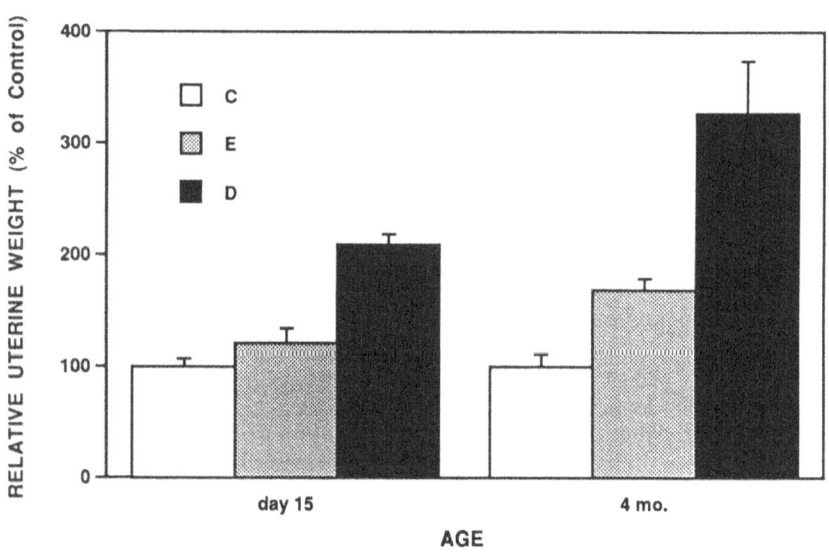

FIGURE 20.6. Comparison of neonatal DES versus E_2 treatment effects on uterine growth in prepubertal and adult hamsters. Animals were injected on the day of birth with 50 μL of corn oil vehicle either alone (control, open bars) or containing 100 μg of E_2 (cross-hatched bars) or DES (solid bars). For the prepubertal animals sacrificed at 15 d of age, uterine horns plus the rudimentary oviducts and ovaries were weighed. The adult animals that were sacrificed at 4 months of age had been ovariectomized and also received a Silastic E_2 implant at 21 d of age. Tissue weights in the two neonatal estrogen treatment groups were compared to mean tissue weight in the respective control group (set equal to 100%) and are expressed as mean ± SEM, n = 3.

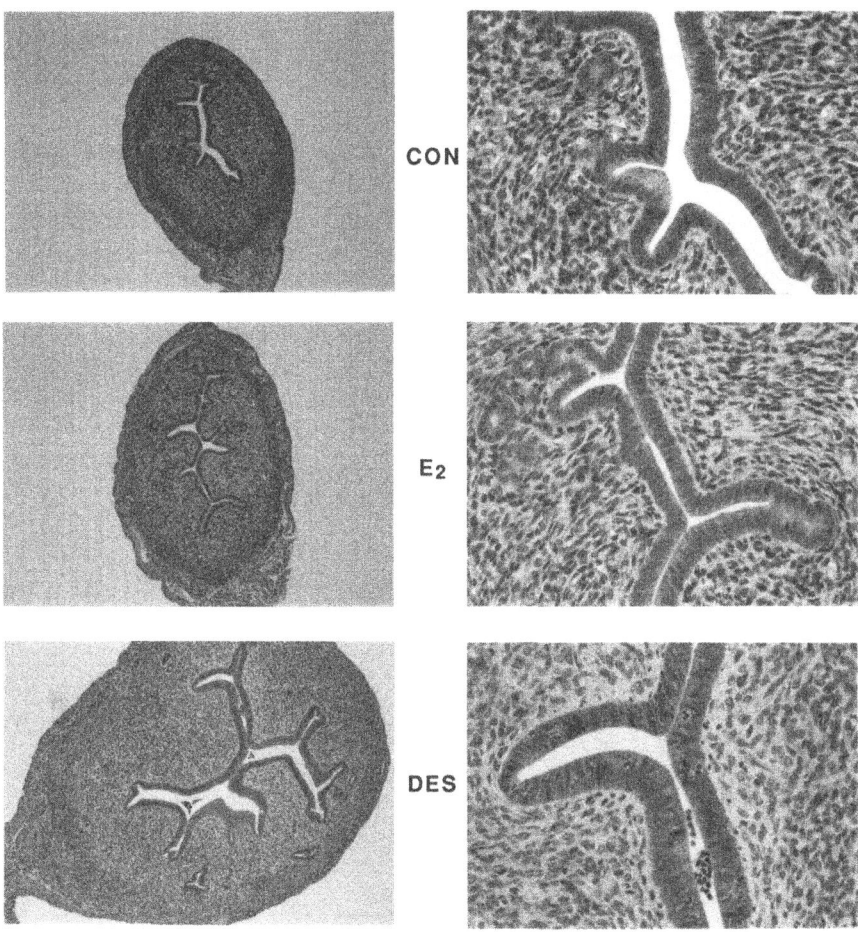

FIGURE 20.7. Comparison of neonatal DES versus E_2 treatment effects on uterine histology in 15-day-old hamsters. Animals were injected on the day of birth with 50 μl of corn oil vehicle either alone (control, CON) or containing 100 μg of either E_2 or DES. Cross sections of paraformaldehyde-fixed and paraffin-embedded uterine horns from 15 day old animals were stained with hematoxylin/eosin and photographed at 40X (left panels) and 400X (right panels) magnification.

development of endometrial glands. Higher magnification (right panels) shows that the uterine lumen in control and neonatally E_2-treated animals was lined by a simple cuboidal/columnar epithelium. In contrast, the epithelium in neonatally DES-exposed uteri was relatively hypertrophic (taller pseudostratified columnar cells) and often heavily infiltrated with inflammatory cells. Preliminary evidence indicates that the population of infiltrating cells seen in DES-exposed uteri from prepubertal animals and from E_2-stimulated adults contained a high percentage of eosinophils (unpublished observations).

The micrographs in Figure 20.8 illustrate that post-pubertal E_2 stimulation promoted a distinct histopathological profile in the uteri of DES-exposed hamsters. Low magnification (left panels) shows that both neonatal treatments enhanced the size of the uterus and its endometrial complexity (gland number and organization). However, the DES-exposed uterus also contained cystic glandular structures and extensive areas of "foamy" epithelium. Higher magnification (right panels) shows that the luminal epithelium in control and neonatally E_2-treated animals was somewhat hypertrophic but

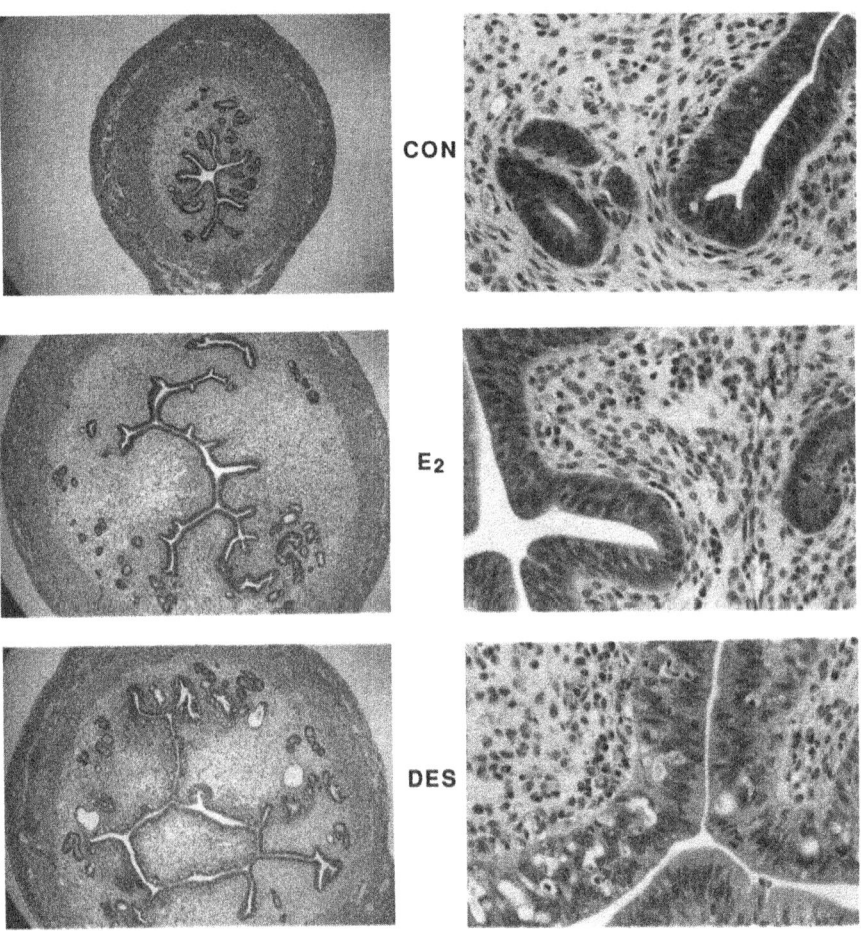

FIGURE 20.8. Comparison of neonatal DES versus E_2 treatment effects on uterine histology in post-pubertal hamsters following nine days of E_2 stimulation. Neonatal treatments, histological processing and photography were as described for Figure 20.7. All animals were ovariectomized and E_2 implanted at 21 d of age and uteri were harvested nine days later (1 month of age).

organization of the pseudostratified columnar cells appeared normal. In contrast, the luminal epithelium in neonatally DES-treated uteri was markedly hyperplastic and/or dysplastic since the cells were extremely tall, disorganized, overcrowded, and poorly demarcated from the underlying stroma. Areas of the epithelium appeared "foamy" because it was riddled with infiltrating cells and fenestrations containing what appeared to be degenerating cells.

In summary, this study showed that neonatal DES exposure altered general morphogenesis of the female hamster reproductive tract to a much greater extent than neonatal E_2 exposure. Furthermore, atypical development of the endometrial epithelial compartment was confined to neonatally DES-exposed animals and this response was promoted by subsequent estrogen stimulation. Thus, we conclude that the synthetic estrogen DES is considerably more potent than the natural estrogen E_2 as a teratogen in the developing hamster uterus.

Endometrial Apoptosis

The degenerating cells shown in the bottom right panel of Figure 20.8 appeared to be undergoing apoptosis, a process known to be hormonally regulated in the endometrial epithelium of hamsters (12), mice (13, 14), and rabbits (15, 16). This was supported by analyses based on a generally recognized biochemical endpoint of apoptosis, internucleosomal cleavage of DNA (17). Figure 20.9 shows the gel electrophoretic separation of genomic DNA extracted from the uteri of control and neonatally DES-treated animals that were ovariectomized and E_2-implanted two months before sacrifice. The characteristic "ladder pattern" of cleaved DNA was much more prevalent in the DES-treated sample. These results are representative of those made in several analyses performed in tissues after various periods of E_2 stimulation.

We have also performed in situ analyses of apoptosis using the TUNEL method (18). Unfortunately, the signal detection system (peroxidase/diaminobenzidine-based) and counterstaining method (hematoxylin) we chose generated an immunohistochemical profile that was not well-suited to black and white photography. However, color photographs (not shown) clearly showed that a brown reaction product was strongly and specifically deposited over the numerous apoptotic bodies within the dysplastic epithelium of DES-exposed uteri. These results indicate that one aspect of altered E_2 responsiveness in the neonatally DES-exposed hamster uterus is a combined increase in cell proliferation and cell death in the endometrial epithelial cell compartment.

Uterine Transplantation Studies

Despite years of study, the mechanistic basis for estrogen-dependent normal, as well as neoplastic, growth remains unresolved. For instance, there is credible evidence for both direct (19–21) and indirect pathways (22–24).

C   D

FIGURE 20.9. Gel electrophoretic analysis of internucleosomal DNA cleavage in E_2-stimulated uteri from control and neonatally DES-treated hamsters. Each lane contained 15 µg of DNA that was separated on an agarose gel and stained with ethidium bromide. Genomic uterine DNA was obtained from control (C) or neonatally DES-treated animals (D) that were ovariectomized and given an E_2 implant at three weeks of age and then killed at three months of age.

Evidence also indicates that either pathway may involve non-steroidal autocrine, paracrine or endocrine factors with either a positive (25–27) or negative influence (28–30) on growth. Thus, the neonatal DES effects we observe in the hamster uterus could be explained by either of two hypotheses:

1. Direct Action-the neonatal DES insult directly alters the cellular physiology and/or composition of the developing uterus such that it thereafter responds abnormally to estrogenic stimulation, or
2. Indirect Action-the neonatal DES insult permanently alters the level or functional activity of a systemic factor(s) involved in estrogen-dependent growth.

Testing these hypotheses depends on a means to experimentally control the developmental environment of the neonatally DES-exposed hamster uterus. Fortunately, we have demonstrated that the uterus from a normal prepubertal hamster can be transplanted to a recipient hamster cheek pouch where it grows, responds to circulating estrogen, and undergoes appropriate morphogenesis (8).

TABLE 20.3. Summary of histopathology segregation for hamster uterine tissues at normal and ectopic sites.

	Epithelial Histopathology	
Uterine tissue site	CON host	DES host
Normal	−	+
Ectopic (Con Donor)	−	−
Ectopic (Des Donor)	+	+

Control and neonatally DES-treated host animals were ovariectomized and given an estrogen implant at three weeks of age. At the same time, the cheek pouches of each host received a uterus from a 7-day-old control (right side) and neonatally DES-exposed (left side) donor animal. At five months of age, host animals were sacrificed and uterine tissues from both the normal and ectopic sites were fixed in 10% neutral buffered formalin and then processed for light microscopy using standard hematoxylin and eosin staining. Tissues were scored for the absence (−) or presence (+) of characteristic histopathology (hyperplasia/dysplasia and apoptosis) in the endometrial epithelial cell compartment. These results are summarized from two independent studies where a total of 31 control transplanted tissues and 28 DES-exposed transplanted tissues taken from 18 control hosts and 18 DES-exposed hosts were examined.

We exploited this system to test the above hypotheses by transplanting the uteri from 7-d old control or neonatally DES-treated donors into the cheek pouches of control or neonatally DES-treated adult hosts. The host animals had also been ovariectomized and E_2 implanted to provide equivalent hyperestrogenic environments. This experimental approach yielded the results summarized in Table 20.3. For the normally situated uteri, characteristic histopathological lesions (hyperplasia/dysplasia and apoptosis) developed in the endometrial epithelial cell compartment of DES-exposed, but not control, hosts, as expected. This histopathologic profile was never fully encountered in ectopic tissues that were transplanted from control donors into either control or DES-exposed hosts. However, the histopathological profile was expressed to a comparable degree in ectopic tissues that were transplanted from DES-exposed donors into either control or DES-exposed hosts. We interpret these results as unequivocal evidence for the direct action hypothesis (atypical estrogen responsiveness is an inherent property of the neonatally DES-exposed uterus) and against the indirect action hypothesis (atypical estrogen responsiveness is not a consequence of any DES-induced change in the host endocrine environment). This suggests that the neonatal DES insult acts as an initiation event and subsequent estrogen exposure acts as a promoter for the development of uterine lesions that can progress to endometrial cancer.

Altered Protooncogene Expression

Recent studies have shown that neonatal estrogen exposure leads to persistent induction of certain estrogen-regulated genes such as peptide growth factors/cytokines and their receptors (31–33). Growth factor or estrogen stimulation of quiescent cells also increases the expression of a set of "immediate early genes," including the proto-oncogenes c-*jun* (34, 35), c-*fos* (36, 37), and c-*myc* (38). In fact, considerable evidence now supports the concept that differential expression of these genes can drive either normal cellular proliferation and differentiation or the development of neoplasia (39). Somewhat paradoxically, however, other studies indicate that the enforced or inappropriate expression of c-*myc, c-jun,* and c-*fos* may induce apoptosis (40–42). The emerging consensus from these and other studies (43, 44) is that the cellular decision between cell proliferation and apoptosis depends on a precise integration of various regulatory inputs, among which the above proto-oncogenes represent only a subset.

   Another family of genes appears to be more direct regulators of apoptosis. Its original member, the *bcl-2* gene, codes for a 26 kD protein (Bcl-2) that can extend cell survival by inhibiting apoptosis (45). Another member of the *bcl-2*-related gene family, *bcl-x*, codes for two distinct cDNAs, $bcl\text{-}x_L$ and $bcl\text{-}x_s$, whose protein products can either block cell death ($Bcl\text{-}x_L$) in a manner analogous to Bcl-2 or allow apoptosis to proceed ($Bcl\text{-}x_s$) by countering the effects of Bcl-2 (46). Another member of this gene family encodes a 21-kD protein termed Bax that forms homodimers and heterodimers with Bcl-2 or $Bcl\text{-}x_L$ in vivo (47). When Bax predominates apoptosis is accelerated, indicating that the *bax* gene product may be a direct effector of apoptosis (41). In fact, it has been proposed that the fate of a cell during development is decided by the ratio or balance of death repressors (Bcl-2 and $Bcl\text{-}x_L$) to death inducers (Bax and $Bcl\text{-}x_s$) in that cell at any given time (41, 46, 48, 49).

   Several groups have shown that transforming growth factor-beta (TGF-β) isoforms also play an important role in regulating apoptosis (50–52). Other studies showed that when eosinophils are recruited to sites of developing tumors, they predominantly associate with the malignant epithelium, and most of the tumor-associated eosinophils express the cytokine, TGF-α (53–55). Lastly, studies of hormonally regulated apoptosis in the hamster uterine epithelium have shown that it involves phagocytosis of apoptotic bodies by macrophages and possibly neutrophils (12).

   Since we are interested in the mechanistic links between neonatal DES treatment and subsequent effects in the hamster uterus (altered E_2 responsiveness, putative eosinophilic infiltration, combined hyperplasia/apoptosis, and progression to neoplasia), we have analyzed expression dynamics for the genes and regulatory factors discussed above. Our immunohistochemical results have been particularly informative and are summarized in semiquantitative form in Table 20.4. This table deals specifically with the endometrial epithelial compartment where DES-induced changes were most conspicuous.

   For the experimental series reported in Table 20.4, control and neonatally DES-treated animals were ovariectomized and E_2 implanted at 3

TABLE 20.4. Summary of immunohistochemical findings in the endometrial epithelial cell compartment.

Antigen	Signal observations	
	Cellular localization	DES-induced intensity change
Proto-oncogenes		
c-Myc	nucleus	↑
c-Jun	nucleus	↑
c-Fos	nucleus	↑↑
Apoptosis-relevant		
Bcl-2	diffuse cellular	↓
Bcl-x	diffuse cellular	↑
Bax	diffuse cellular	↑↑
Growth factors		
TGF-α	apoptotic bodies	↑↑
TGF-β_1	apoptotic bodies	↑↑
TGF-β_2	apoptotic bodies	↑↑

Control and neonatally DES-treated animals were ovariectomized and given an estrogen implant at three weeks of age. Uterine tissues harvested from 40-day-old animals were fixed in 4% formalin, embedded in paraffin, subjected to heat-induced antigen retrieval, and reacted with primary antibodies that were detected by the avidin brotin complex (ABC) method. For each antigen, signal intensity in the endometrial epithelial cell compartment of DES-exposed uteri was compared to that in control uteri and was scored as being either increased (↑) or decreased (↓) to either a moderate (single arrow) or dramatic (double arrow) extent.

weeks of age and uterine tissues were harvested about 3 weeks later. The DES-induced changes are representative of those observed in other independent analyses of uteri from E_2-stimulated adult animals. For the proteins encoded by some of the protooncogenes that are implicated in the regulation of both cell proliferation and apoptosis (c-Myc, c-Jun, c-Fos), levels were enhanced (especially pronounced for c-Fos). As expected for these members of the nuclear class of proto-oncogenes (56–58), the reaction product in positive cells was localized over their nuclei. For the proteins encoded by the genes that are implicated in the positive regulation of apoptosis (Bax and Bcl-x), levels were also enhanced. For the protein implicated in the blockage of apoptosis (Bcl-2), levels were reduced. The diffuse intracellular localization of the reaction product for the latter three proteins is consistent with evidence that they associate with various intracellular membranes (59–62). For the growth factor isoforms studied, intense reaction product was deposited specifically over apoptotic cells in the epithelial compartment. These results suggest that neonatal DES treatment induces imbalances in the estrogen-regulated expression pattern of specific genes that regulate both cell proliferation and cell removal in the endometrial epithelial cell compartment.

Concluding Remarks and Future Prospects

This overview describes the important attributes of our experimental system. It is very productive for the induction and study of endometrial tumors, and the cheek pouch transplantation approach provides a unique and facile means to exert experimental control over host and tissue exposures. Further exploitation of this system should help to resolve the basic mechanisms by which estrogens induce normal, as well as neoplastic, growth of certain target tissues. We have also determined that the endometrial epithelial cell compartment is a site for both apoptosis and hyperplasia/neoplasia in the neonatally DES-exposed hamster uterus. This is noteworthy in view of the increasing evidence that a balance between cell proliferation and apoptosis plays a role in various developmental processes, the progression of neoplastic diseases, and the success or failure of some traditional cancer therapies (63–65). Since apoptosis may normally serve to eliminate mutated cells that develop as a result of extremely rapid proliferative activity (66), perhaps frank neoplasms erupt at sights where apoptotic activity is either lost or overwhelmed. In fact, we have preliminary evidence that this may be so in our system. Further probing of this topic is now one of our primary research efforts.

Acknowledgments. These studies were supported in part by an institutional Pilot Study Program funded through the American Cancer Society and the Arkansas Cancer Research Center, grants from the National Institute of Child Health and Human Development (HD28074) and the National Cancer Institute (CA60250), by the Flossie West Memorial Trust Foundation, and by the United States Food and Drug Administration.

References

1. Herbst AL, Scully RE, Robboy SJ. Prenatal diethylstilbestrol exposure and human genital tract abnormalities. Monogr Natl Cancer Inst 1979;51:25–35.
2. Herbst AL, Ulfelder H, Poskanzer DC. Adenocarcinoma of the vagina: association of maternal stilbestrol therapy with tumor appearance in young women. N Engl J Med 1971;284:878–81.
3. Marselos M, Tomatis L. Diethylstilbestrol: I. Pharmacology, toxicology and carcinogenicity in humans. Eur J Cancer 1992;28A:1182–9.
4. Marselos M, Tomatis L. Diethylstilbestrol: II. Pharmacology, toxicology and carcinogenicity in experimental animals. Eur J Cancer 1993;29A:149–55.
5. Leavitt WW, Evans RW, Hendry WJ. Etiology of DES-induced uterine tumors in the Syrian hamster. In: Leavitt WW, ed. Hormones and Cancer. New York: Plenum Publishing Corporation, 1982:63–86.
6. Evans RW, Chen TJ, Hendry WJ, Leavitt WW. Progesterone regulation of estrogen receptor in the hamster uterus during the estrous cycle. Endocrinology 1984;107:383–90.
7. Hendry WJ, Leavitt WW. Binding and retention of estrogen in the uterus of

hamsters treated neonatally with diethylstilbestrol. J Steroid Biochem 1982;17: 479–87.

8. Hendry WJ, Branham WS, Sheehan DM. The hamster cheek pouch as a convenient ectopic site for studies of uterine morphogenesis and endocrine responsiveness. Differentiation 1992;51:49–54.

9. Hendry WJ, Leavitt WW. Altered morphogenesis of the immature hamster uterus following neonatal exposure to diethylstilbestrol. Differentiation 1993;52:221–7.

10. Newbold RR, Bullock BC, McLachlan JA. Diverticulosis and salpingitis isthmica nodosa (SIN) of the fallopian tube. Am J Pathol 1984;117:333–5.

11. Saunders FJ. Effects of sex steroids and related compounds on pregnancy and on the development of the young. Physiol Rev 1968;48:601–43.

12. Sandow BA, West NB, Norman RL, Brenner RM. Hormonal control of apoptosis in hamster uterine luminal epithelium. Am J Anat 1979;156:15–36.

13. Terada N, Yammamoto R, Takada T, et al. Inhibitory effect of progesterone on cell death of mouse uterine epithelium. J Steroid Biochem 1989;33:1091–6.

14. Pollard JW, Pacey J, Cheng SVY, Jordan EG. Estrogens and cell death in murine luminal epithelium. Cell Tissue Res 1987;249:533–40.

15. Nawaz S, Lynch MP, Galand P, Gerschenson LE. Hormonal regulation of cell death in rabbit uterine epithelium. Am J Pathol 1987;127:51–9.

16. Rotello RJ, Hocker MB, Gerschenson LE. Biochemical evidence for programmed cell death in rabbit uterine epithelium. Am J Pathol 1989;134:491–5.

17. Schwartzman RA, Cidlowski JA. Apoptosis: the biochemistry and molecular biology of programmed cell death. Endocr Rev 1993;14:133–51.

18. Gavrieli Y, Sherman Y, Ben-Sasson SA. Identification of programmed cell death in situ via specific labeling of nuclear DNA fragmentation. J Cell Biol 1992;119:493–501.

19. Stack G, Gorski J. Direct mitogenic effect of estrogen on the prepuberal rat uterus: studies on isolated nuclei. Endocrinology 1984;115:1141–50.

20. Holinka CF. Proliferation and responsiveness to estrogen of human endometrial cancer cells under serum-free culture conditions. Cancer Res 1989;49:3297–301.

21. Silberstein GB, Horn KV, Shyamala G, Daniel CW. Essential role of endogenous estrogen in directly stimulating mammary growth demonstrated by implants containing pure antiestrogens. Endocrinology 1994;134:84–90.

22. Shafie SM. Estrogen and the growth of breast cancer: new evidence suggests indirect action. Science 1980;209:701–2.

23. Richards J, Imagawa W, Balakrishnan A, Edery M, Nandi S. The lack of effect of phenol red or estradiol on the growth response of human, rat, and mouse mammary cells in primary culture. Endocrinology 1988;123:1335–40.

24. Alkhalaf M, Propper AY, Adessi GL. Proliferation of guinea pig uterine epithelial cells in serum-free culture conditions: effect of $17-\beta$ estradiol, epidermal growth factor and insulin. J Steroid Biochem Mol Biol 1991;38:345–50.

25. Gardner RM, Kirkland JL, Ireland JS, Stancel G. Regulation of the uterine response to estrogen by thyroid hormone. Endocrinology 1978;103:1164–72.

26. Lippman ME, Dickson RB, Gelmann EP, et al. Growth regulation of human breast carcinoma occurs through regulated growth factor secretion. J Cell Biochem 1987;35:1–16.

27. Ignar-Trowbridge DM, Nelson KG, Bidwell MC, et al. Coupling of dual signaling pathways: epidermal growth factor action involves the estrogen receptor. Proc Natl Acad Sci USA 1992:4658–62.

28. Mizejewski GJ, Vonnegut M, Jacobson HI. Estradiol-activated á-fetoprotein suppresses the uterotropic response to estrogens. Proc Natl Acad Sci USA 1983;80: 2733–7.

29. Soto AM, Sonnenschein C. Cell proliferation of estrogen-sensitive cells: the case for negative control. Endocr Rev 1987;8:44–52.

30. Knabbe C, Lippman ME, Wakefield LM, et al. Evidence that transforming growth factor-β is a hormonally regulated negative growth factor in human breast cancer cells. Cell 1987;48:417–28.

31. DiAugustine RP, Petruez P, Bell GI, et al. Influences of estrogens on mouse uterine epidermal growth factor precursor protein and messenger ribonucleic acid. Endocrinology 1988;122:2355–63.

32. Gardner RM, Verner G, Kirkland JL, Stancel GM. Regulation of uterine epidermal growth factor (EGF) receptors by estrogen in the mature rat and during the estrous cycle. J Steroid Biochem 1989;32:339–43.

33. Murphy LJ, Ghahary A. Uterine insulin-like growth factor: regulation of expression and its role in estrogen-induced uterine proliferation. Endocr Rev 1990;11:443–53.

34. Weisz A, Cicatiello L, Persico E, Scaloma M, Bresciani F. Estrogen stimulates transcription of c-jun protooncogene. Mol Endocrinol 1990;4:1041–50.

35. Chiappetta C, Kirkland JL, Loose-Michell DS, Murthy L, Stancel GM. Estrogen regulates expression of the jun family of protooncogenes in the uterus. J Steroid Biochem Mol Biol 1992;41:113–23.

36. Loose-Mitchell DS, Chiappeta C, Stancel GM. Estrogen regulation of c-fos messenger ribonucleic acid. Mol Endocrinol 1988;2:946–51.

37. Jouvenot M, Pellerin I, Alkhalaf M, Marechal G, Royez M, Adessi GL. Effects of 17 beta-estradiol and growth factors on c-fos gene expression in endometrial epithelial cells in primary culture. Mol Cell Endocrinol 1990;72:149–57.

38. Lau LF, Nathans D. Expression of a set of growth-related immediate early genes in BALB/c 3T3 cells: coordinate regulation with c-fos or c-myc. Proc Natl Acad Sci USA 1987;84:1182–6.

39. Luscher B, Eisenman RN. New light on Myc and Myb. Part 1. Myc. Genes Dev 1990;4:2025–35.

40. Colotta F, Polentarutti N, Sironi M, Mantovani A. Expression and involvement of c-fos and c-jun protooncogenes in programmed cell death induced by growth factor deprivation in lymphoid cell lines. J Biol Chem 1992;267:18278–83.

41. Williams GT, Smith CA. Molecular regulation of apoptosis: genetic controls on cell death. Cell 1993;74:777–9.

42. Hermeking H, Eick D. Mediation of c-Myc-induced apoptosis by p53. Science 1994;265:2091–3.

43. Gewirtz DA. DNA damage, gene expression, growth arrest and cell death. Oncol Res 1993;5:397–408.

44. Ueda N, Shah SV. Apoptosis. J Lab Clin Med 1994;124:169–77.

45. Hockenbery DM, Zutter M, Hickey W, Nahm M, Korsmeyer SJ. Bcl-2 protein is topographically restricted in tissues characterized by apoptotic cell death. Proc Natl Acad Sci USA. 1991;88:6961–5.

46. Boise LH, Gonzalez-Garcia M, Postema CE, et al. bcl-x, a bcl-2-related gene that functions as a dominant regulator of apoptotic cell death. Cell 1993;74:597–608.

47. Yang E, Zha J, Jockel J, Boise LH, Thompson CB, Korsmeyer SJ. Bad, a heterodimeric partner for Bcl-x$_L$ and Bcl-2 displaces Bax and promotes cell death. Cell 1995;80:285–91.

48. Oltvai Z, Milliman C, Korsmeyer SJ. Bcl-2 heterodimerizes in vivo with a conserved homolog, Bax, that accelerates programmed cell death. Cell 1993; 74:609–19.
49. Barinaga M. Cell suicide: by ICE, not fire. Science 1994;263:754–6.
50. Rotello RJ, Lieberman RC, Purchio AF, Gerschenson LE. Coordinated regulation of apoptosis and cell proliferation by transforming growth factor $\beta1$ in cultured uterine epithelial cells. Proc Natl Acad Sci USA 1991;88:3412–5.
51. Moulton BC. Transforming growth factor-β stimulates endometrial stromal apoptosis in vitro. Endocrinology 1994;134:1055–60.
52. Bursch W, Oberhammer F, Jirtle RL, et al. Transforming growth factor-$\beta1$ as a signal for induction of cell death by apoptosis. Br J Cancer 1993;67:531–6.
53. Elovic A, Galli SJ, Weller P, et al. Production of transforming growth factor alpha by hamster eosinophils. Am J Pathol 1990;137:1425–34.
54. Chiang T, McBride J, Chou MY, Nishimura I, Wong DTW. Molecular cloning of the complementary DNA coding for the hamster TGF-α mature peptide. Carcinogenesis 1991;12:529–32.
55. Ghiabi M, Gallagher GT, Wong DTW. Eosinophils, tissue eosinophilia, and eosinophil-derived transforming growth factor á in hamster oral carcinogenesis. Cancer Res 1992;52:389–93.
56. Vogt PK, Bos TJ. *jun:* oncogene and transcription factor. Adv Cancer Res 1990;55: 1–35.
57. Distel RJ, Spiegelman BM. Protooncogene c-*fos* as a transcription factor. Adv Cancer Res 1990;55:37–55.
58. Koskinen PJ, Alitalo K. Role of *myc* amplification and overexpression in cell growth, differentiation and death. Semin Cancer Biol 1993;4:3–12.
59. Lithgow T, van Driel R, Bertram JF, Strasser A. The protein product of the oncogene *bcl*-2 is a component of the nuclear envelope, the endoplasmic reticulum, and the outer mitochondrial membrane. Cell Growth Differ 1994;5:411–7.
60. Krajewski S, Tanaka S, Takayama S, Schibler MJ, Fenton W, Reed JC. Investigation of the subcellular distribution of the Bcl-2 oncoprotein: residence in the nuclear envelope, endoplasmic reticulum, and outer mitochondrial membrane. Cancer Res 1993;53:4701–14.
61. Krajewski S, Krajewska M, Shabaik A, Miyashita T, Wang HG, Reed JC. Immunohistochemical determination of in vivo distribution of Bax, a dominant inhibitor of Bcl-2. Am J Pathol 1994;145:1323–36.
62. Krajewski S, Krajewski M, Shabaik A, et al. Immunohistochemical analysis of in vivo patterns of Bcl-X expression. Cancer Res 1994;54:5501–7.
63. Wyllie AH. Apoptosis and the regulation of cell numbers in normal and neoplastic tissues: an overview. Cancer Metastasis Rev 1992;11:95–103.
64. Kerr JF, Winterford CM, Harmon BV. Apoptosis: its significance in cancer and cancer therapy. Cancer 1993;73:2013–26.
65. Thompson CB. Apoptosis in the pathogenesis and treatment of disease. Science 1995;267:1456–62.
66. Preston GA, Lang JE, Maronpot RR, Barret JC. Regulation of apoptosis by low serum in cells of different stages of neoplastic progression: enhanced susceptibility after loss of a senescence gene and decreased susceptibility after loss of a tumor suppressor gene. Cancer Res 1994;54:4214–23.

Author Index

Subject Index

PROCEEDINGS IN THE SERONO SYMPOSIA USA SERIES

Continued from page ii

The manufacturer's authorised representative in the EU is Springer
Nature Customer Service Centre GmbH, Europaplatz 3, 69115 Heidelberg,
Germany. If you have any concerns regarding our products, please
contact ProductSafety@springernature.com

Printed and bound by CPI Group (UK) Ltd, Croydon, CR0 4YY
24/04/2026
02096348-0007